Windows on the Wild®

Biodiversity Basics
An Educator's Guide
to Exploring the
Web of Life

Dear Educator,

A Hollywood scriptwriter couldn't have come up with anything better: A mysterious microscopic organism goes out of control and creates havoc in nine rivers in the eastern United States. In 1997, this tale of terror actually took place. And what was this tiny troublemaker? A species of plankton called *Pfiesteria piscicida*. Although it was identified in the late '80s, *Pfiesteria* went largely unnoticed for more than a decade. But then something happened that turned the normally nontoxic predator into a mass killer. In just a few months, *Pfiesteria*-infested waters killed more than a billion fish and caused some people who had been in contact with the water to experience memory loss and skin rashes. No one knew what made the organism change into such a toxic pest. Although researchers are still searching for answers, many believe that the millions of tons of hog and chicken waste that have been pouring into rivers in the East for the past half century might be changing the ecology of our freshwater river systems. These changes could be the cause of imbalances that led to the *Pfiesteria* outbreaks.

In the same year, an exhibition of Ansel Adams's spectacular photographs of nature toured the nation's museums. Thousands of people flocked to see his dramatic black and white images. Some people went back two and three times to wonder at Adams's skill at capturing the beauty of the natural world.

The year also brought the first international agreement to reduce global warming, a potentially worldwide threat to biodiversity. In December 1997, representatives of more than 150 countries met in Kyoto, Japan, to discuss the global impacts of climate change and what we should do as an international community to reduce the threat.

As these examples indicate, our lives are intimately linked to biodiversity—the variety of life on the planet. We depend on it, appreciate it, and are threatening it by our actions. We also have the power to protect it. Loss of biodiversity is one of the most pressing environmental issues of our time. And at the same time, biodiversity offers an exciting framework for exploring the world and our place in it. As an educational theme, it promotes interdisciplinary teaching that links science, writing, economics, politics, civics, geography, ethics, and more. And as a moral issue, biodiversity challenges all of us to clarify our beliefs and values about how to balance conserving species and wild spaces with meeting human needs.

Biodiversity Basics is designed to provide ideas about how to integrate biodiversity into teaching—at many levels, in both formal and nonformal settings. The activities target students in grades 6, 7, 8, and 9, but many work well with younger and older age groups too. Although this module is not intended to be a complete biodiversity textbook, you can use it to build a unit, a course, or a year-long, interdisciplinary program. We also encourage you to use current articles from newspapers and magazines as well as other resources to add variety, depth, and different points of view.

As you explore the issue of biodiversity, let us know what you think by filling out the feedback form on page 477. And check out our Web site at <www.worldwildlife.org> to find out more about workshops, resources, and other biodiversity education opportunities.

Have fun and good luck!

Judy Braus

Judy Braus, Director of Education

WWF CONSERVATION PROGRAM
Executive Staff

Kathryn S. Fuller
President and Chief Executive Officer

James P. Leape
Senior Vice President, Conservation Program

Diane W. Wood
Vice President, Research and Development

Bruce W. Bunting
*Vice President, Asia, Conservation Finance,
and Species Conservation*

Eric Dinerstein
Chief Scientist

William M. Eichbaum
*Vice President, U.S. Conservation
and Global Threats*

Ginette Hemley
Vice President, Species Conservation

Twig Johnson
Vice President, Latin America and the Caribbean

Richard N. Mott
Vice President, Living Planet Campaign

Henri Nsanjama
Vice President, Africa and Madagascar

DEVELOPMENT TEAM

Director and Editor
Judy Braus

Senior Environmental Education Specialist
Betty Olivolo

Managing Editors
Nicole Ardoin, Andrew Burnett

Environmental Education Specialists
Jeff England, Eddie Gonzalez, Christy Vollbracht

Administrative Assistants
Catherine Fuller, Scott Sophos, Heather Zehren

Senior Writers
Judy Braus, Jody Marshall, Sara St. Antoine,
Christy Vollbracht, Luise Woelflein

Senior Editors
Gerry Bishop, Dan Bogan, Jody Marshall, Luise Woelflein

Contributors
Dan Bogan, Randy Champeau, Peggy Cowan,
Joe Heimlich, Diane Hren, Mary Jo Larson, Rob Larson,
Terry Lawson-Dunn, Diane Lill, John Ramsey,
Mary Schleppegrell, Julie Velasquez Runk, Heather Zehren

Interns
Lani Asato, Michelle Belt, Kimberley Marchant,
Katherine Munson, Ian Signer, Jennifer Steinberg,
Ethan Taylor, Jennifer Taylor, Dawn Turney

Design
Greg Moraes Studios

Illustrators
Bettye Braus, Blaise P. Burke, Meryl Hall,
Kimberley Marchant, Devorah Romanek

Marketing and Outreach Consultant
Barb Pitman

ACKNOWLE

In developing this module, we have been fortunate to work with some of the most talented, dedicated, and thoughtful professionals in the field. During the past three years, hundreds of individuals have taken the time to write activities, edit copy, pilot materials, review them for accuracy, and field-test activities across the country. We'd especially like to thank the members of the *Windows on the Wild* Advisory Board, Council, and Evaluation Steering Committee, as well as the *Biodiversity Basics* Working Group.

The following acknowledgement list includes many of the individuals who helped produce this module. Unfortunately, we weren't able to include everyone who reviewed copy and participated in the workshops, focus groups, and writing sessions—to those individuals, we offer our sincere thanks. We'd also like to recognize our many colleagues throughout World Wildlife Fund who helped with content review, fact checking, and moral support during the production of this module.

Several organizations and individuals deserve special thanks. First, we'd like to recognize Eastman Kodak Company for its generous support of *Windows on the Wild* and its ongoing commitment to environmental education. Special thanks also go to the following colleagues for their terrific suggestions, energy, and support as this program evolved: Dr. Joe Heimlich, The Ohio State University; Dr. John Fien, Griffith University, Australia; Dr. Danie Schreuder, University of Stellenbosch, South Africa; Diane Wood, World Wildlife Fund; Francis Grant-Suttie, World Wildlife Fund; Dr. Gordon Orians, University of Washington; and Gerry Bishop, National Wildlife Federation. And finally, we'd like to give a huge thank you to Dr. Randy Champeau, Director of the Wisconsin Center for Environmental Education at the University of Wisconsin, Stevens Point, for his vision and support in getting this project off the ground and for providing insightful direction throughout the development of this module.

BIODIVERSITY BASICS Curriculum Working Group

Randy Champeau
Director, Wisconsin Center for Environmental Education, and Professor of Resource Management, University of Wisconsin, Stevens Point

Peggy Cowan
Education Specialist, Alaska Department of Education

Joe Heimlich
Associate Professor, Environmental Education, School of Natural Resources, The Ohio State University

Rob Larson
Curriculum Coordinator, Willamette Primary School, West Linn, Oregon

John Ramsey
Associate Professor of Science Education, College of Education, University of Houston

WINDOWS ON THE WILD Advisory Board and Council

Janet Ady
Chief, Division of Education Outreach, National Education and Training Center, U.S. Fish and Wildlife Service

Lynne Baptista
Curator of Education, Baltimore Zoo

Rich Block
Executive Director, Santa Barbara Zoological Gardens

Dan Bogan
Education Specialist and Teacher, Green Fire, Inc.

Jeffrey Bryant
Education Director, Monterey Bay Aquarium

Bruce Carr
Director, Conservation Education, American Zoo and Aquarium Association

Randy Champeau
Director, Wisconsin Center for Environmental Education, and Professor of Resource Management, University of Wisconsin, Stevens Point

Dwight Crandell
Executive Director, St. Louis Science Center

Carmel Ervin
Secondary Education Specialist, National Museum of Natural History, Smithsonian Institution

Paul Grayson
Vice President, Indianapolis Zoo

Steve Hage
Environmental Science Educator, School of Environmental Studies, Minnesota

Joe Heimlich
Associate Professor, Environmental Education, School of Natural Resources, The Ohio State University

Bob Hoage
Chief of Public Affairs, National Zoological Park

Nancy Hotchkiss
Director of Education, Zoological Society of Florida

Lou Iozzi
Professor, Department of Natural Resources, Rutgers University

David Jenkins
Associate Director for Interpretive Services, National Zoological Park

Douglas Lapp
Executive Director, National Science Resources Center, Smithsonian Institution

David Love
Executive Vice President, WWF-Canada

Thane Maynard
Director of Education, Cincinnati Zoo

Kathy McGlauflin
Vice President of Education Programs, Project Learning Tree

Annie Miller
Educator, Jefferson Junior High School, Washington, D.C.

Terry O'Connor
Curator of Education, Woodland Park Zoo

Mary Schleppegrell
Assistant Professor, Linguistics, and Director, English as a Second Language, University of California, Davis

Samuel Scudder
Educator, Hart Junior High School, Washington, D.C.

Talbert Spence
Vice President, Education Division, National Audubon Society

Cynthia Vernon
Director of Communications, Brookfield Zoo

Cherie Williams
Education Specialist, The Seattle Aquarium

Keith Winsten
Curator of Education, Brookfield Zoo

WINDOWS ON THE WILD Evaluation Steering Committee

Randy Champeau
Director, Wisconsin Center for Environmental Education, and Professor of Resource Management, University of Wisconsin, Stevens Point

Paul Hart
Professor, Faculty of Education, University of Regina, Saskatchewan

Joe Heimlich
Associate Professor, Environmental Education, School of Natural Resources, The Ohio State University

Lou Iozzi
Professor, Department of Natural Resources, Rutgers University

Tom Marcinkowski
Acopian Chair, Graduate Program in Environmental Education, Florida Institute of Technology

Martha Monroe
Assistant Professor and Extension Specialist, School of Forest Resources and Conservation, University of Florida

Phyllis Peri
Environmental Education Resources Coordinator, Wisconsin Center for Environmental Education, University of Wisconsin, Stevens Point

Mary Schleppegrell
Assistant Professor, Linguistics, and Director, English as a Second Language, University of California, Davis

Danie Schreuder
Professor of Biology and Environmental Education, University of Stellenbosch, South Africa

Trudi Volk
Professor, Curriculum and Instruction, Southern Illinois University at Carbondale

EXPERT REVIEWERS

Allen Allison
Director for Research and Scholarly Studies, Bishop Museum

Eric Anderson
Associate Professor and Coordinator of Wildlife, University of Wisconsin, Stevens Point

Pedro Barbosa
Professor, Department of Entomology, University of Maryland

Rich Block
Executive Director, Santa Barbara Zoological Gardens

David Blockstein
Senior Scientist, Committee for the National Institute for the Environment

Robin Braus
President, MacDonald and Company

Herb Broda
Assistant Superintendent, Wooster Public Schools, Ohio

Caitlin Brune
Special Assistant to the Senior Vice President, World Wildlife Fund

Scott Burns
Director of Marine Conservation Program, World Wildlife Fund

Carol Cala
Manager, Technical Issues, Eastman Kodak Company

Steven Chambers
Fisheries Specialist, Department of Endangered Species, U.S. Fish and Wildlife Service

Randy Champeau
Director, Wisconsin Center for Environmental Education and Professor of Resource Management, University of Wisconsin, Stevens Point

Robert Chipley
Director of Communications, Conservation Science Division, The Nature Conservancy

Jason Clay
Senior Fellow, Social Science and Economics, World Wildlife Fund

Theo Colborn
Senior Scientist, World Wildlife Fund

Bob Cooper
Curator of Zoology, Rochester Museum and Science Center

Peggy Cowan
Education Specialist, Alaska Department of Education

Dwight Crandell
Acting President, St. Louis Science Center

Peter DeBrine
Senior Program Officer, World Wildlife Fund

Eric Dinerstein
Chief Scientist, World Wildlife Fund

Diane Elam
Fish and Wildlife Biologist, Sacramento Valley Branch, U.S. Fish and Wildlife Service

John Fien
Associate Professor, School of Environmental Studies, Griffith University, Australia

Trish Flaster
Ethnobotanist, Botanical Liaisons

Marie Foegh
Department of Surgery, Georgetown University Medical Center

Donna Ford-Werntz
Associate Professor and Herbarium Curator, Department of Biology, West Virginia University

Andrea Freed
Adjunct Faculty,
Woodring College of Education,
Western Washington University

Steve Gough
Economist, University of Bath

Sharon Haines
Manager of Natural Resources,
International Paper

Don Hall
Professor, Entomology,
University of Florida

Jim Hasler
Manager of Environmental Affairs,
The Clorox Company

Carl Haub
Conrad Taeuber Chair of Population
Information, Population Reference
Bureau

Joe Heimlich
Associate Professor, Environmental
Education, School of Natural Resources,
The Ohio State University

Susan Helms
Research Associate, Tellus Institute

Gary Hevel
Public Information Officer,
Entomology, National Museum of
Natural History, Smithsonian Institution

Bill Hilton
Education Director,
Hawk Mountain Sanctuary

Kathleen Hogan
Education Development and Research
Specialist, Institute for Ecosystem Studies

Leslie Hudson
Environmental Education Consultant

Malcolm Hunter
Professor, Department of Wildlife
Ecology, University of Maine

Georgia Jeppesen
Coordinator, Florida Biodiversity Middle
School Project, Florida Gulf Coast
University

Fatima Jackson
Professor, University of Maryland
Bioanthropology Research Lab

Paulette Johnson
Director, Pennsylvania Center for
Environmental Education

Stephen Kellert
Professor, School of Forestry and
Environmental Studies, Yale University

Diane Lill
Environmental Education Consultant

Rich Liroff
Senior Program Officer,
World Wildlife Fund

Frankie Long
Educator, Sutton Middle School,
Atlanta, Georgia

Lori Mann
Environmental Education Consultant

Peter Martin
Head of Education, WWF-UK

Ed McCrea
Executive Director,
North American Association
for Environmental Education

Kathy McGlauflin
Vice President of Education
Programs, Project Learning Tree

Ari Michelsen
Professor, Department of Economics,
Washington State University

Katherine Munson
Coastal Zone Management Planner,
Worcester County, Maryland

Melissa Nelson
Executive Director,
Cultural Conservancy

Fran Nolan
Curator of Education,
North Carolina Zoological Park

Fiona Norris
Collections and Education Manager,
Botanical Research Institute of Texas

Elliott Norse
President, Marine
Conservation Biology Institute

David Olson
Senior Scientist,
World Wildlife Fund

Marlar Oo
Program Coordinator, Research
and Development Program,
World Wildlife Fund

Gordon Orians
Professor Emeritus, Department of
Zoology, University of Washington

VieVie Pickett
Coordinator, Office of Environmental
Education, Florida Gulf Coast University

Dan Polhemus
Research Entomologist, National
Museum of Natural History,
Smithsonian Institution

Maria Bober Rasmussen
Senior Engineer, Issues Management,
Eastman Kodak Company

Sorrayut Ratanapojnard
Faculty Member, Faculty of Science,
Mahidol University, Bangkok, Thailand

Joe Salanitro
Microbiologist, Shell
Development Company

Danie Schreuder
Professor, Biology and Environmental
Education, University of Stellenbosch,
South Africa

Fulai Sheng
Senior Program Officer,
World Wildlife Fund

John Shores
Environmental Specialist, Peace Corps

Katie Sieving
Associate Professor of Avian Ecology,
Wildlife Ecology and Conservation
Biology, University of Florida

ACKNOWLE

Bora Simmons
Associate Professor, Curriculum and Instruction, Northern Illinois University

Sheri Sykes Soyka
Associate Director, Project Learning Tree

Eleanor Sterling
Program Director, Center for Biodiversity, American Museum of Natural History

Ron Tipton
Director of U.S. Ecoregion Conservation, World Wildlife Fund

Keith Tomlinson
Administrator, Meadow Lark Gardens

John Wagner
Biologist, The Field Museum

Arjen Wals
Professor, Department of Agricultural Education, Wageningen Agricultural University

Shirley Watt-Ireton
Director of Special Publications, National Science Teachers Association

Brenda Weiser
Executive Director, National Envirothon

Larry Weiser
Director, Center for Economic Education, University of Wisconsin, Stevens Point

Chris Wille
Director of Conservation, Rainforest Alliance, Costa Rica

Cherie Williams
Education Specialist, Seattle Aquarium

David Wood
Educator, Sidwell Friends School, Washington, D.C.

Paul Zeph
Executive Director, Iowa State Office of the National Audubon Society

Pilot Educators

Elizabeth Adams
Falls Church High School, Falls Church, Virginia

Afzal Aziz
Falls Church High School, Falls Church, Virginia

Cliff Bueneman
Elliot Junior High School, Washington, D.C.

Sara Carlson
Mercer Middle School, Seattle, Washington

Phyllis Collie
Frost Middle School, Fairfax, Virginia

Judy Esten
Marsellar Middle School, Manassas, Virginia

Chuck Gastin
Southern Middle School, Lusby, Maryland

Romanita Harrod
Paul Junior High School, Washington, D.C.

Vernon Howell
Francis Junior High School, Washington, D.C.

Diane Hren
The Langley School, McLean, Virginia

Alan Kaplan
Tilden Nature Center, Berkeley, California

Anne Koch
Notre Dame High School, San Jose, California

Bill Kraegel
Lathrop E. Smith Center, Rockville, Maryland

Mayday Levine
Merritt Elementary School, Washington, D.C.

Frankie Long
Sutton Middle School, Atlanta, Georgia

Atiba Madyun
Kramer Middle School, Washington, D.C.

Shaun McCarthy
Westwood Middle School, Gainesville, Florida

Annie Miller
Jefferson Junior High School, Washington, D.C.

Debbie Palmer
Swanson Middle School, Arlington, Virginia

Margaret Pennock
Sidwell Friends School, Washington, D.C.

Neil Rockwell
Mercer Middle School, Seattle, Washington

Barbara Running
Romig Middle School, Anchorage, Alaska

Karen Schmeltzer
Creston Middle School, Creston, Ohio

Marilyn Schwille
Neelsville Middle School, Germantown, Maryland

Carlas Starling
Falls Church High School, Falls Church, Virginia

Patricia Travis
Southern Middle School, Lusby, Maryland

Jose Veras
Monroe Middle School, Rochester, New York

David Wood
Sidwell Friends School, Washington, D.C.

Peggy Soong Yaplee
Mercer Middle School, Seattle, Washington

TABLE OF CONTENTS

CHAPTER 2: Why Is Biodiversity Important?

CHAPTER 3: What's the Status of Biodiversity?

"Biological diversity is the basis for sustainability." —Jerry Franklin, scientist

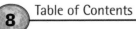

☼ =Outside Activity   ⚲ =Challenging Activity

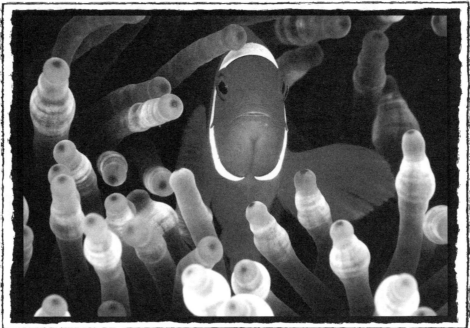

©Mike Stevens/Tony Stone Images

*"What better gift can we give the
next generation than a passion for science
and discovery, and a commitment to
caring for the natural world?"*

**–Julie Packard, President,
Monterey Bay Aquarium**

Welcome to Windo
and Biodiversity B

"The decisions we

make in the next

decade and beyond

may well determine

the fate of our planet—

and of ourselves."

**–United Nations
Environment Programme**

World Wildlife Fund

ws on the Wild®

asics

Welcome to Windows on the Wild®

"Biodiversity represents the very foundation of human existence. Besides profound ethical and aesthetic implications, it is clear that the loss of biodiversity has serious economic and social costs. The genes, species, ecosystems, and human knowledge that are being lost represent a living library of options available for adapting to local and global change. Biodiversity is part of our daily lives and livelihood and constitutes the resources upon which families, communities, nations, and future generations depend."

—Global Biodiversity Assessment,
Summary for Policy Makers
United Nations Environment Programme

 **B**iodiversity! Although the term may seem intimidating to some, you couldn't choose a more engaging and stimulating topic—or one as all-encompassing and important for our future.

Biodiversity is the variety of life on Earth. It's everything from the tiniest microbes to the tallest trees, from creatures that spend their entire lives deep in the ocean to those that soar high above the Earth's surface. It's also the word used to describe the wealth of habitats that house all life forms and the interconnections that tie us together. All of Earth's ecosystems and the living things that have evolved within them—including the fantastic range and expression of human cultures—are part of our planet's biodiversity.

Biodiversity Basics— An Educator's Guide to Exploring the Web of Life is part of World Wildlife Fund's environmental education program called *Windows on the Wild*, or *WOW*. *WOW* uses the topic of biodiversity as a "window" to help learners of all ages explore the incredible web of life. *WOW* also explores the complexity of biodiversity— and looks at it within scientific, social, political, cultural, and economic contexts.

We believe that biodiversity is an important and powerful issue that draws learners in. As a theme, it cuts across many disciplines and provides real-world contexts and issues that promote critical and creative thinking skills, citizenship skills, and informed decision making. Biodiversity also illustrates the complexity of environmental issues and makes plain that there are many perspectives as well as much uncertainty.

The diversity of life on Earth shapes and nourishes every facet of our existence. But because these connections are seldom obvious, we humans have often pursued our short-term interests with limited regard for the well-being of other species and the places they live. At the same time, social and economic inequities have forced some people to exploit resources to meet their basic needs. As a result, biodiversity is rapidly declining. If we want to ensure the long-term health of the planet, we need to develop an informed and motivated citizenry that understands what biodiversity is and why it's important. And we need citizens who have the skills and confidence to rise to the challenge of protecting biodiversity and who feel empowered to do so. Education, we believe, is one of the best tools we have for achieving this goal.

An Overview of Biodiversity Basics

Biodiversity Basics is the first curriculum module in the *Windows on the Wild* environmental education middle school program. As you read through it, you may be surprised by the variety of topics it covers—everything from genetics to chewing gum to population growth. That's because biodiversity touches nearly every aspect of our lives. It's exciting and complex. And it can spark your students' creativity and interest.

Biodiversity Basics is packed with ideas and information designed to help you explore both the ecological and social dimensions of biodiversity with your students. You'll find a comprehensive overview of biodiversity, followed by more than thirty teaching activities. We've also included a section on developing effective unit plans—one that we hope will give you a variety of ideas about how to link activities into broader units. In the back of the module, you'll find appendices that include resource ideas, planning charts, our biodiversity education framework, and other resources to enrich your teaching.

We've designed *Biodiversity Basics* for use with students in grades 6, 7, 8, and 9, but we've found that many of the activities work well with younger and older age groups too. We've also designed the module for use in both schools and nonformal education settings, such as museums, nature centers, zoos, aquariums, botanical gardens, science centers, and other community education institutions. Many of the activities can also be adapted for use in home teaching.

Our goal for this program is not to teach your students *what* to think about biodiversity, but to introduce them to some fascinating topics, raise some challenging questions, and guide them to explore, analyze, evaluate, and discuss these issues from an informed position. We're aware that biodiversity issues are both controversial and complicated. Your students may bring many diverse perspectives to the biodiversity-related issues you introduce. But with careful guidance, these different points of view can contribute to a dynamic learning environment in which students clarify their own thinking, learn how to listen to others, and gain new insights about these intriguing issues.

In short, we hope this module opens your students' minds to the wondrous diversity of life around them. We also hope it engages them in

BIOFACT

Wild Dentistry—Many kinds of roots used by people in Africa as "chewing sticks" to clean teeth have been shown to contain fluoride and other cavity-fighting ingredients, natural antibiotics, and even, in some cases, substances that combat sickle-cell anemia.

thoughtful dialogue about their place on the planet—and about the future of the world we all share.

Biodiversity Basics is built on a set of underlying principles about education. As you read through the activities in this module, you'll see many familiar strategies and approaches—from constructivist education, which values prior experiences and knowledge, to innovative assessment strategies, group learning, problem-solving, interdisciplinary teaching, and experiential learning. (For more about the teaching strategies we've used in *Biodiversity Basics*, see "Putting the Pieces Together" on pages 378–409.)

Education should challenge students to think critically and creatively about their world—to question how and why we do things, and how we might do them differently. It should promote positive

> *"In the delicate world of relationships, we are tied together for all time."* **–John Steinbeck, writer**

change (both personally and within communities), help students envision a better society, increase respect and tolerance for others, and build effective citizenship skills and stewardship. In *Biodiversity Basics*, we've emphasized four overlapping themes that we believe can help create a more sustainable society: futures education, community action and service learning, education that examines "sustainability," and creating a sense of wonder. We've touched on each below.

The Building Blocks of Biodiversity Basics

Futures Education: Looking Ahead

When kids watch movies about the future, they often see a world gone awry. In fact, many writers and moviemakers center their fictional future breakdown on environmental disasters. So how can we teach about biodiversity loss and other environmental issues without making students cynical, or even terrified, about the future they'll inherit? The answer may lie in *futures education*— education that encourages students to envision a positive future and the role they can play to make such a future happen.

Thinking constructively about the future may be more important now than ever before. For the first time in history, the way we choose to live is affecting global natural systems, from the atmosphere to the oceans. Rather than letting your students start to feel like victims of an inevitable future disaster, you can encourage them to see themselves as active participants in creating a more livable future.

In this module, you'll find a number of activities designed to help your students look forward. These activities encourage them to imagine futures that will contribute to their own quality of life, as well as to the well-being of both local and global communities. (For more information on futures education, contact the World Futures Society at 7910 Woodmont Avenue, Suite 450, Bethesda, MD 20814.)

Devoted Dads—Male emperor penguins may be the most "devoted" dads in the world. While their mates are off at sea in search of food, the males stand in darkness and howling Antarctic winds for two full months, holding their single egg on top of their feet. By the time the females finally return to take over the egg-sitting, the starving males have to walk up to 100 miles across the ice to find open water in which to feed. At a slow, half-mile-per-hour pace, they may have to march for 200 hours without a stop.

Learning from the Community

In schools and communities around the world, educators are finding that one of the best ways to prepare students for their future role as active, voting citizens is to get them involved in local issues. By addressing a real community need, students can learn about the political process, environmental issues, careers, project planning, and what it means to be a responsible citizen.

Biodiversity lends itself to *service learning* and community investigations. Every community faces environmental challenges that affect the well-being of both people and wild species: pollution, rapid development, transportation problems, park planning, and so on. By getting involved in a biodiversity-related project, your students will invest energy in their community and see that they can help to improve its condition.

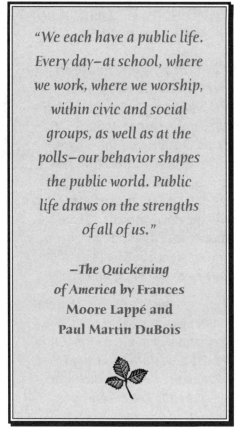

"We each have a public life. Every day—at school, where we work, where we worship, within civic and social groups, as well as at the polls—our behavior shapes the public world. Public life draws on the strengths of all of us."

—The Quickening of America by Frances Moore Lappé and Paul Martin DuBois

Service Learning— A Closer Look

More and more schools across the country are embracing service learning as a way to engage students in community activities that solve real-life problems, apply academic skills, and help others in the process. The National and Community Service Act of 1990 defines service learning as "thoughtfully organized service experiences that meet actual community needs ... and that help students learn and develop through active participation." Many educators predict that service learning will continue to grow and that, by encouraging students to take part in projects that focus on the environment, health, the arts, the elderly, politics, and other important community issues, we will produce a nation with more caring, committed, and skilled citizens.

At best, service learning and community action are learner-centered and teacher-facilitated, with constant opportunities for both students and facilitators to reflect on the problem, approach, and achievements of the project. This cycle of action, reflection, and revised action is known as *action research*. Many educators already rely on action research to improve their teaching. And educators report increased motivation and maturity among their students as the students become more involved in and thoughtful about their learning process. (For more about service learning, see *Enriching the Curriculum Through Service Learning*, edited by Carol W. Kinsely and Kate McPherson, Eds., Association for Supervision and Curriculum Development, 1995. For related insights, read *Environmental Education for Empowerment: Action Research and Community Problem Solving* by William B. Stapp, Arjen E. J. Wals, and Sheri L. Stankorb, Kendall/Hunt Publishing, 1996. For tips on getting your students involved in local issues, see "Getting Involved!" on pages 358–377.)

Environmental Education and Agenda 21

In 1977, at the first meeting of its kind, representatives from more than 70 countries gathered in Tbilisi, Georgia (in the former U.S.S.R.), to discuss education and the environment. The goal was "to create new patterns of behavior of individuals, groups, and society as a whole towards the environment." Since then, environmental education has been evolving to meet the challenges facing the environment and society.

Today, many countries have national environmental education programs and are trying to implement Agenda 21—the detailed plan resulting from the 1992 Earth Summit, which was sponsored by the United Nations Conference on Environment and Development (UNCED). Held in Rio de Janeiro, Brazil, the UNCED Summit embraced education as a key to our efforts to build a more sustainable society. To find out more about new thinking in environmental education, see the resources on page 452.

"The frog does not drink up the pond in which he lives."

—Indian Proverb

Education for Sustainability

Many of the activities in *Biodiversity Basics* explore pathways to sustainability—meeting the needs of the present without compromising our ability to meet the needs of the future. In the process, the activities examine the relationships between ecological integrity, economic prosperity, and social equity. And, because these three goals often come into conflict, the module also works to develop students' ability to negotiate, listen, compromise, persuade, and analyze.

Thinking in terms of sustainability—and finding ways to balance economic issues, social equity, and ecological integrity—also requires thinking beyond our immediate needs and interests. The activities in this module encourage students to consider the perspective of other individuals, communities, and cultures, and to look forward to assess the way actions today will affect the lives of people and other species in the United States and around the globe in the future. These activities also challenge students' thinking about fairness, individual and community responsibility, and other concerns that are critical to our understanding of sustainability. At the same time, the activities in *Biodiversity Basics*, and many of those listed in the resources section, also encourage students to work on developing a set of personal ethics—a framework by which they can make decisions.

One View of Sustainable Development

"Sustainable development meets the needs of the present without compromising the ability of future generations to meet their own needs. Choosing to be sustainable in businesses, schools, government institutions, and in our individual lives demands a national commitment to the nation's economic prosperity, ecological integrity, and social equity." (President's Council on Sustainable Development, April 1995)

Environmental Justice for All

The Environmental Justice movement is an active effort to make sure people of all races, cultures, and incomes have equal opportunities to live and work in a healthy environment. By encouraging students to examine issues from all sides and to develop an ethical framework for making decisions and taking action, you can help them understand the role that social equity plays in creating a more sustainable world. (For more information, contact EPA's Office of Environmental Justice.)

Biodiversity Basics introduces the theme of sustainability to help students, educators, and the public create a more positive vision of the future. To facilitate this process, the activities encourage creative thinking and problem-solving skills—both of which are vital to taking action toward a more sustainable and equitable future.

Creating a Sense of Wonder

Increasingly, the wonders of the natural world are lost on many of us. Each day the saga of human affairs dominates the media and demands our attention, leaving little room for any awareness of what's happening right outside our doors—in the soil beneath our feet, the trees lining the block, the sky overhead.

At WWF, we believe that it's important to nurture a perspective that includes an understanding of natural processes and rhythms. After all, how can people be expected to value or protect something that they've had little or no exposure to and have little or no understanding of?

When people become aware of the fantastic phenomena that routinely take place in the natural world, it's probably safe to say that most experience a sense of awe and appreciation. The intricate dance a honey bee performs when communicating information about a food source, the unexpected appearance of a rainbow, the 22,000-mile annual migration of the arctic tern, the delicate craftsmanship of a bird's nest—our planet is full of wonders such as these. Some have argued that we're "hardwired" to respond to these things with a sense of wonder, or at least with curiosity.

In *Biodiversity Basics* we provide many opportunities for educators to draw out students' own natural curiosity and sense of wonder. We can think of no other subject that's better suited to stimulating these natural predispositions—and no better way to help students understand biodiversity. But there are also important educational advantages to presenting the wonder and "gee-whiz" of biodiversity. Provoking curiosity leads students to ask questions and think creatively, to explore, and to challenge previous knowledge.

To help your students learn more about their local environment, we encourage you to spend as much time outside with them as you can—and to make use of the many natural areas and outdoor educational institutions that exist in your communities, from local nature centers to zoos to city and rural parks.

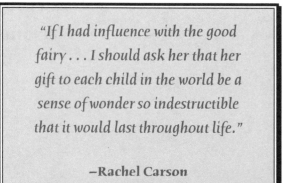

"If I had influence with the good fairy . . . I should ask her that her gift to each child in the world be a sense of wonder so indestructible that it would last throughout life."

—Rachel Carson

Space Matters—To survive, a single harpy eagle requires nearly 39 square miles of rain forest. An eyelash mite spends its entire life in one of the follicles of a person's eyelashes.

Exploring the Web

Biodiversity

is "life in all

its dimensions

and richness and

manifestations."

–Michael Soulé,
conservation biologist

# of Life

An Introduction to Biodiversity

You'll find it everywhere— from the largest ocean to the smallest puddle, from the densest forest to the driest desert, from the richest marsh to the most barren mountaintop. It marches in vast parades across the African savanna and lurks in the merest pinch of city-lot soil. It's a living bank account, providing the "capital" that underlies all human enterprise and prosperity. And its beauty, abundance, and complexity continually fascinate and inspire us—as it has since the first flicker of human consciousness more than 1.5 million years ago.

What it is, is *biodiversity**—the wonderful variety of life on Earth.

We humans are late additions to an incredibly complex and interdependent web of life that has been evolving on this planet for at least 3.5 *billion* years. And we're just beginning to understand our place in it. We're beginning to realize how dependent we are on biodiversity for most if not all of our needs—and how a significant loss of biodiversity could seriously undermine our long-term economic, intellectual, physical, and emotional well-being.

We're learning how a number of species, including humans, may depend on certain *keystone* species. Imagine that a species of wild bee has become extinct. The plants that depend on the bee for pollination and the animals that rely on the plants for food could be adversely affected by this loss—and so might the farmers whose crops need the bees for pollination. Also at risk are local economies that rely on uninterrupted and bountiful harvests.

Or consider the Pacific yew, a small understory tree found only in the dwindling old-growth forests of the American Northwest. The yew is a unique natural source of a potent anticancer drug called taxol. Researchers around the world are trying to create a synthetic version of taxol, which has

* Biodiversity is short for biological diversity, the variety of life on Earth, reflected in the variety of ecosystems and species, their processes and interactions, and the genetic variations within and among species.

Keys to Survival

Certain species, called keystone species, play critical roles in ecosystems. Species such as sea otters, beavers, elephants, and mycorrhizal fungi, for example, affect the survival and abundance of many other species in the community where they live. If a keystone species is removed from or added to a natural community, it can change the makeup of the community and sometimes even the community's physical environment.

provided what some call one of the most significant advances in cancer therapy. But if the last of the forests had fallen before the discovery of natural taxol, there might currently be no model for the creation of a synthetic version. And cancer research might not have come as far as it has to date.

What other useful compounds remain undiscovered in wild species living in threatened habitats around the world? Could an obscure grass, now hanging onto survival in North America's remaining patches of tallgrass prairie, contain the genes needed to make corn resistant to a new disease? Or could a fungus found only in a tropical rain forest hold the key to a new antibiotic? What we do know is that we're just beginning to understand the wealth to be found in the life all around us.

If our well-being is so dependent upon biodiversity, why aren't we more aware of it? The answer may be partly due to our loss of intimacy with other living things. As we have moved from walking the land to hurtling past it on paved roads, from hunting and growing our food to buying it at a store, from dipping water from a stream to turning on a faucet, we've lost contact with the natural foundations on which our lives are built. Most of us can now get along just fine without knowing about weather patterns, soil conditions, water sources, or the migration patterns of game.

Some say it's OK to lose our knowledge of nature, that it's not relevant to modern life. But others say our ignorance is catching up with us. They suggest that we may be oblivious to the

incremental—but important—changes in the health of our natural environment, and that we're thus letting biodiversity slip away without realizing the value of what we're losing.

There are mounting fears that an accelerating decline in biodiversity could undermine the ecological systems that support us and all other living species. That's why many scientists, economists, philosophers, business leaders, educators, and health officials are now urging us to think more carefully about how we interact with other living things on the planet. They are challenging us to ask questions such as these:

- How much biodiversity are we losing, and what are the effects of this loss in both the short term and long term?
- How can we use limited natural resources more efficiently and equitably?
- Which communities are most affected by pollution and resource scarcity and why?
- How do the varieties of food crops we eat and the methods we use to grow them threaten biodiversity?
- How do our consumer habits affect biodiversity and resource distribution around the world?
- How is natural resource management tied to poverty, conflict, and other social and political issues?
- What's the status of global fish populations; of world forests; and of the habitats of songbirds, salamanders, and sea turtles?
- What type of education will best prepare tomorrow's citizens and leaders to deal with biodiversity issues?
- What are our moral and ethical obligations, if any, to preserve biodiversity?
- Does a close association with other species and natural systems provide for some essential emotional, intellectual, or other human need?
- Do species have an inherent right to exist?

By asking these types of questions, say scientists and others, we'll start to see more clearly the invisible connections that bind our lives to biodiversity. As we come to understand these connections, perhaps we'll gain a new perspective on our place in the natural world. And in the process, we may learn not only how to better protect the living things on which we depend, but also how to create a more sustainable and responsible society.

> "Biological diversity is the actual stuff of life, encompassing life's myriad adaptations. . . . We do not even know how many species there are in the world, or how they relate to one another, much less how they fit together in the vast tapestry we call life."
>
> —Kenny Ausubel, writer

Our planet is literally crawling, not to mention swimming and flying, with life. Scoop up a handful of soil from your backyard, and you might just scoop up an impressive collection of biodiversity along with it—everything from microorganisms to seeds and spores to insects, mites, and worms.[1] There could be thousands of species right there in the palm of your hand—and many of these might well be unknown to science. The Earth is so rich with life that scientists don't even know for sure how many different kinds of organisms may exist. So far, they've identified and named more than 1.7 million species, including nearly 270,000 species of plants, 950,000 species of insects, 19,000 species of fish, 10,500 species of reptiles and amphibians, 9,000 species of birds, and 4,000 species of mammals. The rest includes mollusks, worms, spiders, fungi, algae, and microorganisms. But scientists think that millions more species, mostly microorganisms and invertebrates, are yet to be discovered.[2]

What Is Biodiversity?

They're All Important

Many scientists use the term "hotspots" when referring to those areas of the world that are not only rich in biodiversity, but also unique and threatened. These hotspots include ecosystems that have the most variety of species, such as rain forests and coral reefs. But they also include ecosystems that have a large number of endemic species and landscapes—those that are found nowhere else on Earth. Islands in particular, both large and small, are rich in endemic species.

While scientists are justifiably concerned about saving unique ecosystems or those with the most diversity, many also agree that it is important to conserve the biodiversity remaining in our most settled or disturbed regions, such as vacant city lots, suburban backyards, and heavily cultivated farmland. The value that biodiversity provides does not decrease in areas where human activity has already taken its toll. In fact, where biodiversity has been lost, the remaining species may become that much more precious. And where we have done the most damage, perhaps ethics demand that we are even more obligated to preserve what is left and restore what was once there.

Three Levels of Life

The sheer variety of *species* on Earth—from microscopic bacteria to blue whales the length of a city block—is pretty impressive on its own. But biodiversity isn't limited to the numbers and kinds of organisms. It also includes Earth's *ecosystems*: its savannas, rain forests, oceans, marshes, deserts, and all the other environments where species evolve and live. And it includes *genetic* diversity, which refers to the variety of genes within a species. All three are critical to understanding the interconnections that support all life on the planet.

Rich, Unique, and Threatened

Certain ecosystems around the world harbor especially large numbers of species. The most familiar of these are tropical rain forests. The forests of New Guinea, for example, are home to about the same number of bird species as the United States and Canada combined—yet the island covers less than 3 percent of that area.[3] Other incredibly diverse ecosystems include coral reefs, large tropical lakes, and parts of the deep-ocean floor.

In general, scientists consider ecosystems with naturally large numbers of species to be among the most important ones to focus on in the effort to conserve biodiversity. But the number of species is just one measure of an ecosystem's importance. Another factor is the uniqueness of an area—from the types of species that live there to the physical landscapes within it. Still another factor is whether the ecosystem performs a key function, such as flood control or water purification. Scientists also consider whether an ecosystem is threatened when determining its biological importance.

BIOFACT

DOWNTOWN DIVERSITY—Even the most bustling cities can be rich with species. In one recent biological survey of a park in downtown Washington, D.C., about 1,000 different species of plants and animals were identified in a single 24-hour period![4]

The Global 200: Setting Conservation Priorities

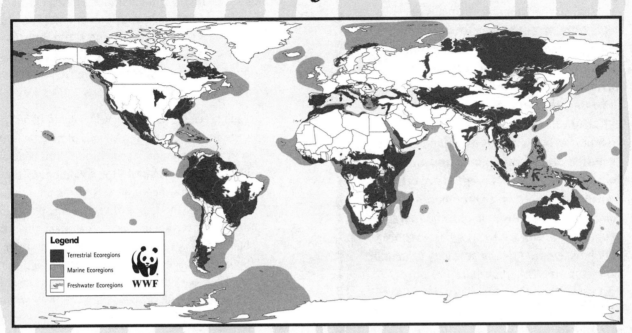

Legend
- Terrestrial Ecoregions
- Marine Ecoregions
- Freshwater Ecoregions

WWF

The above map is an example of how scientists at WWF and its partners around the world have considered a variety of factors—from richness (the number of species) to ecological importance (flood control, water purification, and so on) to uniqueness (species and landscapes that are found nowhere else)—to determine conservation priorities. From the windswept tundra of Alaska's North Slope to the warm tropical forests of the Congo Basin, this map highlights more than 200 of the richest, rarest, and most distinct natural areas on the planet. Together, these areas are part of a comprehensive assessment of the world's biodiversity that WWF is calling the Global 200.

The Global 200 represents a science-based approach to setting priorities for conservation. At its core is a simple concept: If we conserve the broadest variety of the world's habitats, we can conserve the broadest variety of the world's species, along with the ecological and evolutionary processes that maintain the web of life. The goal is to implement a global strategy to protect these biologically distinct areas and the biodiversity they harbor.

Many of these bio-rich areas are in trouble: Nearly half of the Global 200 terrestrial ecoregions are critically endangered and only a quarter are still relatively intact. In addition, many of the marine and freshwater ecoregions are also endangered and face serious threats.

Many scientists believe that the Global 200, which was developed with input from hundreds of experts worldwide, represents an important blueprint for long-term conservation action. WWF is now working with its partners to develop strategies for protecting these ecoregions, with a focus on the need to address biological, social, economic, cultural, and political factors.

For more about this strategy, see "Mapping Biodiversity" on page 252 and the enclosed map of the Global 200. And for more about how other organizations around the world are trying to protect biodiversity, see "How Can We Protect Biodiversity?" on pages 50–61 and the resources on pages 452–469.

Culture and Nature

The diversity of human cultures is an important part of biodiversity. All humans belong to a single species (Homo sapiens), but within our species are thousands of different cultures. And there's evidence to show that the natural world is nearly as important in influencing the development of different peoples as it is in influencing the evolution of different plants and animals. For example, cultural practices are often heavily influenced by environmental conditions. Nomadism among the Tuareg people of the Sahara Desert, for instance, may have evolved as the best way to acquire sparse or unevenly distributed resources. And prehistoric Hawaiians developed a natural cultivation system that protected forests and watersheds. They planted ti and kukui tree crops in the moist upper reaches of the valleys and extensive taro terraces in the flooded, open lowlands. It was a well-integrated mountain-to-ocean flow-through system that made clever use of topography and water resources.

Many of our physical features have also developed in response to environmental conditions. One of the most obvious of these is skin pigmentation: In general, those of us who live (or whose ancestors lived) in areas with more intense solar radiation, such as equatorial regions, have more of the protective skin pigment called melanin than those with ties to temperate or arctic areas. And where we live influences our resistance to certain diseases.

The Gene Scene

Though not as obvious as Earth's impressive array of species and ecosystems, genes—the basic units of heredity—are an equally important aspect of biodiversity. Many plant and some animal species have as many as 400,000 genes—giving rise to enormous possibilities for genetic variation.[5] Dr. Edward O. Wilson, a biologist at Harvard University, describes the importance of genetic diversity in this way:

> Each species is like an encyclopedia of genetic information, containing billions of genetic letters that give it a unique "code of life." This code allows each species to adapt to the conditions of the ecosystem in which it lives. For example, over hundreds of thousands of years, some plants have developed certain chemicals that make them taste bad, which keeps insects from devouring them. Some animals have developed sharp claws, thick fur, keen eyesight, and other adaptations that help them survive. All of these traits are the result of coded messages in our genes that get passed from one generation to the next. And when a species goes extinct, all that valuable information is lost.[6]

> *"A typical mammal such as the house mouse . . . has about 100,000 genes. . . . The full information contained therein, if translated into ordinary-size letters of printed text, would just about fill all 15 editions of the Encyclopaedia Britannica published since 1768."*
>
> —Edward O. Wilson, biologist

For a vivid portrait of genetic diversity, just look around. It's evident in the different colors of apples in the supermarket, in the different eye and hair colors of mammals, and in the different songs and behavior patterns of birds, to name a few examples. The total genetic diversity within a species or a population of that species is its gene pool.

We are still learning about the importance of genetic diversity to the long-term survival of individual species—including how the loss of genetic diversity affects small, isolated populations. But we know that most species probably require a great amount of variety in their gene pools to survive over the long run. It is this variety that allows enough individuals to adapt to unexpected changes in their environment and thus to perpetuate the species. Armed with this knowledge, geneticists and conservation biologists are trying to measure and maintain the genetic variation of some endangered species, including the red-cockaded woodpecker, cheetah, white rhinoceros, golden lion tamarin, and many endangered plants.

Besides being relevant to the survival of wild species, the genetic aspect of biodiversity also has direct links to humans. For example, we know from experience that genetic diversity in agriculture is important in preventing crop failure (see page 33).

©Frans Lanting/Minden Pictures

The Big Picture

It doesn't make sense to think of the three levels of biodiversity—species, ecosystems, and genes—as separate categories, because each level influences the others in significant ways. And if one level is disrupted, the effects can ripple through the others.

The Florida Everglades provides a good example of this "ripple effect." This unique ecosystem once covered more than 3,600 square miles and supported a tremendous diversity of wading birds, mammals, reptiles, insects, and other species.[7] But beginning in the 1930s and 1940s, the Army Corps of Engineers and other agencies built a system of canals, levees, and pumping stations to decrease the risk of flooding from major tropical storms in south Florida and to allow for more farming. As more and more people moved into the area, the Everglades ecosystem shrank and much of the slowly flowing "river of grass" was converted into a series of water management areas connected by canals. Changing the water flow and reducing the size of the wetlands put pressure on many of the species that lived there such as wood storks and Everglade kites. And for some species, such as the Florida panther (a relative of the cougar), the fragments of suitable habitat were so small that only a few individuals could survive in them. With fewer and fewer breeding partners, the panthers began to interbreed, thus reducing the genetic variety within the species. Scientists became concerned that the animals might not be able to survive, so they started importing a closely related subspecies of cougar from Texas. They hope that the "new" cats will add critical genetic diversity to the remaining panther population.

Coming to Terms with Terms

It's easy to get confused about the meanings of ecological terms such as habitat, community, ecosystem, ecoregion, and biome. Since we'll be using all five terms throughout this module, here's a brief clarification of each one:

*The word **habitat** refers to a place that provides whatever a species needs to survive, such as nutrients, water, and living space. "Black bear habitat" and "sagebrush habitat" mean places that support those specific organisms.[8]*

*When talking about an area and all the species that live and interact in it, we use the word **community**. When talking about the sum of interactions between living things in a community and the nonliving components of that community—such as bodies of water, soil, and energy sources—we use the word **ecosystem**. Ecosystems can be very large, such as a desert or rain forest ecosystem. (In fact, some people consider the entire Earth to be an ecosystem.[9]) They can also be very small, such as a pond or rotting log ecosystem.*

*An **ecoregion** is a geographically distinct area (like the Chesapeake Bay, the Sonoran Desert, or the Florida Everglades) that is characterized by climate, vegetation, terrain, altitude, soil type, and the types of communities that exist there. More than 100 ecoregions are recognized in the United States and Canada, and more than 1,000 in the world.[10]*

*Another word used to describe different natural areas of the world is **biome**. A biome is a large region characterized by the type of plant that dominates the area. Temperate grasslands, for example, are a biome found in many parts of the world. Within each biome, there are usually many different ecoregions and thousands of different ecosystems. (Some scientists also use the phrase "major habitat type" in place of biome.)[11]*

Countless Connections

Intricate connections don't exist only among the levels of biodiversity; they also exist among species. These connections are a fascinating and important part of biodiversity. For example, some plants and their pollinators have evolved to be extremely well suited to one another: A long-billed bird pollinates a flower with a long tube and a night-roving moth pollinates a night-blooming cactus. And some insect parasites have evolved to live in only one host, such as a species of wasp that lays its eggs on only one type of caterpillar. Such close partnerships and inter-relationships between species mean that if one species becomes extinct, the other could too.

We're just beginning to understand the remarkable interactions among species. We're also beginning to realize that such interactions may be important in ways we don't yet understand—not only to the species but also to their entire ecosystem. This "argument from ignorance," as writer David Takacs calls it, is made often in discussions of biodiversity.[12] Those who make it claim that our ignorance is one reason the loss of biodiversity is such a serious problem. They believe that, since we don't really know what the ramifications are of losing species and the interactions they are a part of, protecting species and ecosystems is a prudent strategy for all nations.

> "Scientists have only recently discovered a world whose astounding diversity of life holds the promise of new medicines to treat our ailments, a sustainable supply of delectable, healthful foods, and a wealth of understanding unlikely to be matched anytime soon in the sterile vastness of outer space. This new world is called the sea."
>
> **—Elliot Norse, President,**
> **Marine Conservation Biology Institute**

A Watery World

Diversity doesn't end at the water's edge. In fact, an incredible variety of living things inhabits the world's lakes, rivers, wetlands, and oceans. A single coral reef, for example, can support more than 3,000 species of fish and invertebrates, such as giant clams, sea urchins, sea stars, and shrimp. And acre for acre, there can be more life in a healthy wetland than in almost any kind of habitat. Take mangroves, for instance. These rich wetlands, which are found along many of the world's coasts, are nurseries for some of our most commercially important species, including shrimp, crabs, oysters, and fish. They also provide a variety of ecosystem services including flood control, water purification, and food and shelter for a host of aquatic creatures.

Raining Cats

Malaria once infected nine out of ten people on the island of Borneo, now shared by the countries of Brunei, Malaysia, and Indonesia. In 1955, the World Health Organization (WHO) began spraying dieldrin (a pesticide similar to DDT) to kill malaria-carrying mosquitoes. The program was so successful that the dreaded disease was almost eliminated from the island. But other unexpected things happened. The dieldrin killed other insects, including flies and cockroaches inhabiting the houses. The islanders applauded. But then small lizards that also lived in the houses died after gorging themselves on dead insects. Then cats began dying after feeding on the dead lizards. Without cats, rats flourished and began overrunning the villages. Now people were threatened by sylvatic plague carried by the fleas on the rats. The situation was brought under control when WHO parachuted healthy cats onto parts of the island. On top of everything else, roofs began to fall in. The dieldrin had killed wasps and other insects that fed on a type of caterpillar that either avoided or was not affected by the insecticide. With most of its predators eliminated, the caterpillar population exploded. The larvae munched their way through one of their favorite foods, the leaves used in thatching roofs. In the end, the Borneo episode was a success story; both malaria and the unexpected effects of the spraying program were brought under control. But it shows the unpredictable results of interfering in an ecosystem.

Story adapted from: **Living in the Environment, Fifth Edition.** G. Tyler Miller, Jr. (Wadsworth Publishing Company, 1988).

If you were to ask a dozen biologists or sociologists to explain why a high level of biodiversity is important, you'd probably get a dozen different perspectives. But you'd hear one message loud and clear: The quality of our lives depends on it. In this section, we'll explore the many ways biodiversity enriches our lives, and why we're so dependent on it. In the process, we'll focus on some of the complex ecological interactions that make up life on Earth.

"[If our civilization is to have an economy of meaning and purpose], then it must have two agendas. It must serve and nurture the aspirations of the poor and uneducated. And it must also, as its underlying goal, seek to reconstruct, know, or revive genotypes, species, ecosystems, forests, vernal pools, allelomorphs, subspecies, grasslands, seral stages, reserves, natives, gradients, corridors, and habitat blocks."

**—Paul Hawken,
writer, businessperson**

Why Is Biodiversity Important?

The Power of Plants

There's almost no end to the role plants play in the workings of our planet. For example, besides the essential role of giving off oxygen and thus maintaining a breathable atmosphere, plants keep us cooler by releasing moisture through their leaves and by providing shade. They also remove the main "greenhouse gas," carbon dioxide, from the atmosphere. (See page 47 for more about the role of carbon dioxide in global warming.)

But that's not all. Scientists are finding more and more examples of the importance of plant communities to overall ecosystem vitality. In some parts of Africa and the Middle East, excessive cutting of trees for fuel wood and the destruction of grasses and shrubs by goats and cattle is converting more and more land to desert. And in parts of the northwestern United States, loss of tree cover has caused massive mud slides and flooding.[13] These examples show that when plant communities are damaged or destroyed, so too are important biological services and functions that people and other living things depend on.

Life Support Systems

Biodiversity does more than provide a variety of products and resources; it also keeps the planet livable for us and for all other species. Biodiversity helps maintain the atmosphere, keep the soil fertile, purify water, and generally keep the world running smoothly.

Want to see some of these services in action? Pull a weed from your backyard or a vacant lot and use a magnifying glass to take a close look at the soil clinging to the roots. Here, an entire community performs many of the services that help maintain much of the terrestrial life on the planet. Springtails, nematodes, earthworms, and other invertebrates feed on dead plants and animals, decomposing their remains and, in the process, providing the nutrients that other organisms depend on. A battalion of bacteria too small to see converts nitrogen gas—an element essential to life—into a form that the plant can use. And a network of fungi, often visible as thin white threads, forms links with the plant's roots, helping them to take up the nutrients produced by the decomposition process. The plant's roots transfer the nutrients (including the nitrogen from the bacteria) to its leaves and other tissues. And the leaves use the sun's energy to create food in a process called photosynthesis.

BIOFACT

SALTY SOURCE OF OUR SEAFOOD—Salt marshes are probably the greatest nurseries in the world: More than three-quarters of the seafood we eat begins life in these habitats. And each year, more than 800,000 ducklings hatch in salt marshes along the southeast coast of the United States alone.

Eat, Drink, and Be Diverse

It's especially easy to see what biodiversity looks like when you walk into a grocery store. In most produce sections, you'll see several varieties of oranges, apples, corn, and potatoes. A "variety" is a distinct genetic lineage within a species, and the genetic differences usually make one variety physically distinct from another. For example, red delicious, golden delicious, and Granny Smith are varieties of apples with different appearances, textures, and tastes.

Of about 270,000 plant species on Earth, only about 7,000 species (less than 1 percent of all plants) have been used for food by people throughout history. Today, fewer than 20 of those species make up the bulk of the world's food supply.[14] And of all the varieties within these species, very few are commonly used. For example, the National Academy of Sciences found that 50 percent of the land planted with wheat in the 1970s contained only nine varieties, 75 percent of the land planted with potatoes contained only four varieties, and 50 percent of the cotton acreage included only three varieties. And in India, where 30,000 varieties of rice were once grown, fewer than 10 varieties now make up 75 percent of the country's rice production.[15]

Intensive agricultural production of just a few crop varieties, called industrial farming, is a common practice in the United States, as well as in much of the rest of the world. These varieties are attractive to many farmers because they have been bred to produce an optimum yield on mechanized farms. In addition, seed companies generate the most profits by producing and selling just a few varieties. Unfortunately, these high-yield varieties (which often

lack genetic diversity) tend to be more susceptible to destruction by disease, droughts, floods, and insect pests. Thus, more pesticides and water often have to be used to produce them. In addition, relying on only a few varieties can create serious problems: When a pest or disease attacks an individual plant, it can quickly spread through a large field of plants that are all the same variety. Major crop failures, such as the southern leaf blight outbreak in the U.S. corn crop in the early 1970s and the peanut crop collapse in 1990, can be attributed to a lack of genetic diversity.[16]

> *"We are shaped by the Earth. The characteristics of the environment in which we develop condition our biological and mental being and the quality of our life. [If only] for selfish reasons, therefore, we must maintain variety and harmony in nature."* **–René Dubos, bacteriologist**

For the reasons outlined above, many farmers and agronomists are beginning to recognize the value of growing and maintaining different varieties of crops and seeds. Many are also trying to find economical alternatives to industrial farming.

Native American and other traditional farming communities are often invaluable sources of genetic diversity for agricultural plants. In these communities, many varieties of native seeds have survived and have been grown and nurtured for generations.

Protected wild places can also be good sources of genetic diversity for crops, since wild relatives of our domesticated plant varieties often still exist in these places. By crossbreeding these wild species with our most popular domesticated crop plants, agronomists can increase crop yields and improve nutritional content, as well as enhance disease and pest resistance. Biodiversity protection, then, has important implications for agriculture: If we protect the places that harbor wild plants, we'll be better able to ensure a continued, reliable source of new genes for agriculture.

The Link to Human Health

- *Understanding why bears do not lose bone mass during their long winter dormancy could lead to treatments for osteoporosis, a condition that costs the U.S. economy more than $10 billion in direct health care costs and productivity loss each year. But time may be running out for researchers: Many bear species are endangered.*

- *In Ghana, where the tropical forest has been reduced to 25 percent of its original size, approximately 75 percent of the human population depends on wild game to supplement its diet. Forest depletion, along with subsequent reduction in game populations, has resulted in a sharp increase in malnutrition and disease.*

- *More than 4.5 billion people—some 80 percent of the world's population—still rely on traditional plant- and animal-derived medicines as their primary source of health care.*

- *Some people have theorized that the loss of tropical forests in Africa, along with other habitat disruptions, is a contributing factor in the emergence of serious viruses and other contagious disease agents in recent years.*

Adapted from Biodiversity, Science, and the Human Prospect by the Center for Biodiversity and Conservation at the American Museum of Natural History, 1997.[17]

A Diverse Prescription

Open your medicine cabinet, and you'll likely see a number of products derived from wild plants and animals. In fact, more than 25 percent of the medicines we rely on contain compounds derived from or modeled on substances extracted from the natural world.[18] About 120 useful prescription drugs are derived from 90 species

of plants alone. And many of our best drugs for treating cancer are derived from plants.[19] The Pacific yew tree, once burned as trash in forestry operations, produces compounds found to be effective in treating ovarian, breast, and other types of cancers.[20] Animal-derived products are also important in medical treatments. For example, calcitonin, a hormone used for the treatment of osteoporosis, and protamine sulphate, an important medicine used in open heart surgery, both come from salmon. But microorganisms may well be the most well-represented species in medicine cabinets around the world: More than 3,000 antibiotics, including penicillin and tetracycline, were originally derived from these tiny life forms.

As each of the above examples points out, all species—even those that seem "worthless," like the Pacific yew—have the potential to provide us with useful or even life-saving products. Many people feel that this alone is a strong argument for conserving biodiversity.

Searching for plants with medicinal qualities has become a major pursuit of some of the world's largest drug manufacturers. Fortunately, they're finding that they don't have to start their searches from scratch: Human communities that have

A CLEAN WAY TO CUT COSTS—New York City recently avoided spending $6–8 billion on filtration facilities by investing $1.5 billion to protect the watershed lands around its upstate reservoirs. Keeping land undeveloped around water sources may be the key to clean, inexpensive drinking water.

traditionally used plants for medicinal purposes can provide a wealth of botanical knowledge. Scientists estimate that about 24,000 plant species are used as remedies by traditional cultures throughout the world.

When researchers interview shamans and other healers, they learn which plants are considered medicinal, and for what purposes. Then they try to isolate the active compounds in the laboratory. Many biologists and sociologists are working to ensure that local communities earn income and other benefits from their medicinal plants and their knowledge of these plants—something that hasn't always happened in the past. Ensuring compensation is especially important when local plants are used to create new drugs that may be highly profitable to the pharmaceutical industry.

Biodiversity and the Bottom Line

A huge number of products derived from wild species help to boost all levels of our economy. In fact, many businesses and manufacturers have found that biodiversity can turn a hefty profit. Here are a few examples:

- A species of wild tomato discovered in Peru in 1962 was worth an estimated $8 million a year to commercial makers of tomato paste. This newly discovered species had genes that produced twice the sugar content of any other tomato. After 20 years of breeding, using genes from the wild species, a new hybrid tomato was produced that was much better suited for canning purposes than the original canning tomato.[21]

- Sales of prescription drugs that contain ingredients extracted or derived from wild plants totaled more than $15 billion in the United States in 1990.[22]

- One of the most important new products of the multi-billion-dollar biotechnology industry is an enzyme that can withstand high temperatures. This enzyme has the potential to advance the development of new medical procedures, space travel technology, and fire-resistant materials. It's derived from a bacteria that's found only in certain hot springs of Yellowstone National Park.[23]

- Each year, more than 350 million people visit our national parks, wildlife refuges, and other public lands managed by the Department of Interior. This visitation generates more than 400,000 jobs and more than $28 billion of economic activity.[24]

- Certain types of bacteria make nitrogen available for use by crops, pastures, forests, and natural vegetation. Economists estimate that the value of this activity is $33 billion annually.[25]

- More than 40 crops produced in the United States, valued at approximately $30 billion, depend on insect pollination.

- Bees, butterflies, birds, bats, and other animals pollinate 75 percent of the world's staple crops and 90 percent of all flowering plants.[26]

- The dollar value of services provided by ecosystems throughout the world is estimated to be $33 trillion per year. (The value of all human-produced goods and services per year is about $18 trillion.)[27]

The Nature Connection

As our pace of life becomes more hectic, more of us are finding that the natural world—with all its diversity—offers irreplaceable opportunities to relax and rejuvenate ourselves. Many psychologists and biologists go so far as to say that for most of us there's a deeply ingrained *need* to connect

A Sampling of Free Services from the Wild

- *Through photosynthesis, green plants remove carbon dioxide from the atmosphere and replenish the supply of oxygen.*

- *A diverse group of bacteria convert atmospheric nitrogen into a form that plants can use.*

- *Phytoplankton in the oceans are the foundation for marine food chains and help regulate global atmospheric cycles.*

- *Dead and decaying organisms are decomposed and recycled by fungi, bacteria, insects, and other scavengers and decomposers.*

- *The flow and quality of water within watersheds are regulated by the biodiversity within the area's soils, waterways, and upland and wetland living communities.*

Adapted from Biodiversity, Science, and the Human Prospect by the Center for Biodiversity and Conservation at the American Museum of Natural History, 1997. [30]

with nature and other species. They argue that this need may be more important than we realize to our mental well-being and to our success as a species—especially as we become more urbanized and disconnected from natural things. This argument is called the *biophilia hypothesis.* Biophilia means "affinity for living things." (Some say that the enormous popularity of pets around the world is evidence of biophilia. In the United States alone there are more than 58 million pet dogs and 66.2 million pet cats.[28]) Although there is as yet little experimental or clinical evidence to support the idea of biophilia, a growing number of people believe that this hypothesis will lead us to learn more about our needs as biological beings.[29]

The desire of many of us to connect to the natural world is very easy to see. Visit a park, nature center, zoo, or wildlife refuge on a beautiful day and you'll see people hiking, fishing, bird watching, boating, and enjoying other forms of nature-based recreation. Drive through the mountains of New England during the fall, and you may find yourself in a traffic jam caused by people flocking to see the brilliant colors of maple, dogwood, and sumac trees. Or just walk through a neighborhood on a summer Saturday and count the number of people tending flowers and vegetable gardens. Although social scientists are still in the early stages of learning about the links between nature and the human mind, we do know that wildlife and wildlands provide a sense of wonder, fascination, and rejuvenation for millions of people, old and young, around the world.

All Species Count

Many people believe that biodiversity should be preserved not just because it is valuable to us in some way, but simply because it exists. Some people who hold this opinion believe that each species should be respected and protected because it is the product of many thousands or millions of years of evolution, and we have no right to interrupt the evolutionary process. They also argue that we have no right to destroy something we did not create and that future generations deserve a natural world that is rich and varied. Because we have the power to destroy species and ecosystems, they say, humans have a moral obligation to be careful stewards of the Earth.

Our children will inherit the planet with whatever biodiversity we pass on to them. The decisions we make as individuals and as a society

today will determine the diversity of genes, species, and ecosystems that remain in the future. Many of these decisions are not easy, especially when they involve balancing the immediate needs, rights, and desires of individuals and communities with the measures necessary to protect nature for the long term. Because people have different values and desires, conflicts over the conservation of biodiversity often develop. Understanding what biodiversity is and how different people value it is an essential first step to designing strategies for long-term conservation.

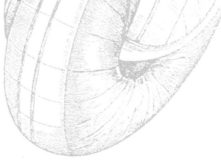

Biotechnology and Biodiversity

The pursuit of profits through biotechnology raises many questions—from who owns the economic rights to products derived from wild species to whether releasing new species into areas where they aren't naturally found would harm native species already living there.

Many experts feel that these questions have not been addressed adequately and that much more work and thinking must take place to ensure that people and natural environments are protected.

> "Biodiversity is out there in nature, everywhere you look, an enormous cornucopia of wild and cultivated species, diverse in form and function, with beauty and usefulness beyond the wildest imagination. But first we have to find these plants and animals and describe them before we can hope to understand what each of them means in the great biological—and human—scheme of things."
>
> —Hugh H. Iltis, botanist

Many scientists believe that we are now beginning to witness a spasm of extinction—the greatest since the dinosaurs and many other organisms died out some 65 million years ago. Some scientists estimate that the current extinction rate is more than 1,000 times the natural rate of extinction. They're very concerned about the long-term effects of such rapid and massive extinction—not only on natural communities, but also on the planet's ability to sustain large numbers of humans.

It's no secret that human activity is the cause of most of the biodiversity loss we're experiencing.[31] Such a connection seems straightforward enough: Our population and rate of resource consumption are growing, causing increasingly serious problems of pollution and habitat destruction. But the relationships between human activity and biodiversity loss are often subtle and complex. This chapter explores some of these relationships, focusing not only on their biological and ecological aspects, but also on their political and social dimensions.

What's the Status of Biodiversity?

The Nature of Change

There's nothing unnatural about the process of extinction or, for that matter, about disturbance to Earth's ecosystems. More than 99 percent of all species that ever existed have become extinct over a period of more than 3.5 billion years of evolution, and ecosystems have expanded or receded in the wake of climate changes and natural disturbances.[32] For example, areas of tropical forest in Africa and South America gradually gave way to savanna during climate changes associated with the ice ages. And as this gradual process occurred, many forest animals and plants died out while others evolved into species more suited to the changing habitat.

Sometimes relatively rapid rates of extinction, called spasms, are brought on by catastrophic natural events. Many scientists believe that such an event—in the form of an asteroid slamming into the Earth—might have caused the demise of the dinosaurs.[33] They theorize that the asteroid's impact may have created a massive cloud of atmospheric dust and vapor that blocked the sun long enough for many plants to die. This event set off a chain of extinctions (including that of the dinosaurs) that forever changed the composition of Earth's fauna and flora.

Five mass extinctions have punctuated the history of life on Earth.[34] There are some important differences, however, between these extinctions and the one many scientists say we're facing today. One difference, according to conservation biologists, is that today's extinction spasm is occurring more rapidly than those of the past—especially for plant species. Another difference scientists point to is that the potential for species and ecosystems to recover from this episode of extinction is probably far lower than with past episodes. That's because, in the past, certain species have escaped extinction by surviving in pockets of habitat that happened to be relatively unaffected by natural catastrophic events.[35] (Certain tropical areas and islands like New Zealand and New Caledonia are good examples of such pockets.) Species that survived a global extinction episode in these refuges were able to gradually recolonize surrounding areas when conditions became more favorable. But today, many potential refuges are being destroyed or degraded by human activity. And many scientists worry that even if conditions do eventually improve, we have already lost too much habitat and genetic diversity to allow ecosystems to recover.[36]

Of course, the overarching difference between extinctions of the past and those of the present can be summed up in one word: humans. Never before on our planet has one species been responsible for such a massive decline of global biodiversity.[37]

Losing Ground

The loss of habitats—the places where organisms live and get the nutrients, water, and living space they need to survive—is the primary reason biodiversity is in decline.[38] When people cut down a forest, fill in a wetland, trawl a seabed, or plow a prairie, they change the natural habitat of the species that live there. Not only can such changes kill or force out many animals, microorganisms, and plants; they also can disrupt complex interactions among species.

Often, habitats aren't so much destroyed outright as they are chiseled away little by little. When not carefully planned, roads, housing developments, agricultural fields, shopping centers, and other modifications fragment natural areas, forming a patchwork of habitat islands in a sea of development. For some species, this "sea" is a barrier that isolates them from sources of food or water. It can also isolate species from others of their kind, resulting in inbreeding and a loss of genetic diversity.

Habitat fragmentation can cause other problems too. For example, when a highway is built through a forest, areas that were once in the forest's interior become exposed to more light, wind, and temperature fluctuations. Animals that are adapted to conditions inside forests often can't tolerate these changes and are forced to move elsewhere, if they can. And plants, which can't just pick up and move, may die out.

In the long run, habitat destruction can disrupt human communities as well as those of plants and animals. Dams on rivers in the Pacific Northwest have produced inexpensive electricity and have redirected water for agriculture—but they've also interrupted salmon migrations, drastically lowering the number of salmon that have made it to their spawning grounds. As a result of the dams and other changes, wild salmon populations have dropped dramatically, causing economic hardship for those who depend on the salmon for their livelihood.[39]

©Paul Chesley/Tony Stone Images

"The hemorrhaging of so much life on Earth has principally resulted from widespread habitat destruction. The reasons for this loss are many: large-scale farming, mining, forestry, grazing, water impoundment and diversion, various forms of urbanization and industrialization, road and highway construction, and more. All these forces . . . have been largely the consequence of increasing human numbers, technology, energy use, and per capita consumption of space and materials."

—Stephen R. Kellert, writer, professor

Vanishing Along with Their Habitats

While we have a pretty good idea about how many mammals, birds, and other vertebrates are threatened or have already become extinct, we have no idea how many species of invertebrates and microorganisms have been or are being lost. What we do know, mainly through recent studies in the tropics, is that vast numbers of unidentified species of invertebrates and microorganisms are adapted to and therefore dependent on localized habitats.[40] Since these habitats are being wiped out by logging, farming, and other human activities, many scientists believe that scores of species, unknown to science, are being extinguished in the process.

Here are some facts that help put habitat loss into focus.

- *Less than 1 percent of North America's original tallgrass prairie ecosystem remains.[41]*

- *More than one-half of the original wetlands in the United States have been lost or severely degraded in the last 300 years as a result of draining and filling.[42]*

- *More than 95 percent of the lower 48 states' original primary forests are gone. The largest areas of primary forest are in the Pacific Northwest, where about 10 percent of the original forests remain.[43]*

- *The Yellowstone River is the only large U.S. river (longer than 600 miles) that is not severely altered by dams.[44]*

- *Between 90 percent and 98 percent of the nation's rivers are degraded enough to be unworthy of federal designation as wild or scenic.[45]*

- *In Arizona, about 90 percent of river and stream (riparian) habitat has been destroyed by dams, conversion to farmland, excessive pumping of groundwater, cattle grazing, and urban development.[46]*

A Growing Concern

In 1998, the number of people on Earth reached an estimated 5.9 billion—more than twice as many as in 1950. And about 88 million more people are added to the planet each year. This annual addition, equivalent to the population of Germany, amounts to almost a quarter of a million births every day. Most experts think that the human population is likely to stop growing sometime around 2200, when the number of people in the world is expected to reach at least 10 billion.[47]

With a population of around 268 million, the United States is the world's third most populated country. (China and India are first and second, respectively.) The U.S. Census Bureau projects that the nation's population will rise to more than 390 million by 2050, which is equivalent to adding a city with the population of Chicago (approximately 2.7 million people) to nearly every state.[48]

There is little doubt that our growing population will result in continuing habitat loss and will put enormous pressure on Earth's natural resources. Resources, such as certain minerals and other nonrenewable resources, could someday become depleted. While some people argue that new technologies could indefinitely extend the use of nonrenewable resources and allow the use of alternative ones, many others feel that technology can only delay the time when the Earth loses its ability to sustain us.

Renewable resources, such as trees, won't necessarily run out, but unmanaged use of them can lead to other problems. For example, increasing demands for paper and other wood products could lead to increased plantings of monocultures—stands of a single species of tree. Planting monocultures is an efficient way to ensure a steady supply of wood for wood products, and may thus allow large areas of high biodiversity, such as tropical rain forests, to be spared. But monocultures can be less resistant to disease than more diverse forests, which often means that they need to be sprayed with more pesticides. They also may require high levels of chemical fertilizers to remain productive. And perhaps most importantly, they generally support very little biodiversity.

World Population Growth

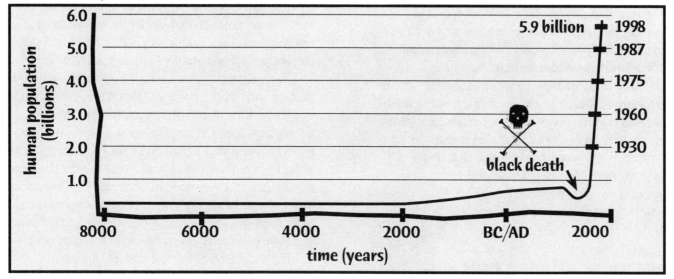

Human Need and Human Greed

Population growth alone doesn't account for the increasing consumption of natural resources that is largely responsible for biodiversity's decline. Patterns of affluence and poverty also have a huge impact. For example, those of us living in affluent, industrialized nations, such as the United States, consume a disproportionate amount of the Earth's fossil fuels, forests, and other natural resources. And in less-industrialized areas of the world, people struggling to survive often have little choice but to overuse the few resources available to them.[49]

A few statistics can shed light on the connection between economic status and biodiversity. First, consider the consumption patterns of affluent nations, using the United States as an example. The United States uses 3 times as much iron ore, 3.6 times as much coal, and 12 times as much petroleum as does India, but it has one-third the population. And a typical Chinese household uses less than 0.03 percent of the energy consumed in the average American home.[50] While many people point out that resource use helps fuel economic growth, others state that the consumption rates of affluent nations run the risk of depleting resources more rapidly than natural processes can replace them. Also, under current consumption patterns, affluent nations are benefiting disproportionately from our planet's

limited resources—sometimes at the expense of developing nations. Consequently, some people argue that affluent nations have a responsibility to not only dramatically reduce their consumption of natural resources, but also to help less industrialized nations achieve their economic goals without making the same mistakes as the developed nations have made.[51]

Now consider the connection between poverty and biodiversity loss. Of the 5.9 billion people in the world, about 4.7 billion live in developing nations. And in many of these nations, a handful of wealthy individuals owns a disproportionate percentage of the arable land and other natural resources.[52] The result is that large numbers of poor people are often forced to sustain themselves on land that's not well suited to the task. This situation often degrades the land to such an extent that it can no longer support humans and certain other species.

Cycle of Destruction

The destruction of tropical rain forests in South America is one example of how poverty and overpopulation can lead to biodiversity decline. For decades, poor farmers living in Brazil and other South American nations have been chipping away at their forests by clearing land to grow crops. (Much of the good farmland in these nations is unavailable to them because it's owned by an affluent minority.) But rain forest soil is typically shallow and nutrient poor. It doesn't hold nutrients for long; those that are not quickly absorbed by tropical vegetation are leached out of the soil and into streams by the heavy rainfall. Thus, farmers trying to grow crops in rain forest soil can use the area for only a few years before the soil is depleted. And when this happens, the farmers must move on and clear another patch of forest.

Up to a point, the land can accommodate this kind of farming: Historically, nearby species were able to recolonize small patches of spent farmland once the farmers moved on. But the numbers of poor farmers has greatly increased, and so has the number of acres of degraded land. Tropical forests have not been able to regenerate in many areas because the disturbance has been too severe and widespread. As a result, large areas that once supported lush rain forests are now incapable of supporting much more than a few hardy grasses.[53]

Many similar examples of poor land use exist in the United States. In the Great Basin region of the western United States, for instance, too many cattle have overgrazed the land to the point where, in certain areas, native grasses and other species can no longer thrive. Non-native weeds have invaded the area, resulting in decreased diversity of native plants and animals. And in California's Central Valley, over-irrigation has caused salt to build up in the soil. This has resulted in decreased soil productivity. Some areas, in fact, have become incapable of supporting much life at all.

Calculating Human Impact

According to population scientist Paul Ehrlich, an equation can help us understand the impact that human population growth has on the global environment. The equation is $I = PAT$, which stands for: Impact = Population x Affluence x Technology.[54]

This equation tells us why, for example, affluent nations such as the United States have such a huge impact on the environment, even though their populations aren't as large as those of less affluent nations. (The A and T multipliers for each person in an affluent nation are high.) It also tells us why the large populations of developing nations, such as China and India, are so damaging. (When the P multiplier is so high, even small amounts of technology and affluence can have enormous environmental consequences.)

The total environmental impact of human populations can be lowered, says Ehrlich, by decreasing any of the equation's variables. But he adds that any lowered impact would be offset if one or more of the other variables were to increase.

Others feel that the benefits of affluence and technology should not be overlooked. For example, Mark Sagoff of the University of Maryland argues that affluence generates the very kinds of research and high technology that can help preserve biodiversity.[55] Genetic engineering, for example, could lead us to highly productive crops that require little use of pesticides and chemical fertilizers, while advances in solar cell technology could allow the cheap and efficient conversion of sunlight into electricity, thus helping to conserve fossil fuels and limit pollution.

Here's a closer look at each variable in the equation:

I = impact of any human group on the environment

P = number of people

A = a measure of the average person's consumption of resources (also an index of affluence)

T = an index of the environmental disruptiveness of the technologies that provide the goods people consume

Born Consumers

From birth, people from highly industrialized nations are rampant consumers, using far more resources per person than residents of other nations. Most highly industrialized nations consume far more than their fair share of resources. For example, highly industrialized nations, which make up approximately 20 percent of the world's population, account for the lion's share of total resources consumed:

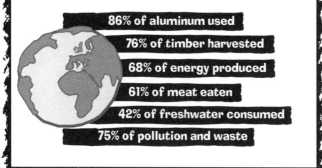

- 86% of aluminum used
- 76% of timber harvested
- 68% of energy produced
- 61% of meat eaten
- 42% of freshwater consumed
- 75% of pollution and waste

Environmental Injustice

Many people point out that problems related to environmental destruction and biodiversity loss often hit hardest in poor communities around the world. For example, toxic waste dumps in the United States are most often located in low-income (and often minority) communities. Toxic waste is also frequently shipped to developing countries, whose governments benefit from the disposal fees paid by the waste producers—often without informing local communities about the hazards of these materials.[56] In effect, then, communities and nations with greater wealth often transfer the environmental costs and risks of their activities to those who are less well off.

Wildlife for Sale

Trade in wild animals and plants is a major threat to biodiversity worldwide. Annually, such trade—which encompasses not only wild species themselves, but also products made from them—is worth more than $10 billion. In the United States, which is the world's largest consumer of wild species, we buy as many as 12,000 primates, 2.5 million orchids, 200,000 live birds, 2 million reptiles, 250 million tropical fish, and millions of wildlife products each year.[57]

Although most wildlife and wildlife products that are available for purchase in department stores, pet stores, or specialty catalogs are legal imports, about one-quarter of this trade is illegal. Driven by consumer demand, people unlawfully take thousands of rare and endangered plants and animals from their native habitats each day. Wildlife trade threatens populations of plants and animals around the world, and is especially devastating to those species—such as rhinos, tigers, and pandas—that are already vulnerable because of habitat loss, pollution, and other problems affecting their environment.

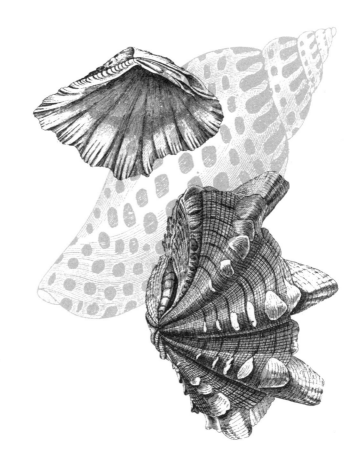

CO$_2$: An 'Innocent' Gas Goes Bad

At first glance, carbon dioxide (or CO$_2$) seems harmless enough. After all, it's nontoxic to people and other living things except in extremely high concentrations. But CO$_2$ could turn out to be one of the most dangerous pollutants of our time. That's because it's one of the greenhouse gases–gases that act like the glass in a greenhouse to keep heat trapped near the Earth's surface.

Ironically, this "greenhouse effect" is, in moderate doses, essential to life. Carbon dioxide occurs naturally in our atmosphere–and without a certain amount of this and other greenhouse gases, Earth would be a frigid and potentially lifeless rock. But people have changed the levels of atmospheric greenhouse gases, particularly CO$_2$. We emit these gases into the atmosphere whenever we burn oil, coal, and other fossil fuels, as well as when we burn forests to make way for agricultural land. All of these activities have increased since the beginning of the industrial era, and so have atmospheric concentrations of CO$_2$.[58]

Scientists don't yet know exactly what effects the growing levels of atmospheric CO$_2$ will have. But many computer models now predict that increased concentrations of CO$_2$ and other greenhouse gases will cause an increase in the overall global temperature, a rise in the sea level, and a change in climate patterns. Most data shows that such changes are already under way.[59]

Even a small increase in the overall global temperature could adversely affect biodiversity–especially since the change will probably happen too quickly for many species to adapt to the new conditions or migrate to more suitable habitats.[60] For example, a rise in the sea level and coastal flooding could destroy essential feeding grounds for migratory birds. And the melting of ice and permafrost in the Arctic tundra would change the habitat of polar bears, walruses, and other Arctic species. In parts of Antarctica, rising temperatures are already melting sea ice, thus limiting the areas where Adelie penguins can breed.[61]

Global warming is likely to most severely affect endangered species and ecosystems, which already have a tenuous hold on existence. But it could benefit certain other species and ecosystems. For example, some species might be able to take advantage of the warming temperatures by spreading into areas that were previously too cool for them. Of course, what's good for one species isn't necessarily a boon to others, or to the ecosystems involved. For example, tropical mosquitoes could be one of the beneficiaries of global warming. And as these insects spread into new areas, they'd most likely bring with them the malaria, yellow fever, and other tropical diseases that they routinely carry.

Although most scientists studying the issue agree that global warming is occurring, people are having a harder time agreeing on what to do about it. The majority say we should cut CO$_2$ emissions now, across the globe. Others say we should do more research before implementing CO$_2$-cutting policies that could jeopardize the economic health of some nations and bring hardship or at least inconvenience to millions of people. But many believe that if we cut CO$_2$ now, we won't necessarily face economic hardships. In fact, some say we could reap many economic and health benefits. The debate is likely to continue for sometime to come–but in the meantime, more and more scientists, economists, insurance companies, and others warn that there will be severe consequences if we let current practices continue much longer.[62]

Carbon Dioxide Concentrations in the Atmosphere

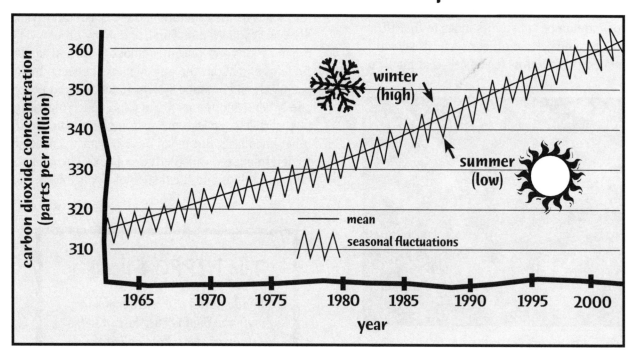

Oil in the Ocean; Gases in the Air

Whether they're renewable or nonrenewable, it's safe to say that the more resources our growing population consumes, the more pollution we're likely to create. Pollution's effects on biodiversity can be obvious, such as an oil spill that causes populations of sea life to plummet in the area of the spill, or forests and lakes that have been damaged by acid rain.[63] But pollution can also cause problems that, though not visibly dramatic or immediately apparent, are every bit as deadly as an oil spill. Insecticides, for example, can cause reproductive failure in fish, birds, and mammals—and research is starting to show that these substances have the potential to cause problems in people.[64] Other insidious pollutants include gases released into the atmosphere by industry (such as in electric power production and manufacturing) and by everyday activities—the most damaging of which are probably driving cars and using electricity.[65]

Many predict that the public debate about global warming and our dependence on fossil fuels will heat up in the next decade—especially as we learn more about how global warming affects biodiversity.

Chemical Connections

A growing number of studies indicate that DDT, DDE, PCBs, and certain other chemicals known as *endocrine disruptors* (chemicals that mimic or

interfere with normal hormones in living things) may be affecting biodiversity in subtle ways by causing a host of developmental, reproductive, and behavioral problems. For example, researchers have found gulls in the Great Lakes region with both male and female reproductive organs. Gulls feed on fish, and research has shown that fish in some areas of the Great Lakes contain high levels of endocrine-disrupting chemicals. Scientists have demonstrated a connection between the birds' diet and the presence of abnormal sex organs.[66]

Terns off the Massachusetts coast and salmon in the Great Lakes have also been found with both male and female organs. And male alligators living in a number of pesticide-contaminated lakes in Florida have been found with genitals so stunted that the animals were unable to reproduce. In addition, scientists have found that the fetuses of some animals are particularly sensitive to chemicals at certain times.[67]

The Environmental Protection Agency, the National Wildlife Federation, the U.S. Fish and Wildlife Service, World Wildlife Fund, and many other organizations are concerned about these and similar findings and about the possible links between toxic chemicals and immune system functioning in people and other living things. There's also concern about how certain chemicals may act in combination with other chemicals. But much more research is needed to determine the varied effects of toxic chemicals on living things and the degree of risk involved.

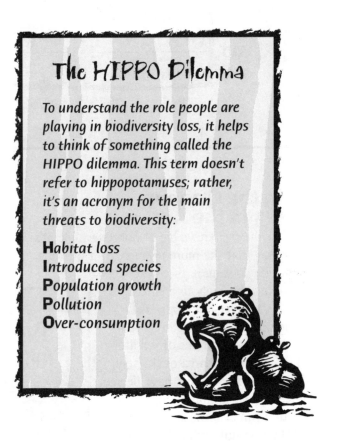

The HIPPO Dilemma

To understand the role people are playing in biodiversity loss, it helps to think of something called the HIPPO dilemma. This term doesn't refer to hippopotamuses; rather, it's an acronym for the main threats to biodiversity:

Habitat loss
Introduced species
Population growth
Pollution
Over-consumption

Meeting Human Needs

Many biologists, economists, historians, sociologists, and others agree that any plans to preserve biodiversity on a global scale must go hand in hand with addressing such critical issues as human poverty, hunger, and health. They warn that failing to do so will ultimately limit the effectiveness of even the best-laid conservation plans.[68]

The Sum of the Parts

Many scientists believe that certain negative trends, such as the increasing incidence of both infectious and non-infectious diseases, mounting tensions between different groups over land and resources, and erratic weather patterns can be traced to our species' collective impact on the planet. [69]

Aliens Are Among Us!

Another contributor to biodiversity decline—and one that is sometimes an indirect result of population growth and consumption—is introduced species. When people introduce organisms into new areas, either intentionally or accidentally, the organisms can take a toll on native plants and animals. Also called *alien*, *invasive*, or *exotic* species, introduced species often have no natural predators or diseases in their new homes. And native species often have no defenses against the introduced species or the diseases they carry. As a result, introduced species may thrive—often at the expense of native species. Examples of damaging species that have been introduced into the United States include starlings, kudzu, the chestnut blight fungus, purple loosestrife, Japanese honeysuckle, zebra mussels, gypsy moths, fire ants, and wild pigs.

Occasionally, an introduced species will have a beneficial effect on its new home. For example, the introduction of dung beetles to Australia—which had no dung beetles of its own—has mitigated the problem of too much cow manure on ranches. Overall, however, introduced species probably cause more harm than good. And their impact is often not limited to their effects on native plants and animals. Zebra mussels (a kind of freshwater mollusk) are a prime example of this. Accidentally introduced into the Great Lakes in the 1980s when ballast water was drained from a foreign ship, zebra mussels spread rapidly throughout much of the northeastern and midwestern United States. Today they're clogging the intake valves of pumping stations, power plants, and industrial facilities; impeding navigation and boating; and interfering with sport fishing. They're also damaging local ecosystems and contributing to the decline of native freshwater species.

Most experts agree that the accidental introduction of species should be minimized and that deliberate introductions (such as the release of non-native insect predators to control pests) should be undertaken only after careful planning and study of the possible effects. But even under the best conditions, no one can predict with certainty what the impact of introducing an exotic species will be.

The Fall of Native Fish

The decline of many freshwater fish species in North America can be traced to the introduction of non-native species. For example, a study of the Salt River in Arizona in 1900 identified fourteen native and no non-native fish species. Another sample, in 1920, revealed seven native species and two introduced fish species. By 1950 the native fish had disappeared altogether. In their place were some twenty invasive fish species.[70]

One of the greatest challenges we face in protecting biodiversity is how to balance the needs of the present without jeopardizing those of the future. We're finding that there's no one way to address this challenge —in part because there's no one reason that we're losing biodiversity. Ensuring the survival of species, genes, and ecosystems will require a combination of approaches, and the collective thinking of people from all disciplines and backgrounds.

"If we want to pass on a living planet to our children and our children's children, then we need to embrace the idea that our actions matter, that the decisions we make—as individuals and as institutions— reverberate throughout the web of life and endure through time."

—Kathryn S. Fuller, President, World Wildlife Fund

How Can We Protect Biodiversity?

Tourists to the Rescue?

In some areas, ecotourism may help protect biodiversity by creating incentives to protect it. Organized "ecology-sensitive" trips to natural areas have been on the rise for the past few years, as have the number of people participating in such trips. From bird-watching tours in Arizona to elephant research trips in Zimbabwe, ecotourism often boosts local economies by creating a demand for accommodations, food, and native guides while taking care to minimize the impacts on local ecosystems. [71]

Any form of tourism has its drawbacks, though. For example, more visitors to natural areas can result in more pollution and disturbance to animals and plants. And if enough of the income generated from ecotourism doesn't end up in the hands of local people, they have little financial incentive to continue to protect the areas that tourists come to see.

Is the Sustainable Obtainable?

Increasing pollution and consumption of natural resources are clearly contributing to biodiversity decline and could eventually threaten large segments of the human population. (People in areas where natural resources have already been seriously depleted, such as Nicaragua and Honduras, are now suffering the consequences.[72]) Nobody can avoid using resources and creating waste—it's a necessary part of being alive. But decreasing our impact on the planet can go a long way toward ensuring that humans and other species can thrive indefinitely.

Attaining such a state of relative equilibrium with our environment is often described as achieving *sustainability.* A society that reaches sustainability is one that is able to persist for many generations without producing significant amounts of pollution, depleting natural resources, and causing a decline in biodiversity. According to many experts, most of the world's societies are far from sustainable.[73] They also point out that efforts to achieve sustainability must address not only the conservation of biodiversity and other natural resources, but also issues of economic security and social equity.[74] Obviously, we have a long way to go in understanding what a sustainable world would look like and what is required to get us there.

Different Points of View

Many organizations and governments have attempted to develop projects or policies that could help us take steps toward sustainability. These projects or policies have often dealt with transportation, water use, community design, and energy use. For example, the President's Council on Sustainable Development has offered a set of goals designed to promote economic health, biodiversity protection, and equality and justice for all people living in the United States.[75]

BIOFACT

PROTECTING WILD SPACES—Today only about 5 percent of the Earth's land area is designated as parks or reserves, and much of that is weakly protected.[76] Many conservation organizations would like to see at least 10 percent of the world's natural areas, including marine ecoregions, protected for the future.

But not everyone believes that it's necessary to develop policies specifically designed to promote sustainability. Some economists, for example, are concerned that such policies will unreasonably limit or direct economic growth and interfere with the free market system. They feel that special policies are unnecessary because, in their view, market forces will naturally encourage the development of new technologies that will solve our environmental problems—including those related to biodiversity loss. Proponents of this view believe that the combination of human ingenuity and unfettered markets can prevent us from ever running out of resources.[77]

Other experts feel that it's wishful thinking to expect that ingenuity and market forces will solve future environmental problems—especially if we deplete scarce and nonrenewable resources. They feel that the decisions we make today about natural resources will have serious economic consequences for future generations, and that we're being reckless to use up the "natural capital" that fuels the market in the first place. In addition, many people feel that it's unwise to place so much faith in a market system that does such a poor job of recognizing the value of species and ecosystems beyond the role they play in providing raw materials for human consumption.

Land for the Future

Many people point out that, in addition to focusing on long-term sustainability, we must also engage in shorter-term projects that address the immediate needs of dwindling species and ecosystems. One such approach is to set aside more land for wildlife. However, protecting large amounts of land from all human use is often impractical or impossible, particularly as the global population continues to expand. That's why increasing attention is being paid to improving habitats where large and growing numbers of people already live. (See "Bringing Back the Habitats" on page 55.)

As urban areas around the world continue to expand, leaving room for other species and ecosystems will require careful planning. Landscape architects and land-use planners are already thinking about what elements are essential to ecologically sensitive design. Such elements include zoning that allows for multiple but compatible uses of the land around reserves, creating corridors between habitats to allow for wildlife migration, and paying more attention to entire watersheds and other natural boundaries when development decisions are being made. Equally important is paying attention to the people who will be affected by planning decisions. For land-use plans to work, landowners, citizens, planners, and others who share an interest in these areas will have to work together.

THE FAR SIDE   By GARY LARSON

"Someday, son, this will all be yours"

BIOFACT

Bunches of Birds—There are about 800 species of birds in all of North America. There are about 850 species of birds in Costa Rica—a Central American country that's half the size of Tennessee.

Genetic Engineering and the Web of Life

Many scientists are using genetic engineering and other forms of biotechnology to harness the natural biological processes of animals, plants, and microorganisms. They see this technology as having tremendous benefits. Others, however, are concerned that genetic engineering could have serious consequences—including negative effects on the planet's biodiversity.[78]

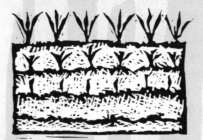

There's no doubt that genetic engineering, which involves the manipulation of genes, is transforming the ways we use living things. Traditional methods of breeding food plants to get the desired combination of traits can take decades. But through genetic engineering, desirable traits can be quickly incorporated into a plant. And the new gene can come from an unrelated species. For example, an insect-repelling gene from a common soil bacterium has been successfully incorporated into potato plants. These "new" plants now have extra protection against certain insects.

But experts disagree about the extent to which we should pursue genetic engineering and about its effects on the natural world and biodiversity.

Some scientists believe that genetic engineering will be environmentally positive because engineered plants could, for example, require fewer chemical pesticides and less fertilizer. Proponents also point out that genetic engineering could allow us to develop plants that are drought resistant, able to withstand poor soil, and contain more nutrients. "Engineered" organisms, they say, can also increase profits for farmers, business people, and others.

Other scientists point out, though, that "super" varieties of crops developed through the genetic engineering industry could be used to the exclusion of other varieties—thus accelerating the loss of genetic diversity in crops. Many feel that we should avoid such loss, since genes that seem worthless to us today may be the key to the health and viability of future crops. Another concern is that genetically engineered species could gain a foothold in wild ecosystems and could displace or otherwise harm native species.

BIOFACT

SUPER SILK—Scientists are working to produce a material with the strength of spider silk. Theoretically, one strand of some types of silk, made as thick as a pencil, would be strong enough to stop a Boeing 747 in flight.

Bringing Back the Habitats

Around the world, scientists and others have begun to recognize the far-reaching impacts that humans have had on the natural world. They are also discovering that many ecosystems no longer can provide the many services that they once did because human activities have severely damaged them. In an effort to recover these ecosystems and the critical services they provide, scientists have begun an important effort called *ecological restoration*. Ecological restoration is the process of returning a damaged ecosystem to a condition as close as possible to what it was before it was disturbed. This process has become an important part of the efforts to protect biodiversity.

Almost every region of the United States is undergoing some type of restoration effort. In many areas, scientists, school and youth groups, and others are working to restore our nation's rivers and watersheds to a more natural state—an effort that, in several cases, has resulted in the removal of outdated dams that impede the passage of migrating fish species. In the Midwest, several groups are working to restore parts of a vast prairie ecosystem that has all but been destroyed. They are not trying merely to replace all of the species that should be in a prairie, but also to restore the complex interactions among the species and the nonliving aspects of grassland ecosystems, such as natural fires.

The federal government has also taken responsibility for restoring habitats in areas where its activities have damaged ecosystems. For example, the Department of Defense Environmental Restoration Program has taken an aggressive approach to cleaning up environmental contamination that has resulted from military activities.

In the Florida Everglades, the largest restoration project ever attempted is under way. This unique wetland is home to Everglades National Park, one of our most endangered national parks. The task force developed to lead this restoration effort is made up of engineers, scientists, anthropologists, and government and tribal representatives. The challenge they face is to attempt to restore, to the extent possible, the natural flow of water through the Everglades system and to ensure that it remains as free of pollutants as possible.[79] The federal government has pledged more than $1.5 billion to help the task force achieve its goals.

From the rivers of the Pacific Northwest, where salmon are starting once again to make their incredible journeys up rivers no longer obstructed by dams, to East Coast beaches, where complex dune systems have taken the place of paved parking areas, we are slowly learning what it takes to return our natural areas to a condition that's close to the way they were before we changed them. But restoration is a long-term experiment. Because ecosystems are so complex, it will be many years before we can know how effective our efforts have been. Right now, we know only that many species are struggling to survive in habitats created or altered by humans, and that their survival may well depend on our attempts to undo the damage we've caused.

Rapid Research

Only a small percentage of the 1.7 million identified species on Earth have been studied in any detail. [80] In some areas of the world, habitat destruction is occurring so fast that species are disappearing before we even know they exist. Conservation biologists agree that we need to find out as much as possible about not only what species are out there, but also about how species depend on their habitats, how people can manage habitats to ensure healthy populations, and which habitats and ecosystems are most at risk. As biologist Gordon Orians says, "We can't manage what we don't understand." With these points in mind, many of the world's top biologists are forming teams to rapidly assess the biological diversity of lesser-known regions all over the globe. Because there's so much work to be done, biodiversity research will need more support—and more scientists to conduct the research itself—in the years to come.

Legal Action

Many endangered species in the United States owe their continued survival to a law called the Endangered Species Act. Hundreds of waterways are cleaner than they once were because of the Clean Water Act. And without the international cooperation encouraged by the Montreal Protocol, Earth's ozone layer would be more degraded by CFCs (chlorofluorocarbons) than it is now.[81] These and many other laws and treaties are an extremely important part of the effort to protect biodiversity. (For more about environmental legislation, see page 415 of the Appendix.)

In fact, many experts believe that the most important thing we can do to protect biodiversity in the short term is to implement effective laws and policies that protect natural resources worldwide, and then to enforce them. But others worry about the limits to regulation and legislation. For example, many landowners feel that current regulations can sometimes work against biodiversity. They point to how the presence of an endangered species or the existence of wetlands on private property can limit an owner's use of his or her land and can discourage stewardship and support for biodiversity protection. Instead of regulation, they would like to see stewardship encouraged through other means, such as tax incentives to protect biodiversity on private land. Many would also like to see Congress cancel programs that they feel encourage the overuse of resources, such as farm subsidies and below-cost timber sales.

Although legislation is often controversial, most would agree that taking legal action is important. However, it's just one tool in the conservation toolbox. Others feel that nonpolicy options, such as fostering "green markets" and promoting environmental education, may be as effective in protecting biodiversity—or even more so—than enacting and signing more laws. But whichever methods of conserving biodiversity we choose, the challenge we face is how to balance the short-term economic interests and rights of individuals with long-term conservation—and how to ensure, in the process, that future generations will inherit a healthy environment.

Conservation Biology —A New Field for the Future

Investigating the effects of humans on biodiversity and developing practical strategies to protect it is the work of conservation biologists. Conservation biology is a new, interdisciplinary science designed to explore human impacts on biodiversity and to develop practical solutions to the loss of biodiversity. Michael Soulé, a renowned conservation biologist, sums it up as "the science of scarcity and diversity." [82]

conservation biologists studying rhinos in Nepal

In their efforts to protect species, conservation biologists usually emphasize the long-term protection of entire natural communities rather than focus on individual species. This emphasis means considering conservation priorities and human needs. Determining the best strategies for protecting rare species, designing nature reserves, and reconciling conservation concerns with the needs of local people are a few of the activities you might find conservation biologists taking part in. Accomplishing these tasks requires working with many other experts, including anthropologists, botanists, economists, geographers, educators, and zoologists.

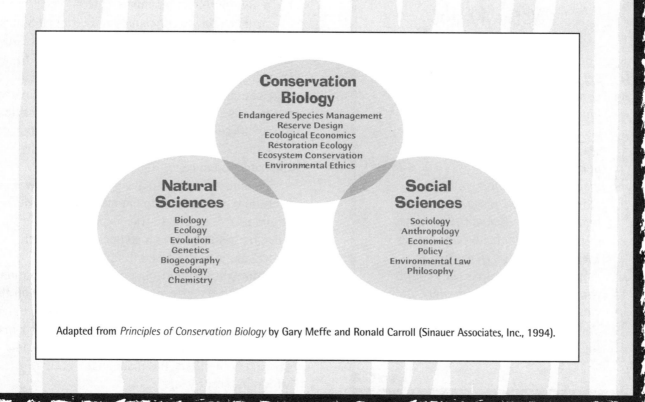

Conservation Biology

Endangered Species Management
Reserve Design
Ecological Economics
Restoration Ecology
Ecosystem Conservation
Environmental Ethics

Natural Sciences

Biology
Ecology
Evolution
Genetics
Biogeography
Geology
Chemistry

Social Sciences

Sociology
Anthropology
Economics
Policy
Environmental Law
Philosophy

Adapted from *Principles of Conservation Biology* by Gary Meffe and Ronald Carroll (Sinauer Associates, Inc., 1994).

The Corporate Connection

Some corporations are taking unprecedented action to integrate stewardship of nature into their decision making. This is partly due to the desire of the public to use processes or products that are less harmful to the environment. But it is also because, in many cases, conserving energy or using alternative industrial processes simply saves money. For example, several large chemical companies have begun to use some of their toxic wastes as a source of inexpensive raw materials for other products—a practice that not only helps prevent pollution but also increases corporate profits. The federal government also sponsors a "Green Chemistry" award for individuals or companies that develop new techniques for producing less waste or nontoxic waste.[83]

Besides awards, laws are also encouraging entrepreneurs to develop environmental technologies and services. Recycling units were developed in response to laws that banned the venting of certain chemicals from air conditioners. And in some places companies get tax breaks when they provide

incentives, such as shower facilities, that are designed to encourage employees to walk or cycle to work instead of driving.

Through a variety of programs, U.S. companies are also working with federal, state, and local governments to set and reach environmental goals. For example, the U.S. Environmental Protection Agency's "Partners for the Environment" program encourages businesses, trade associations, citizens, and others to decrease energy use and to reduce toxic wastes and other pollution. Cooperative programs such as these are part of an increasing trend toward finding new mechanisms for addressing the pollution that companies create.[84]

Regional power companies in the United States are taking part in yet another innovation in pollution control. They are exercising pollution "offsets" to compensate for some of the pollution they create. One way companies do this is by contributing money to the conservation of tropical rain forests. By helping to save these forests, the companies are maintaining ecosystems that can absorb some of the CO_2 the companies produce.

Employees are also playing a role in the way firms do business, advocating everything from programs for recycling office paper to programs for reducing pesticide use on company land. Such efforts, along with those of industry leaders, governments, and non-governmental organizations, are reducing the effects of industrial activity on the natural world and its biodiversity. Although we have a long way to go in curbing industry-related damage to the environment, the business community, in working with others, has made many positive efforts in recent years.

Natural Light Makes the Grade

Scientists have been comparing natural light to electric light in work environments to see how light affects performance and profits. For example, in Raleigh, North Carolina, researchers increased the number of windows in certain schools and left others without windows; then they compared the results. They found that in the schools with more windows, the electricity bills went down while student performance increased.[85]

The Pros and Cons of Captive Breeding

Once whooping cranes were in big trouble. Today, thanks to captive breeding efforts, they're doing a lot better. Captive breeding can ensure a species' continued existence, and in some cases captive-bred animals can eventually be reintroduced into the wild.

The red wolf is another example of a species that has been aided by captive breeding efforts. Nearly wiped out due to persecution and habitat loss, red wolves were bred in captivity and reintroduced into suitable habitat in eastern North Carolina. Other species that are being bred in captivity and released in the wild include the black-footed ferret, the Arabian oryx, and the California condor. There's even a captive breeding program for a species of snail from Tahiti.

But captive breeding isn't easy. Successful reintroduction requires detailed knowledge of the animal's habitat needs and social patterns. Captive-born animals simply may not know how to survive in the wild and therefore must be taught before they can be returned to their native habitat.

And maintaining or increasing a species' numbers is only part of a successful captive breeding program. The other essential ingredient is habitat protection. It does little good to breed animals if there's no habitat for them to live in.

Although there are many captive breeding success stories, this strategy won't work with more than a tiny percentage of threatened species. One reason for this is that zoos, public aquariums, and other breeding facilities simply don't have enough space or resources for all the species that need help. Another reason is that reintroduction is often not feasible.

Even if scientists and zoo managers had the resources to breed more species, they may not have the kind of information necessary to successfully release them. Despite these limitations, captive breeding is making an important, though limited, contribution to the preservation of biodiversity.

Saving Seeds

To help preserve the world's diversity of food crops, medicinal plants, and other plants, people all over the world are storing seeds and other plant parts in *gene banks*. Scientists hope that they'll be able to save rare plants by cultivating them—or maybe even cloning them—from the seeds and preserved parts. For plants in danger of becoming extinct in the wild, this approach could be a real safety net.

Of course, as with captive breeding, preserving plant parts is not a substitute for preserving the habitat of endangered and threatened plants. And maintaining seed and gene banks is very expensive.

California condor

Stewardship, Citizenship, and Democracy

One of the most important things that we can do to conserve biodiversity is to get involved—in our roles as parents, community members, educators, landowners, voters, employees, employers, politicians, and business leaders. For many, that means changing the way we educate our children and ourselves about what it means to be a citizen in a democracy. As Frances Moore Lappé and Paul Martin DuBois say in *The Quickening of America*, "Democracy requires a lot more of us than being intelligent voters. It requires that we learn to solve problems with others—that we learn to listen, to negotiate, and to evaluate. To think and speak effectively. To become partners in problem solving."

There are thousands of examples of individuals and communities working together to solve biodiversity problems. They're forming citizen groups to restore habitats, writing letters to elected officials, lobbying on biodiversity issues, taking action as company shareholders, using the media for communication campaigns, raising money for environmental and social organizations, educating fellow employees, conducting workshops on consumer issues, setting up information clearinghouses, volunteering for conservation organizations, and forming community stewardship councils. But many people agree that we still have a long way to go before the majority of U.S. citizens have the confidence, the know-how, the opportunity, and the commitment to bring about changes that ensure the conservation and restoration of biodiversity.

Working Toward the Future

Recent polls of the world's most populous countries show that a majority of people around the world are concerned about environmental quality.[86] This concern has already translated into significant progress toward protecting the environment and biodiversity—from global treaties that focus on reducing pollution to national and international laws designed to prevent the over-exploitation of natural resources.

Nevertheless, we're still facing tremendous challenges. The growing human population and the ever-increasing use of natural resources are linked to environmental degradation the world over. For the first time in history, we are realizing that our collective impact on the environment could be destabilizing the natural systems that make our civilization possible. From the decline of global fisheries to the increase in carbon dioxide emissions, our environmental problems transcend political boundaries and pose threats that will require global cooperation to solve.

WASTE NOT, WANT NOT—Despite the presence of so many elephants, wildebeests, zebras, and other large creatures, the African savanna isn't drowning in dung. That's because more than 2,000 species of dung beetles consume the waste. Their feeding keeps the grasslands cleaner and keeps parasites in check.

Learning the Three Cs

Conserving biodiversity and finding solutions to the intricately connected problems of environmental degradation, social decline, and economic instability will mean feeling, thinking about, and doing things differently from the ways we have before. It will mean fostering more *compassion* for other species and a kind of reverence for living systems too complex for us ever to understand fully. It will mean educating ourselves about the *connections* among all elements of biodiversity and between a healthy natural environment and a healthy human society. And it will mean coming to terms with the *consequences* of our behavior for other people and other species. Conserving biodiversity will also require us to incorporate the concepts of social equity and ecological integrity into how we do business. It will challenge us, in every aspect of our lives, to work toward creating a more sustainable society—one in which human needs are in balance with the needs of other living things. And it will mean developing not only a conservation ethic but also an entire belief system that honors the integrity of the Earth and of ourselves.

©Arthur Tilley/FPG International

"We need to imagine a prosperous commercial culture that is so intelligently designed and constructed that it mimics nature at every step, a symbiosis of company and customer and ecology."

—Paul Hawken, writer, businessperson

Endnotes for Background

1. Suzanne Winckler and Mary M. Rodgers. 1994. *Our Endangered Planet: Soil*. Minneapolis: Lerner Publications Company, p. 8.

2. Edward O. Wilson. 1992. *The Diversity of Life*. New York: Harvard University Press, pp. 134–39.

3. Personal conversation with Bruce Beehler, Counterpart, formerly with the Smithsonian Department of Ornithology, expert on the birds of New Guinea, 27 February 1998.

4. D'Vera Cohn. 1996. Scientists Invade NE Park: 24-Hour Survey Strives To Log Nature's Diversity. *Washington Post*, 6 June. <www.im.nbs.gov/blitz/species.html>.

5. Edward O. Wilson. 1988. *Biodiversity*. Washington, D.C.: National Academy Press, p. 7.

6. World Wildlife Fund. 1994. *WOW!—A Biodiversity Primer*. Washington, D.C.: World Wildlife Fund, p. 4.

7. Personal communication with Ron Tipton, Everglades Specialist, World Wildlife Fund, 27 February 1998. See also William Ross McCluney. 1969. *What You Can Do to Stop the Environmental Destruction of South Florida*. Coral Gables, Fla.: University of Miami Press, pp. 8–15.

8. Malcolm B. Hunter, Jr. 1996. *Fundamentals of Conservation Biology*. Cambridge, Mass.: Blackwell Science, pp. 149–50.

9. James E. Lovelock. 1988. The Earth as a Living Organism. In *Biodiversity*, edited by Edward O. Wilson. Washington, D.C.: National Academy Press, pp. 486–89.

10. Personal communication with David Olson and Colby Loucks, Conservation Scientists, World Wildlife Fund, 18 February 1998.

11. Gary Meffe and Ronald Carroll. 1994. *Principles of Conservation Biology*. Sunderland, Mass.: Sinauer Associates, Inc., p. 87.

12. David Takacs. 1996. *The Idea of Biodiversity*. Baltimore: Johns Hopkins University Press, pp. 83–92.

13. Anders Wijkman and Lloyd Timberlake. 1984. *Natural Disasters: Acts of God or Acts of Man*. Nottingham, United Kingdom: Earthscan Paperback, Russell Press, pp. 53–55. See also Richard Primack. 1995. *A Primer of Conservation Biology*. Sunderland, Mass.: Sinauer Associates, Inc., p. 41. See also Patrick Mazza. 1996. Cascadia's Great Flood of '96: A Clearcut Case of Muddying the Waters. *Cascadia Planet*, 29 February. <www.tnews.com/text/flood3.html>.

14. Edward O. Wilson. 1992, p. 287. See also Edward O. Wilson. 1988, p. 15.

15. Robert E. Rhoades. 1991. The World's Food Supply at Risk. *National Geographic*, April, pp. 74–103.

16. Personal communication with John Baldwin, Extension Agronomist, University of Georgia, 3 March 1998. See also Robert E. Rhoades, 1991.

17. Center for Biodiversity and Conservation. 1997. *Biodiversity, Science, and the Human Prospect*. New York: American Museum of Natural History, p. 5.

18. Edward O. Wilson. 1992, p. 283. See also Stephen R. Kellert. 1997. *Kinship to Mastery*. Washington, D.C.: Island Press, p. 20.

19. Norman Myers. 1997. Biodiversity's Genetic Library. In *Nature's Services: Societal Dependence on Natural Ecosystems*, edited by Gretchen Daily. Washington, D.C.: Island Press, pp. 263–264. See also University of Missouri Chemistry Department. 1997. *The History of Taxol*. <www.missouri.edu/~chemrg/210w97/taxol_bodypage.htm>.

20. Norman Myers. 1997, p. 263.

21. Hugh L. Iltis. 1988. Serendipity in the Exploration of Biodiversity: What Good Are Weedy Tomatoes? In *Biodiversity*, edited by Edward O. Wilson. Washington, D.C.: National Academy Press, pp. 99–103. See also Dorothy Hinshaw Patent. 1994. *The Vanishing Feast.* San Diego: Harcourt, Brace, and Company, pp. 9–10. See also John R. Stommel. 1994. Tomato Fruit Sugar Composition: Impact on Fruit Quality. In *Agricultural Research Service Web Site.* Washington, D.C.: U.S. Department of Agriculture, 17 October. <www.nal.usda.gov/ttic/tektran/data/000003/37/0000033757.html>.

22. Thomas Carr, Heather Pederson, and Sunder Ramaswamy. 1993. Rain Forest Entrepreneurs: Cashing in on Conservation. *Environment* 35, no. 7:35. See also Robert Costanza, Ralph d'Arge, Rudolf de Groot, Stephen Farber, Monica Grasso, Bruce Hannon, Karin Limburg, Shahid Naeem, Robert V. O'Neill, Jose Paruelo, Robert Rasuin, Paul Sutton, and Marjan van den Belt. 1997. The Value of the World's Ecosystem Services and Natural Capital. *Nature* 387:253–260.

23. Michael Satchell. 1996. There's Gold in Them Thar Pools. *U.S. News and World Report*, 2 December. <www.usnews.com/usnews/issue/2spri.htm>. See also Thomas D. Brock. 1994. Life at High Temperatures: Biotechnology in Yellowstone. Yellowstone Association for Natural Science, History and Education, Inc. <www.bact.wisc.edu/bact303/b27>.

24. Personal communication with Carol Anthony, Office of Public Affairs, National Park Service. 14 October 1997. See also U.S. Department of Interior. 1997. U.S. Land Management. In *Draft Strategic Plan Overview.* <www.doi.gov/master2.html>.

25. Gretchen Daily, Pamela Matson, and Peter Vitousek. 1997. Ecosystem Services Supplied by Soil. In *Nature's Services: Societal Dependence on Natural Ecosystems*, edited by Gretchen Daily. Washington, D.C.: Island Press, pp. 124–125.

26. Center for Biodiversity and Conservation. 1997, p. 6.

27. Gretchen Daily et al. 1997, pp. 124–125.

28. Personal communication with Mary Wilson, Public Relations Specialist, Humane Society of the United States, 10 December 1997.

29. Stephen R. Kellert and Edward O. Wilson. 1993. *The Biophilia Hypothesis.* Washington, D.C.: Island Press, pp. 31–35, 74–75.

30. Center for Biodiversity and Conservation. 1997, p. 4.

31. Edward O. Wilson. 1988, p. 13. See also Edward O. Wilson. 1992, pp. 251–59.

32. Edward O. Wilson. 1992, pp. 192, 205–7, 215–17. See also Norman Myers. 1988. Tropical Forests and Their Species: Going, Going . . . ? In *Biodiversity*, edited by Edward O. Wilson, Washington, D.C.: National Academy Press, p. 28. See also David Takacs. 1996. *The Idea of Biodiversity.* Baltimore: Johns Hopkins University Press, p. 52.

33. David Fastovsky and David Weiskampel. 1996. *The Evolution and Extinction of the Dinosaurs.* Cambridge: Cambridge University Press, pp. 404, 421.

34. Gary Meffe and Ronald Carroll. 1994, p. 97.

35. Ernst Mayr and Robert J. O'Hara. 1986. The Biogeographic Evidence Supporting the Pleistocene Refuge Hypothesis. *Evolution* 40(1):55–67.

36. Gary Meffe and Ronald Carroll. 1994, pp. 132, 135.

37. Edward O. Wilson. 1992, pp. 272, 280. See also Gary Meffe and Ronald Carroll. 1994, p. 138.

38. Edward O. Wilson. 1992, p. 253.

39. Confederated Tribes of the Umatilla Indian Reservation. 1996. Salmon Migration and the Dams on the Mainstream Columbia and Snake Rivers. <www.ucinet.com/~ctuir/activity. html#salmon>. See also Mitch Sanchotenas. 1997. Who Benefits From These Dams? *Idaho Post-Register*. 3 April.

40. Edward O. Wilson. 1992, pp. 202–203.

41. National Parks and Conservation Association. 1996. Statement on Tallgrass Prairie and Boston Harbor Islands. April. <www.npca.org/home/npca/testify/ tall-tes.html>.

42. Stephen R. Kellert. 1997, p. 30.

43. World Resources Institute. 1992. *The 1992 Information Please Environmental Almanac.* Houghton Mifflin, Boston. p. 143.

44. Personal communication with Dave Cowan, Communications Associate, Greater Yellowstone Coalition, Bozeman, Mont. 22 April 1998.

45. Reed F. Noss and Robert L. Peters. 1995. *Endangered Ecosystems: A Status Report on America's Vanishing Habitat and Wildlife.* Washington, D.C.: Defenders of Wildlife, p. 61.

46. Thomas L. Fleischner. 1997. Keeping the Cows Off: Conserving Riparian Areas in the American West. In *A Conservation Assessment of the Terrestrial Ecosystems of North America: Vol. I, The United States and Canada* (Forthcoming), edited by Taylor Ricketts, Eric Dinerstein, David Olson, Colby Loucks, William Eichbaum, Kevin Kavanaugh, Prashant Hedao, Patrick Hurley, Karne Carney, Robin Abell, and Steven Walters. Washington, D.C.: World Wildlife Fund, pp. 66–67.

47. Population Division, Department of Economic and Social Affairs, United Nations Secretariat. 1998. World Population Projections to 2150. 1 February. <www.undp.org/popin/wdtrends/execsum.htm>.

48. U.S. Bureau of the Census. 1996. Resident Population Projections of the United States: Middle, Low and High Series, 1996–2050. March. <www.census.gov/population/projections/nation /npaltsrs.txt>. See also U.S. Bureau of the Census. 1998. PopClock Projection. 4 March. <www.census.gov/cgi-bin/popclock>.

49. Nyle Brady. 1988. The Tropical Forestry Action Plan: Recent Progress and New Initiatives. In *Biodiversity*, edited by Edward O. Wilson. Washington, D.C.: National Academy Press, p. 409.

50. World Resources Institute Environmental Education Project. 1988. China's Energy Prospects Teacher's Guide. <www.wri.org/wri/enved/giants/chi-ener.html>.

51. Johan Holmberg. 1992. *Policies for a Small Planet.* London: Earthscan. pp. 297, 334–42.

52. Erik Eckholm. 1982. *Down to Earth: Environment and Human Needs.* New York: W. W. Norton and Company, pp. 24–28.

53. Carl Jordan. 1986. Local Effects of Tropical Deforestation. In *Conservation Biology: The Science of Scarcity and Diversity*, edited by Michael Soulé. Sunderland, Mass.: Sinauer Associates, Inc., pp. 416–18. See also Richard Primack. 1995, p. 220.

54. Gary Meffe and Ronald Carroll. 1994. pp. 543–44, 546–47.

55. Mark Sagoff. 1994. Population, Nature, and the Environment. In *Beyond the Numbers*, edited by Laurie Ann Mazur. Washington, D.C.: Island Press, pp. 37–39.

56. Benjamin Goldman. 1991. *The Truth About Where You Live*. New York: Time Books, pp. 282–83.

57. Kathryn Fuller. 1994. Foreword to *International Wildlife Trade: A CITES Sourcebook*, edited by Ginette Hemley. Washington, D.C.: Island Press, p. vii.

58. Frank Press and Raymond Siever. 1994. *Understanding Earth*. New York: W. H. Freeman and Company, pp. 348–349. See also David M. Gates. 1993. *Climate Change and Its Biological Consequences*. Sunderland, Mass.: Sinauer Associates, Inc., pp. 10–11.

59. David M. Gates. 1993. p. 2.

60. Robert L. Peters. 1992. Conservation of Biological Diversity in the Face of Climate Change. In *Global Warming and Biological Diversity*, edited by Robert L. Peters and Thomas E. Lovejoy. New Haven: Yale University Press, pp. 15–30.

61. *Globewatch*. 1997. Antarctic Penguin Colonies Dwindling, Says Greenpeace. February. <www.mcs.net/~rogers/globe/peng.html>.

62. *BBC News*. 1997. Impact of Global Warming May Be Severe and Wide-Ranging. <www.news.bbc.co.uk/hi/english/special_report/1997/sci/tech/global_warming/newsid_32000/3296/.stm>. See also World Resources Institute. About the CPI. In *Climate Protection Initiative Web Site*. <www.wri.org/wri/cpi/aboutcpi.html>.

63. U.S. Environmental Protection Agency Acid Rain Program. Effects of Acid Rain on Forests. <www.epa.gov/acidrain/student/forests.html>. See also U.S. Environmental Protection Agency Oil Spill Program. 1998. Oil Spill Behavior and Effects. February. <www.epa.gov/superfund/oerr/er/oilspill/effects.htm>.

64. Howard Bern. 1992. The Fragile Fetus. In *Chemically Induced Alterations in Sexual and Functional Development: The Wildlife/Human Connection*, edited by Theo Colburn and Coralie Clement, Princeton, N.J.: Princeton Scientific Publishing, Inc., pp. 1–2, 9–17.

65. John L. Seitz. 1988. *The Politics of Development*. New York: Basil Blackwell, p. 121.

66. Michael Fry and Kuehler Toone. 1981. DDT Induced Feminization of Gull Embryos. *Science* 213:922–24.

67. Howard Bern. 1992, pp. 1–2, 9–17.

68. Jeffrey McNeely, et al. 1990. *Conserving the World's Biological Diversity*. Gland, Switzerland: IUCN, pp. 21, 113–115.

69. Sandra Postel. 1994. Carrying Capacity: Earth's Bottom Line. In *State of the World*, edited by Lester R. Brown. New York: W. W. Norton and Company, p. 3.

70. Stephen R. Kellert. 1996. *The Value of Life*. Washington, D.C.: Island Press, p. 30.

71. Elizabeth Boo. 1990. *Ecotourism: The Potentials and Pitfalls*. Washington, D.C.: World Wildlife Fund, p. xiv–xv. See also David L. Edgell Sr. 1990. *International Tourism Policy*. New York: Van Nostrand Reinhold, pp. 80–83.

72. James C. McKinley, Jr. and William K Stevens. 1998. The Life of a Hurricane, the Death That It Caused. *New York Times*, 9 November. See also Howard LaFranchi. 1998. Poverty and Deforestation Make Storm Effects Worse. *Christian Science Monitor*, 6 November. <www.csmonitor.com/durable/1998/11/06/p652.htm>.

73. Sandra Postel. 1994, p. 4.

74. Johan Holmberg. 1992. *Policies for a Small Planet.* London: Earthscan Publications Ltd., pp. 23–24.

75. President's Council on Sustainable Development. 1997. *Education for Sustainability: An Agenda for Action.* Washington, D.C.: Government Printing Office, pp. 1, 77.

76. John Ryan. 1992. Conserving Biological Diversity. In *State of the World*, edited by Lester Brown. New York: W. W. Norton and Company, p. 15.

77. Julian Simon. 1994. Population Growth Is Not Bad. In *Scarcity or Abundance?* edited by Norman Myers and Julian Simon. New York: W. W. Norton and Company, pp. 23–34.

78. Australian Biotechnology Association. 1996. What is Biotechnology? <www.aba.asn.au/leaf1.html>. See also International Food Information Council. 1988. Food Biotechnology and the Environment. <www.ificinfo.health.org/brochure/bioenv.htm>. See also International Food Information Council. 1988. Food Biotechnology Benefits. <www.ificinfo.health.org/brochure/biobenef. htm>.

79. Courtney LaFountain and Christopher Douglass. 1996. Save the Everglades and 3 Billion. *Journal of Commerce and Commercial*, 13 March. <wuecon.wustl.edu/~douglass/everglade.html>. See also Suzie Unger. 1997. The Restoration of an Ecosystem. In *Everglades National Park Official Web Site.* <www.nps.gov/ever/current/feature2.htm>.

80. Thomas Lovejoy. 1997. Biodiversity: What Is It? In *Biodiversity II: Understanding and Protecting Our Biological Resources,* edited by Marjorie L. Reaka-Kudla, Don E. Wilson, and Edward O. Wilson. Washington, D.C.: Joseph Henry Press, p. 7.

81. Environmental Defense Fund. 1997. "Ozone Protection Anniversary Holds Lessons for Global Warming" press release. 15 September. <www.edf.org/pubs/NewsReleases/1997/sep/e_ montreal.html>.

82. Michael Soulé. 1986.

83. U.S. Environmental Protection Agency Office of Pollution Prevention and Toxics. 1997. The Presidential Green Chemistry Challenge: Quick Reference Fact Sheet. March. <www.epa.gov/opptintr/gcc/fact-2.txt>. See also Donlar Corporation. 1997. The Presidential Green Chemistry Challenge. March. <www.donlar.com/green/index.html>.

84. U.S. Environmental Protection Agency. EPA's New Era of Voluntary Partnership Program. In *Partners for the Environment Web Site.* <www.epa.gov/partners/>.

85. Mike McClintock. 1995. The Difference is Daylight. *The Washington Post: Washington Home.* 5 September, p. 12–17.

86. Environics International. 1998. *Environmental Poll Press Release.* 4 June.

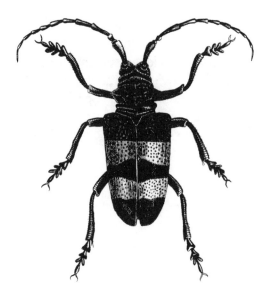

©Don Bonsey/Tony Stone Images

"Imagination is more important
than knowledge. Knowledge is limited;
imagination encircles the world."

–Albert Einstein

Introducing the

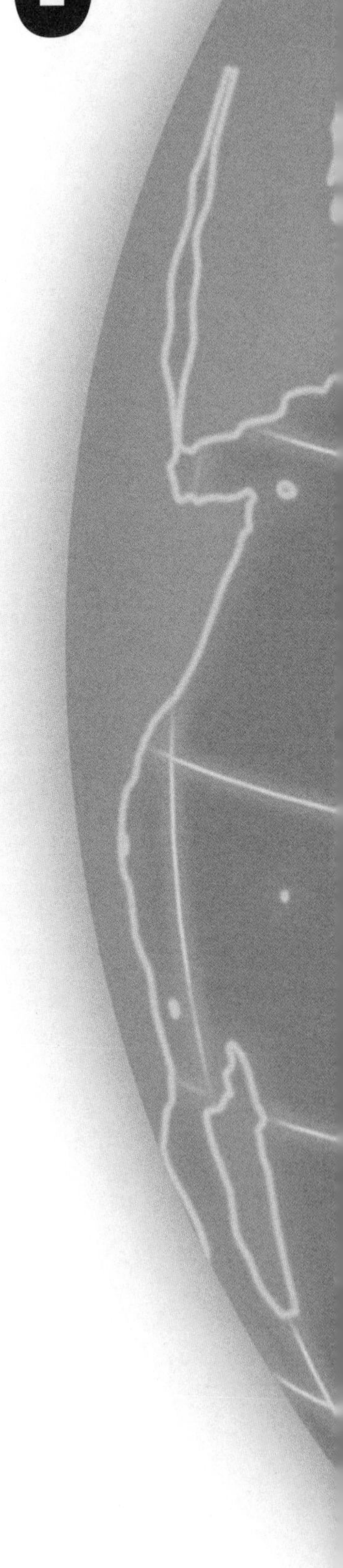

"The real voyage

of discovery

consists not

in seeking new

landscapes, but

in having

new eyes."

–Marcel Proust, writer

Activities

Activities, Activities, and More Activities!

Subjects
Lists specific disciplines. A subject matrix is in the Appendix on pages 442–445.

Skills
Lists the key skills that students will use in the activity. The complete list of skills is in the Appendix on pages 446–450.

Time
Gives an idea of how much time the activity will take based on piloting and educator comments. We have estimated that each session is about 45 minutes.

Connections
Lists related activities in the module that could be used before or after the activity to create more effective lessons and units.

Outdoor and Challenging
Indicates challenge level and where the activity takes place. This symbol ☼ indicates that parts of the activity take place outside. This symbol 💡 indicates a more challenging activity.

Framework Links
Shows specific connections to the biodiversity education conceptual framework on pages 419–428.

Vocabulary
Highlights important words used in the activity that students might not know. Words in bold are defined in the Glossary on pages 412–414.

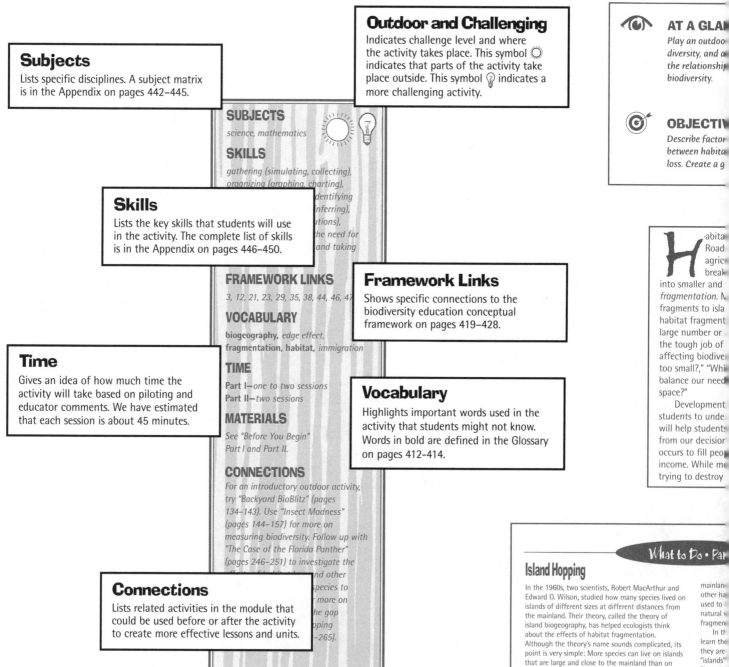

SUBJECTS

science, mathematics

SKILLS

gathering (simulating, collecting), organizing (graphing, charting), ...dentifying ...nferring), ...utions), ...he need for ...and taking

FRAMEWORK LINKS

3, 12, 21, 23, 29, 35, 38, 44, 46, 47

VOCABULARY

biogeography, *edge effect,* **fragmentation, habitat,** *immigration*

TIME

Part I—*one to two sessions*
Part II—*two sessions*

MATERIALS

See "Before You Begin" Part I and Part II.

CONNECTIONS

For an introductory outdoor activity, try "Backyard BioBlitz" (pages 134–143). Use "Insect Madness" (pages 144–157) for more on measuring biodiversity. Follow up with "The Case of the Florida Panther" (pages 246–251) to investigate the ... *and other* ... *species to* ... *more on* ...*he gap* ...*oping* ...*–265).*

AT A GLA...
Play an outdoo... diversity, and a... the relationship... biodiversity.

OBJECTIV...
Describe factor... between habita... loss. Create a g...

Habita... Road... agric... break... into smaller and *fragmentation.* ... fragments to isla... habitat fragment... large number or ... the tough job of ... affecting biodive... too small?," "Whi... balance our need... space?"

Development... students to unde... will help students... from our decision... occurs to fill peo... income. While m... trying to destroy

What to Do • Pa...

Island Hopping

In the 1960s, two scientists, Robert MacArthur and Edward O. Wilson, studied how many species lived on islands of different sizes at different distances from the mainland. Their theory, called the theory of island biogeography, has helped ecologists think about the effects of habitat fragmentation. Although the theory's name sounds complicated, its point is very simple: More species can live on islands that are large and close to the mainland than on islands that are small and far from the mainland.

Why is this theory still important to scientists today? Because it relates directly to the study of the numbers of species in habitat fragments on the

mainla... other ha... used to... natural ... fragmen...
In th... learn the... they are... "islands"... Then the... biogeog... of the th... think ab... planning...

Welcome to the activity section. We've included 34 activities that help teach many of the concepts and skills outlined in the biodiversity education framework in Appendix C (pages 418-429). You'll see that the activities vary in length, depth, and approach. We encourage you to adapt these ideas, combine them with other resources, and devise organizing themes that will best meet your particular objectives. We also encourage you to explore the sample biodiversity units that link the activities based on different themes and approaches. (The unit plans and a description of our educational approach start on page 380.)

The activities are grouped into the same four chapters as the background information (see page 6) to help your students explore these questions: What is biodiversity? Why is biodiversity important? What's the status of biodiversity? How can we protect biodiversity? Each activity follows the format described in the sample below. We also encourage you to check out the resource section (pages 452-469) to find dozens of exemplary educational resources that can help you build the best biodiversity education program possible.

Objectives

Describes what the purpose of the activity is and what types of learning (skills and knowledge) might occur. Objectives describe what students should be able to do after taking part in the activity.

Introduction

Starts off each activity and often includes additional background information that is related to the activity. (Sometimes this background information is contained in boxes outside the normal text. In other cases, it's integrated into the introduction of the activity.)

What to Do

Includes step-by-step directions to conduct the activity. The first part of each step is in boldfaced type. The last step in the directions brings closure to the activity.

WRAPPING IT UP

Assessment

Distribute a sheet of graph paper (with ½-in. to 1-in.

not incorporate any of the concepts regarding fragmentation and species diversity.

Satisfactory—The student's design uses basic ideas to avoid fragmentation.

Excellent—The student's design uses ideas both to avoid fragmentation and to increase species diversity.

Portfolio

Part I has no portfolio documentation. For Part II, use the tables and graphs in the portfolio.

Writing Idea

Have the students write an interview between a

such as these: "So, why are you leaving home?" "Where do you think your travels will take you?" "What are your special habitat needs?" "How could people have reduced the damage this development has caused?" (You might want to have the students do some research in advance to find out about the specific needs of their species.) Afterward, students can share their interviews by taking turns playing the roles of journalists and species being interviewed.

Extensions

• Stage an in-class debate about a current development issue in your area. Have half the students in favor of developing the land and the other half against it. Those in favor

fragments of pine forests, beach-dune systems, or grasslands. Then take your students out on a field trip to investigate some different-sized fragments. Have them think of ways they could investigate the level of biodiversity in the fragments, and then compare the fragments. You might ask a local park ranger, a naturalist, or some other expert to help you organize the trip.

Resources

Environmental Change: Biodiversity (National Science Teachers Association, 1997). (800) 722-NSTA.

The Song of the Dodo: Island Biogeography in an Age of Extinctions by David Quammen (Simon and Schuster, 1996).

The Theory of Island Biogeography by Robert H. MacArthur and Edward O. Wilson (Princeton University Press, 1967).

Visit National Wildlife Federation's Web site at <www.n

Assessment

Suggests strategies for evaluation at the end of most activities. The assessment also includes examples of what excellent, satisfactory, and unsatisfactory results might include. (Some of the introductory activities used to spur interest or to gauge students' understanding do not include assessments.)

Writing Ideas

Encourages creative writing and technical skills. Writing ideas can be integrated into the activity or used as an extension.

Extensions

Provides additional activities that relate to the core activity and can be used to encourage more in-depth investigation or discovery. Some can also be used as assessment strategies.

Resources

Provides more in-depth exploration of a topic. We also recommend that you review the Resources on pages 452–469 to find additional materials that will complement many of the activities.

Windows on the Wild: Biodiversity Basics

What Is Biodiver

The activities in this section introduce the concept of biodiversity and highlight the important roles biodiversity plays in our lives. For background information, see pages 22-29.

sity?

Pete Oxford/ENP Images

"The diversity of life forms, so numerous that we have yet to identify most of them, is the greatest wonder of this planet."

–Edward O. Wilson, biologist

1 | All the World's a Web

World Wildlife Fund

SUBJECTS

science, social studies, language arts

SKILLS

organizing (arranging), analyzing (identifying components and relationships among components), interpreting (relating), applying (creating)

FRAMEWORK LINKS

1, 12, 13

VOCABULARY

biodiversity, ecosystem; *plus any unfamiliar words in the "Web Words" list*

TIME

one session

MATERIALS

flip chart paper, pencils, container for key words, key words on separate cards, audio tape "Life on the Brink" (optional)

CONNECTIONS

This introductory activity works well either before or after "'Connect the Creatures' Scavenger Hunt" (pages 102–107) and "Something for Everyone" (pages 90–93).

 **AT A GLANCE**

Create a "word web" that illustrates the interconnections in nature.

 OBJECTIVES

Define biodiversity and create a word web that illustrates some of the complex connections in the web of life. Discuss at least one way biodiversity affects people's lives.

B iodiversity is the variety of life around us—and much more. It's also everything that living things do—the grand total of interactions of living things among themselves and with their environment. These interactions can be as simple as a moth's dependence on one species of plant for food and the plant's dependence on the moth for pollination. At another level, the moth and the plant also depend on all of the elements that make up their ecosystem—from clean water to the right climate. At still another level, this ecosystem interacts with other ecosystems to form a huge, global system of interacting parts.

This introductory activity is a great way to start a biodiversity unit because it focuses on connections, which are the heart of biodiversity. By making their own word webs using the words provided, students can begin to consider the complex connections that characterize life on Earth. The activity can also give you an idea of how your students are thinking about biodiversity before you start a unit.

Before You Begin

Write each of the *key words* (page 78) on a separate piece of paper, and put all five words into a container. Write the *web words* (page 78) on a chalkboard or sheet of flip chart paper.

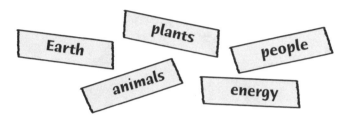

Earth

plants

people

animals

energy

What to Do

1. Review vocabulary and divide the class into groups.

Go over any key words and web words that the students aren't familiar with, and then have someone from each group pick a key word from the container. Tell the groups to write that key word in the center of a piece of paper. Next give them time to create a web using as many of the web words as possible. Encourage them to write in words that describe the connections they're creating. Examples include verbs and phrases such as *influences, affects, benefits, is helped by, can lead to,* and *can cause.* (See pages 76–78 for examples of webs.)

2. Discuss the webs.

Each group should be able to explain the connections that they drew between the key word and the web words, as well as between the different web words. Ask the students if they notice any similarities among different groups' webs, and have them work as a group to identify and write down two or more of these similarities. You might also want to have them write down any differences they notice. Use their ideas to spur discussion.

3. Introduce biodiversity.

Write the word *biodiversity* where everyone can see it, and ask the students for their ideas on its meaning. Use the glossary (pages 412–414) and background information (pages 22–29) to familiarize the students with the word. Explain that biodiversity is the ultimate web because it includes all life on Earth.

4. Create new webs.

Have the groups try their hand at creating webs as before, but this time use the word *biodiversity* as the key word. You can also use the word *nature* if your students are having a difficult time understanding biodiversity. They can add any new web words they might think of. Again, have students share their ideas.

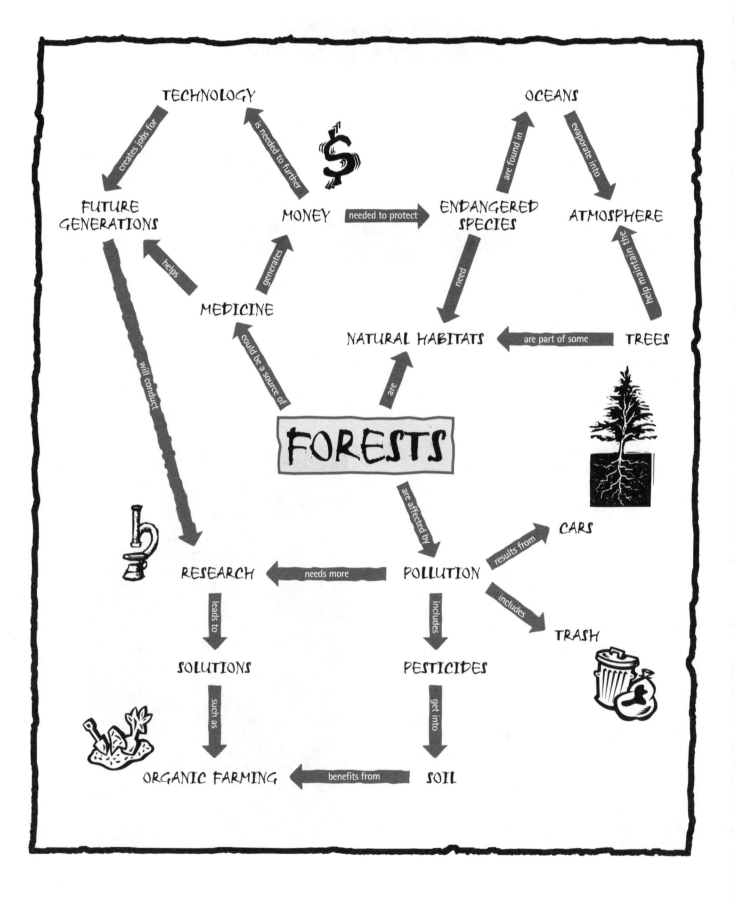

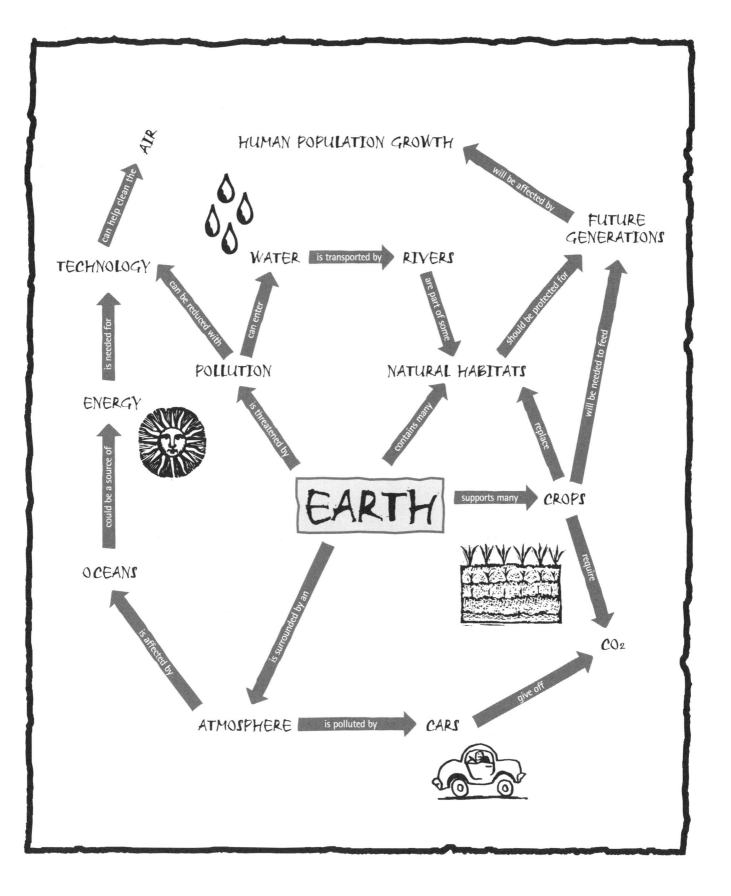

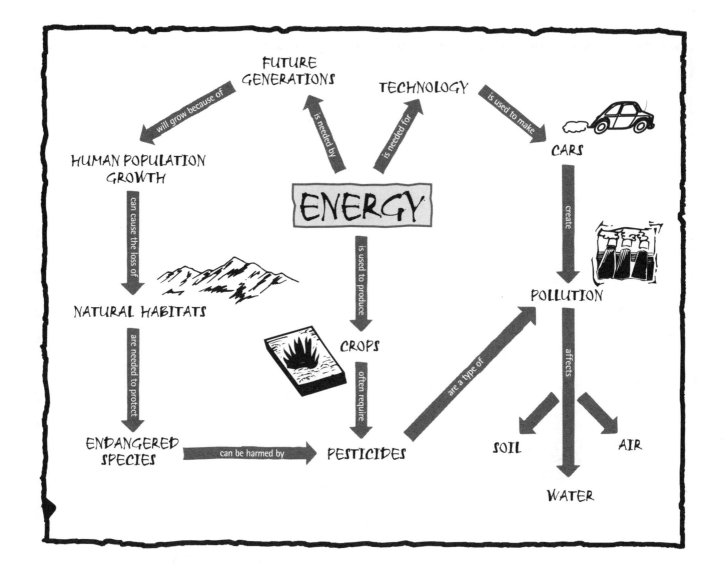

KEY WORDS	WEB WORDS		
	technology	twenty-first century	cars
Earth	natural habitats	pesticides	endangered species
animals	crops	food	organic farming
plants	trash	oceans	atmosphere
people	shopping	money	future generations
energy	soil	water	air
	solutions	human population growth	medicine
	pollution	school	trees

WRAPPING IT UP

Assessment

There is no direct assessment, but this activity can be used as an assessment for other activities. You can have your students create webs at the conclusion of a unit to see if they understand the basic concept of biodiversity and how it is linked to other issues. You can also use this as an assessment for several of the activities and units in this module.

Portfolio

Have each student copy their group's webs onto paper and keep the webs in the portfolio. Each student should date the webs and write any comments related to the activity.

Writing Idea

Have each student choose one of the web words and write a paragraph about how it's connected to biodiversity.

Extensions

- **The Symphony of Life**

    To emphasize how individual species fit together in a complex and interacting web of life, have students compare biodiversity to an orchestra, a rock band, or any other group of musicians who work together. Ask if a band or an orchestra has anything in common with a natural system such as a forest, a coral reef, or even the whole Earth. As students share their ideas, explain that each animal and plant in a natural system plays a part, just as each member of a musical group does. When one of the parts is removed, the system may be able to keep functioning—but perhaps not as well as it used to. For example, if the drums were removed from a rock band, the rest of the band members could keep on playing. But the music wouldn't sound the same.

    Discuss with the students the fact that we may not necessarily notice when one species is removed from a natural system (such as when a species of insect becomes extinct), but the system is changed nonetheless. And if too many parts are removed, the system can't function the way it used to—just as a musical group could not function well if too many of its members were removed.

- **Check It Out!**

    National Public Radio, in conjunction with the National Geographic Society, has produced a series called *Radio Expeditions*. The series uses stories on audiotape to explore a variety of topics. Teacher's kits, featuring an audiotape and a teacher's guide, are available.

The fourth program in the Radio Expeditions *series, called "Life on the Brink," is one of two that explores biodiversity. The companion audiotape in this program compares biodiversity with a piece of music, and the commentary is effectively illustrated with the removal of one type of instrument at a time. Other segments on the tape include an elevator ride from the Earth's molten, lifeless core to its teeming surface, an encounter with Madagascar's lemurs, and a journey upward into the canopy of a tropical rain forest.*

For more information about this program and others in the series, call (202) 414-2726.

Resources

Biodiversity by Dorothy Hinshaw Patent (Clarion Books, 1996).

Biodiversity! Exploring the Web of Life Education Kit (World Wildlife Fund, 1997).

Connections: The Living Planet by Milton McClaren and Bob Samples (Ginn Publishing, 1995).

The Diversity of Life by Edward O. Wilson (Harvard University Press, 1992).

SUBJECTS

science, language arts, social studies

SKILLS

analyzing (discussing), interpreting (inferring, reasoning), applying (synthesizing)

FRAMEWORK LINKS

1, 2, 3

VOCABULARY

biodiversity, ecosystems, genes, migration, species, *plus review "What's Your Biodiversity IQ?" (pages 7–10 in the Student Book) for words that may be unfamiliar*

TIME

one session

MATERIALS

copies of the quiz (pages 7–10 in the Student Book) and copies of the answers that begin on page 82 in this guide

CONNECTIONS

This introductory activity can be followed by any of the other activities in the module.

AT A GLANCE

Take a "gee-whiz quiz" to find out how much you know about biodiversity.

OBJECTIVES

Define biodiversity, discuss facts and issues related to biodiversity, and list reasons why biodiversity is important.

id you know that there are flowers that masquerade as insects, birds that map their migration by the stars, and fungi that find their way into your favorite foods? When it comes to biodiversity, these and other fascinating facts prove that truth really is stranger than fiction.

Have your students take the biodiversity quiz to learn about some of the tantalizing stranger-than-fiction tidbits that biodiversity has to offer. In the process, they'll become familiar with some important biodiversity basics.

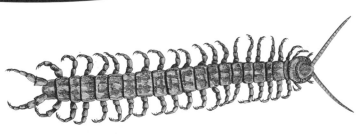

Before You Begin

For each group of three or four students, make a copy of the quiz (pages 7–10 in the Student Book) and the answers (pages 82–87 in this guide).

What to Do

1. Pass out quizzes and go over vocabulary.

Divide the class into groups of three or four students, and give each group a copy of the quiz. Go over any words or terms that may be unfamiliar to the students. For now, however, don't define the word *biodiversity*.

2. Give groups time to complete the quiz.

Assure the students that this is not a real quiz; it's simply a fun introduction to biodiversity issues. Also tell them that their answers won't be graded and that it's OK if many of their answers are wild guesses.

3. Pass out answers to the quiz.

Have the students score their tests. Afterward, discuss each of the questions and answers. How did they do? Were they surprised by any of the answers? Which ones? You may want to suggest that they take the quiz home to test family members and talk about the answers.

4. Develop a class definition of biodiversity and a list of reasons it's important.

Explain to the students that the quiz was designed to point out some interesting facts about the natural world, as well as to introduce the concept of biodiversity. Ask the students what they think biodiversity means, and have them write their ideas on the board. Then use the background information on pages 22–29, along with the glossary, to explain the three levels of biodiversity (genes, species, and ecosystems).

Next have the students use the information on the board and in the quiz to list reasons biodiversity is important. Afterward, combine their group lists to form a single class list. Your class list might include food, clothing, housing, ecosystem services, natural beauty, camping, and hiking.

THE BIO IQ QUIZ SHOW (optional)

You can turn this activity into a quiz show game with you as the host. Give each group a copy of the quiz, and choose a spokesperson for each team. Then start the game by reading a quiz question aloud. Give the groups a minute or so to discuss possible answers. Have team spokespersons raise their hands to indicate their team's readiness to answer the question, and have the groups answer on a first-come, first-served basis. Score one point for each correct answer per question. If the first group misses an answer, other groups may try. Tally the points on the board and see which team comes out on top.

WHAT'S YOUR BIODIVERSITY IQ?—ANSWERS

Each correct answer is worth one point, even if there's more than one correct answer per question.

1. **Which of the following could the fastest human outrun in a 100-yard race?**

   c, e. Lots of animals are quick on their feet (or wings, or scales, or fins), but speed doesn't necessarily count much these days in the race for survival. Cheetahs, for example, are the fastest land animals in the world—but they're also among the world's most endangered.

cheetah

2. **Which of the following actually exist?**

   a, b, c. And you thought science fiction was bizarre!

   a. Certain kinds of ants eat the sugary substances excreted by aphids, which are insects that suck plant juices. The ants actually herd colonies of aphids by moving them from place to place and protecting them from enemies.

   b. Some slime molds have two distinct phases in their life cycle. In the reproductive phase they are stationary, like a plant with a stalk. From this stalk they produce spores. These slime molds may also exist as mobile amoeba-like organisms that feed by engulfing material.

   c. The bee-like flowers of the orchid *Ophrys speculum* attract male bees that are so deceived by the resemblance to female bees that they attempt to mate with the flowers. In the process, the male bees bring pollen from other "females." Other orchids fool certain species of wasps in the same way.

3. **What does a large adult bluefin tuna have in common with a Porsche 911?**

   b, c. Believe it or not, an adult bluefin tuna can fetch as much as $60,000 on the Japanese wholesale market, which is about the cost of a Porsche 911. Why so pricey? Overfishing has made this tuna extremely rare. And you could say that both the car and the fish have big fins.

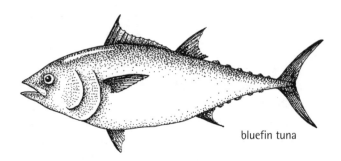

bluefin tuna

WHAT'S YOUR BIODIVERSITY IQ?–ANSWERS (Cont'd.)

4. Which of the following best describes the word biodiversity?

c. The variety of life on Earth includes plants, animals, microorganisms, ecosystems, genes, cultural diversity, and more.

5. U. S. Fish and Wildlife Service agents and U. S. Customs officials in Los Angeles, California, once found which of the following strapped to a man's arms and legs?

a. The man planned to sell the snakes in Taiwan; instead, he was prosecuted and sentenced to six months in jail. While this is one of the more unusual stories of illegal wildlife trade, such trade is far from rare. Annual trade in wildlife and wildlife products is estimated at $10 billion, and up to 25 percent of that trade is illegal. That's a $2.5 billion black market in wildlife trade—one of the largest black markets in the world! If that market didn't exist, the future of African elephants, rhinos, Asiatic bears, and many other species would look a whole lot brighter.

6. Scientists studying bug zappers have learned some interesting facts. Which of the following are among them?

c, d. A recent study at the University of Delaware on bug zappers came up with some "shocking" results. It revealed, for example, that many species of mosquitoes are not attracted to bug zappers at all. Instead, the zappers' blue light attracts harmless insects in droves, many of which are food for birds, bats, and fish. Some of the insects that zappers zap are also important to plants, which need the insects for pollination.

7. Blackpoll warblers are tiny birds that migrate between North America and South America each year. Which of the following statements about them are true?

a, d. Animals that migrate often have remarkable navigational skills. Many use the sun, stars, land patterns, and other means to reach their destinations, which may be thousands of miles and several countries away. And many migrators are able to get where they're going on very little fuel. For example, migrating birds that often travel huge distances and eat very little along the way have incredibly energy-efficient bodies that "burn" body fat for fuel. Some birds, such as the tiny blackpoll warbler, get the equivalent of thousands of miles per gallon of fuel! But being able to get from point A to point B doesn't matter much if the habitat an animal is traveling to has been destroyed. That's one reason why international efforts to conserve habitat are so important.

blackpoll warbler

8. Which of the following can be considered an enemy of coral reefs?

a, b, d. Coral reefs, among the most diverse habitats in the world, are capable of supporting more than 3,000 species of fish

WHAT'S YOUR BIODIVERSITY IQ?–ANSWERS (Cont'd.)

and other marine life. But these lush underwater ecosystems are facing some serious threats. For example, outbreaks of the predatory crown-of-thorns starfish, which relies on coral for food, are a naturally occurring phenomenon. However, human activity has impaired the coral's ability to recover from damage, and may have reduced the number of natural predators of the crown-of-thorns. Other threats to coral reefs include global warming, coral poachers who supply jewelry makers, people who buy coral jewelry, and divers who don't treat reefs gently.

9. What's the most serious threat to biodiversity?

b. So how are we losing habitats? All over the world they're being turned into agricultural land, harvested for wood and fuel, and destroyed or changed to build roads, schools, malls, and other human developments. Because the human population is growing so quickly and consuming so many natural resources, habitat loss is occurring at a rapid pace.

10. The items on the left have been (or are being) developed into important medicines for humans. Match each item with the medicine it inspired by writing the letters in the appropriate blanks.

b bread mold (antibiotic)

c white willow tree (pain reliever)

d vampire bat saliva (medicine to unclog arteries)

a wild yams (anti-inflammatory)

e Caribbean coral (first-aid ointment)

b, c, d, a, e. Biodiversity is like a gigantic pharmacy. Consider plants: More than one-fourth of the drugs commonly used today were originally derived from plants. Animals are a potentially important source of medicines too. In fact, you never know where a future medicine might pop up. Who would have thought that vampire bat saliva could be useful? No wonder researchers are looking to biodiversity to find treatments and cures for cancer, AIDS, and a host of other diseases.

dromedary and Bactrian camel

WHAT'S YOUR BIODIVERSITY IQ?–ANSWERS (Cont'd.)

11. Which of the following are true statements about camels?

b, c, d. Contrary to popular belief, camels store fat–not water–in their humps. And in the cooler months they can actually go for *three* months without drinking. Camels have been important to desert peoples for thousands of years. Camels are a good example of the many creatures around the world that have helped shape human culture.

12. Without fungi, which of the following would you not be able to do?

a, b, c, d. While some forms of fungi may seem less than noble–athlete's foot fungus, for example–the world could not function long without these humble life forms. Fungi and bacteria play a key role in breaking down organic matter and recycling it back into usable nutrients. Without them, dead things would definitely pile up! Besides, without fungi we wouldn't have tasty treats such as mushrooms, yeast bread, or blue cheese.

13. Which of the following statements are true?

c. The potato actually originated in South America. In Peru, some family farmers grow as many as 12 kinds of potatoes. Can you imagine eating purple potato chips or red mashed potatoes? It's sure possible with the thousands of kinds of potatoes out there. Most supermarkets, however, carry only four or five different varieties. And most of the country's baking potatoes are grown in Idaho. (Washington is the second largest producer of potatoes.)

14. Which of the following are real species of animals?

a, b, c. As far as scientists know, werewolves are fiction. The others are just a few examples of some of the many strange and wonderful creatures with whom we share the planet. The vampire moth *(Calpe eustrigata)* uses its sucking mouth parts to pierce the skin of mammals and drink their blood. The Tasmanian devil *(Sarcophilus harrisii)*, the world's largest carnivorous marsupial, has a predominantly black coat of fur and a long, almost hairless tail. The Komodo dragon *(Varanus komodoensis)*, the world's largest monitor lizard, lives in Indonesia, grows up to 10 feet long, and weighs up to 300 pounds!

ostrich

WHAT'S YOUR BIODIVERSITY IQ?–ANSWERS (Cont'd.)

15. If you decided to throw a party to celebrate the diversity of life on Earth and wanted to send an invitation to each species, how many invitations would you need?

d. But scientists have estimated that as many as 100 million species may actually exist—they just haven't gotten around to identifying all of them yet.

16. Which of the following statements about ostriches are true?

b, d. It's a myth that ostriches hide their heads in the sand. What they're actually doing is lying with their necks flat against the ground to hide themselves and their nests from predators.

17. If the number of species on Earth was represented by physical size, which of the following would most accurately illustrate the proportion of insects to mammals?

c. Bugs rule! There are approximately 250 insect species to every mammal species—and that includes only the insects we know about. Scientists think there are millions more species yet to be discovered.

18. Biodiversity includes:

a, b, c, d. Biodiversity describes the incredible variety of life on Earth—and that includes the diversity among genes (which control inherited traits like the color of your eyes), species (from huge whales to tiny soil creatures), and ecosystems (from lush equatorial rain forests to Earth's icy poles).

19. If there were a prize for "the strongest creature for its size," which of the following would win?

c. An ant can carry a load up to 50 times its body weight.

20. Which of the following would people have to do without if there were no bees?

a, b, c, d, e. Bees are worth billions of dollars to the agriculture industry. Each year bees pollinate millions of acres of almond and apple trees, cucumbers, and celery. Other favorite foods we'd miss without bee pollinators include watermelons, avocados, plums, pears, blueberries, cranberries, cherries, carrots, and cantaloupes.

21. Which of the following is an example of an ecosystem service?

a, c, d. Ecosystem services include the "free services" provided by ecosystems around the world—and which most of us take for granted. For example, wetlands help control floods, filter pollutants from water, and provide habitat for all kinds of birds, fish, and other animals. Ladybugs eat aphids, which are common garden pests. And oceans act as a giant thermostat, interacting with the atmosphere and land to control Earth's climate.

WHAT'S YOUR BIODIVERSITY IQ?–ANSWERS (Cont'd.)

22. Some of the world's most fascinating creatures live in really unusual places. Which of the following is sometimes a home for another living thing?

a, b, c, d. Check it out:

a. A very tiny but complex animal was recently discovered living on the mouth of the Norwegian lobster—on what passes for the lobster's lips!

b. Deep within a termite's gut lives a protozoan (a tiny animal) that helps to digest the termite's woody diet.

c. The rhino belongs to a group of hoofed mammals that have bacteria living in their digestive tracts. In the rhino's case, the bacteria live in the colon (large intestine) where they help the rhino digest plant fibers (cellulose), an important part of the animal's diet.

d. Without knowing it, most human beings have mites on their foreheads. Mites are slender creatures with wormlike bodies and spidery heads; they are so small they are almost invisible. One species *(Demodex folliculorum)* dwells in the hair follicles, and the other *(Demodex brevis)* lives in the sebaceous glands.

23. If you had a job that put you in charge of saving all species on the edge of extinction, how many endangered species would you need to save?

d. You'd be pretty busy conserving habitats for 1,082 species. And that's only the number of species of plants and animals listed as threatened and endangered in the United States by the U. S. Fish and Wildlife Service.

Some scientists estimate that up to 27,000 species go extinct each year, and we never even knew that most of them existed.

24. A small population of beluga whales lives in Canada's St. Lawrence River. Which of the following explains why they are threatened?

c. The beluga whale population is in decline, and many scientists think that their health problems are related to the high levels of chemicals in the St. Lawrence River. The St. Lawrence belugas have been known to die from eating too many contaminated fish. Canadian officials who have found belugas washed up on the banks of the river have classified the whales as "toxic waste."

25. Which of the following environments on our planet are too harsh to support life?

e. Amazingly, life has been discovered in all of these harsh environments. Newly identified microorganisms called "extremeophiles" thrive in unimaginable conditions, like boiling sulfur springs and polar ice fields.

hair follicle mite

WRAPPING IT UP

Assessment

If used as an introductory activity, there is no assessment. If used at the end of a unit, the quiz itself can be the assessment tool if you add questions that best reflect your teaching.

Portfolio

Either at the beginning or end of a unit, the quiz can serve as documentation of the students' general knowledge of biodiversity. Have the students note why they answered as they did and record their own definition of biodiversity on the quiz. Use that as portfolio evidence.

Writing Idea

Write each letter of the alphabet on a separate slip of paper. Fold the slips, put them into a container, and have each student pick one. Then have each student write a poem or limerick about an animal, plant, or other life form that starts with his or her letter. As an option, have the students draw or cut out pictures to go along with their writings and put them all together in a book. Possible titles for the book include *Biodiversity A to Z, An Encyclopedia of Biodiversity,* or *A Poetic Look at Biodiversity.* You could also have the students present their poems to younger students.

Extensions

- Have the students read "Ask Dr. B!" on pages 4–10 of *WOW!—A Biodiversity Primer.* Afterward, lead a discussion about the importance of biodiversity and the threats to it.

- Create a bulletin board or other display featuring the class definition of biodiversity, and magazine photos and student artwork that illustrate biodiversity. Encourage students to choose photos and create artwork that portray the different levels of biodiversity. (See pages 22–29 of the background information for more information.) Also encourage students to connect facts from the quiz with their artwork or photos. If the class continues studying biodiversity, the bulletin board can be updated throughout the year to include new knowledge and ideas.

- Have students develop a TV or radio spot of one to three minutes to help people understand the meaning of biodiversity. They can use any of the information on the IQ test to help them. Remind the students that the more creative and entertaining their spots are, the more likely others are to get the message. When they've finished, have them share their efforts with the rest of the class.

- Have students select a quiz question that they answered incorrectly or a question that interests them. Have them research the subject of the question and use the information they find to write a paragraph that either explains the correct answer or gives more details on the subject.

Resources

1996 World Almanac (World Almanac Books, 1996).

The Diversity of Life by Edward O. Wilson (Harvard University Press, 1992).

Encyclopedia of Insects by Christopher O'Toole (Andromeda Oxford Limited, 1995).

WOW!—A Biodiversity Primer (World Wildlife Fund, 1994).

*"Biodiversity, or natural
riches, is a new term that describes
something very old."*

–Alfredo Ortega, writer

SUBJECTS

language arts, science, social studies

SKILLS

gathering (reading comprehension), organizing (matching), applying (synthesizing)

FRAMEWORK LINKS

1, 3, 33, 42

VOCABULARY

biodiversity, genes, habitat, species

TIME

two sessions

MATERIALS

paper and pencils, copies of "Mystery at Leetown High" (pages 11–13 in the Student Book), copies of the "BioForce Fact Book" (pages 15–16 in the Student Book), copies of the "Mystery Notes" (page 14 in the Student Book)

CONNECTIONS

This introductory activity works well before or after "All the World's a Web" (pages 74–79). "Ten-Minute Mysteries" (pages 98–101), "Biodiversity Performs!" (pages 194–197), and "The Spice of Life" (pages 206–213) are good follow-ups to this activity.

 ### AT A GLANCE

Read a mystery story that introduces biodiversity, match mysterious notes from the story with facts about biodiversity, and list ways biodiversity affects people's everyday lives.

OBJECTIVES

Define biodiversity. Identify and list ways biodiversity affects people.

Biodiversity is a big subject—so big that it can be hard to know how to approach it with students. On the other hand, the fact that it *is* so all-encompassing means biodiversity can appeal to all students, no matter what their interests.

In this activity, your students will read a mystery story that will introduce them to some of the many ways that biodiversity affects our everyday lives. They'll try to piece together mysterious messages that will lead them to the amazing truth: Biodiversity impacts what we eat and wear, the medicines we take, the music we listen to, the furniture we sit on, the books we read, the sports teams we root for, and virtually every other part of our lives.

Before You Begin

Make one copy of the story "Mystery at Leetown High" (pages 11–13 in the Student Book) for each student. Make each group of students a copy of the "BioForce Fact Book" (pages 15–16 in the Student Book) and the page of "Mystery Notes" (page 14 in the Student Book).

What to Do

1. Read "Mystery at Leetown High."

Hand out a copy of the story to each student and give the kids plenty of time to read it. (You might want to make the story a homework assignment.)

2. Define and discuss the term *biodiversity*.

Ask the students what they think the word means. Then explain that, as the story points out, biodiversity describes the variety of life on Earth and all its interactions. Point out that biodiversity includes not only the number of different species in the world but also the different kinds of habitats the species live in and the variety of genes that each organism is made up of.

3. Match notes with appropriate biodiversity blurbs.

Along with a copy of the "BioForce Fact Book," hand out a copy of the "Mystery Notes" to each group. Have the students work in small groups to match each note from the story with the biodiversity blurb in the fact book that pertains to it. (Tell the students that some blurbs refer to two or more notes. They can write the number of the appropriate blurb next to each note.) Possible answers are on page 92.

Another option is to forgo having the kids work in groups and simply lead a discussion in which you ask students what they think each note has to do with biodiversity. (If you go this route, you won't need to make copies of the notes and fact book.) Use the fact book blurbs to guide the discussion.

4. Keep a biodiversity journal for a day.

Have the students keep a journal for a day in which they write down all the ways biodiversity affects their lives. To get them started, go over the following examples:

Food—The variety of foods we eat is one daily aspect of biodiversity. Everything we eat has been derived from wild species. For more about food and biodiversity, see page 33 in the background information.

Clothing—From cotton to silk to wool, much of what we wear comes directly from the natural world. Even synthetic fibers are made from the fossilized remains of plants (coal) and animals (petroleum).

Animals—They work for us, feed us, clothe us, entertain us, inspire us, and even become symbols and mascots for our sports teams.

Medicines—Biodiversity is like a medicine chest: It provides us with countless pharmaceuticals, medicinal herbs, and other medical products.

Art—Music, poetry, paintings, and other art forms draw heavily on the natural world for inspiration.

Miscellaneous—Consider wooden desks, chairs, floors, pencils; paper textbooks, magazines, napkins, notebooks; cotton sheets and towels; canvas backpacks and tote bags; herbal shampoos and toothpaste.

You can't get through a day without using, eating, wearing, or otherwise being influenced by biodiversity—whether you realize it or not.

5. Talk about ways biodiversity affects us.

Have students talk about aspects of biodiversity in their lives, based on what they wrote in their journals.

MYSTERY NOTES—ANSWERS

Remember: *Students may have more answers than the ones listed below. Ask them why they chose their various answers. (The numbers in parentheses represent the number of each fact presented in the "BioForce Fact Book.")*

Jehan's Notes

You can't get away from it! (1, 2)

Your very life depends on it! (4, 5, 11, 14)

Ignoring it could be a fatal mistake! (11, 14)

Megan's Notes

Your pizza depends on it! (7)

Without it, nine-nut crackling crunch bars wouldn't even exist! (10)

Leetown High cafeteria food would be even worse without it! (13)

Noah's Notes

Without it, there would be no Detroit Tigers. (3)

The future of the Baltimore Orioles could depend on it! (12)

If it didn't exist, you wouldn't be a member of the Leetown Leopards junior varsity basketball team. (3, 4)

Jamal's Notes

It could provide a cure for cancer! (5, 11)

It's the source of many important prescription drugs. (5)

It's the world's largest pharmacy! (11)

Miscellaneous Notes

Not hundreds. Not even thousands. We're talking millions! (1, 2)

A vampire could save your life! (5, 8)

Aliens are among us! (6)

Don't look now, but there may be something fishy about your eye shadow. (15)

Bugs rule! (8, 9)

WRAPPING IT UP

Assessment

There is no direct assessment.

Portfolio

The biodiversity journal assignment can be kept in the portfolio as a baseline measure of knowledge, attitudes, and awareness.

Writing Idea

Have students create comic books (using Dr. Detecto or some other comic book character) based on some aspect of biodiversity. Possible titles and plots include:

Polly Parrot Tells All

While being illegally smuggled, a parrot learns key words and phrases that Dr. Detecto uses to nail a big time wildlife smuggling ring.

Mystery in Toxic Town

A mysterious fish kill and a rash of illnesses lead Dr. Detecto to track down an illegal toxic waste dumping operation.

Double Identity

Dr. Detecto proves that Barry Noble, an upstanding citizen of Heartland, USA, is actually I. M. Callous, head honcho of an international wildlife poaching ring.

Extensions

- Encourage students to form a biodiversity club. They can use project ideas listed in the story or come up with their own.

- Pass out a copy of *WOW!—A Biodiversity Primer*. Have students read "Ask Dr. B!" on pages 4–10 of the primer. Afterward, discuss their reactions to the article.

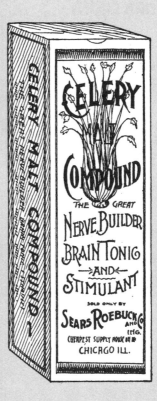

Resources

"Biodiversity—The Fragile Web." *National Geographic* 195, no. 2 (February 1999).

WOW!—A Biodiversity Primer (World Wildlife Fund, 1994).

---

4 | It's a Wild World

SUBJECTS

science, social studies, language arts, art

SKILLS

gathering (researching), applying (creating), presenting (illustrating, writing)

FRAMEWORK LINKS

1, 3

VOCABULARY

bacteria, ecosystem

TIME

one session to two weeks, depending on the activity or actvities chosen

MATERIALS

see individual projects

CONNECTIONS

This introductory activity can be followed by any of the other activities in the module.

 ### AT A GLANCE

Take part in one or more creative activities that introduce the incredible variety of life on Earth.

 ### OBJECTIVES

Define biodiversity, research and discuss amazing biodiversity facts, and describe some of the ways biodiversity affects people's lives.

Imagine coming across the following blurb in a travel brochure:

Fascinating wildlife at every turn. Incredible scenery. Tremendous variety of natural areas to explore. A climate to suit all needs, whether you like it hot, cold, or somewhere in-between. Endless possibilities for fine dining. Opportunities for adventure abound!

If this sounds like a place you'd like to visit, take heart—you're already there! The blurb describes the planet we live on and the amazing collection of life forms and landscapes it harbors, otherwise known as biodiversity.

Here are several ideas for introducing your students to this endlessly fascinating subject and some of the different forms it takes—from beautiful and bizarre animals and plants to the intricate relationships between species to some of the crucial ecological processes we all depend on.

What to Do

1. Introduce or review the term biodiversity.

If this activity is the students' first introduction to biodiversity, use the glossary (page 412–414) and the background information (pages 22–29) to define and lead a brief discussion about it. If the students are already familiar with biodiversity, review the main ideas.

2. Choose a biodiversity project.

Choose one or all of the following suggestions to get the students engaged in discussing biodiversity.

PLANET EARTH: YOUR VACATION DESTINATION

Have your students imagine that, sometime in the far future, the beautiful and lush planet Earth has become an intergalactic vacation spot. As the head of advertising for the Universal Travel Agency, each student should design and create a travel brochure to entice potential alien travelers to come to Earth for a photo-safari eco-vacation. Tell the class to keep in mind that these alien travelers may never have been to Earth before. The student's job is to present the Earth's diverse life forms and ecosystems in an attractive, exciting manner, focusing on particular types of ecosystems or organisms (the ocean or undersea life, for example) or a variety of different types.

YOU READ IT HERE FIRST!

Have students create tongue-in-cheek, tabloid-style newspapers featuring strange-but-true stories that they've researched, along with funny artwork. (To get them started, you might want to have them read *The Natural Inquirer* starting on page 19 of *WOW!– A Biodiversity Primer*.) Here are a few sample titles for articles they could create:

- **Incredible Creature Crawls from the Ooze**
  Lungfish in Africa crawl into the mud when their lakes dry up and breathe through a hole that leads to the surface. When it finally rains, the fish emerge from the mud.

- **Unicorns Live!**
  The narwhal, a type of whale, has a five to ten foot tusk that has reminded people of a unicorn's horn.

- **Daredevil Fish Is Shark's Best Friend**
  The remora, a fish with a sucker on top of its head, attaches itself to a shark and feeds on scraps left behind when the shark tears apart its prey.

- **Aliens Have Landed!**
  The zebra mussel, a European mollusk accidentally introduced into the Great Lakes, is causing big problems for the fishing industry and for native species as it spreads into southern rivers and lakes.

- **Headless Males Litter Countryside**
  During mating, the females of some praying mantis species bite off their mates' heads.

- **Frozen Frog Comes Back to Life**
  Some wood frogs may freeze "solid" in winter but are able to survive.

- **Nighttime Fliers Invade the Skies**
  During the fall and spring, many songbirds, geese, and other birds can be seen migrating. Scientists have learned that they use the stars and other clues to help them navigate at night.

- **Beware of Killer Plants**
  Many species of plants, such as pitcher plants, Venus's flytraps, and sundews, trap and digest insects to get the nutrients they need.

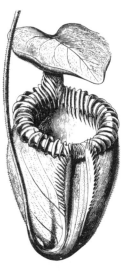

- **Bugs All Over Your Body!**
  Our bodies are home to millions of microscopic mites and bacteria. Many of these organisms are beneficial; some, like eyelash mites, are neither beneficial nor harmful.

Encourage students to further research the subjects of the titles listed here or to find their own stories of weird happenings and strange facts. Magazines such as *Discover* and *Ranger Rick* are great resources for students who are starting to explore the amazing world of biodiversity.

BIODIVERSITY BULLETIN BOARD

This bulletin board project can continue indefinitely. If you change the board periodically to focus on different themes related to biodiversity, your students can discover not only some of the incredible-but-true aspects of the topic but also some of the many ways biodiversity affects our lives.

a) **Choose a biodiversity-related theme for the bulletin board.**
   Have the students decide on a theme and a title for the bulletin board. Possibilities include the following:

- **Biodiversity: The Real World Wide Web**
  a general look at biodiversity and all it involves, from genes to species to ecosystems

- **I Need You: You Need Me**
  amazing relationships and interactions between different species—see background on page 28 for examples

- **Crawling with Creatures**
  a wide-ranging look at some of the different kinds of animals, plants, and other organisms that Earth supports, from the bizarre to the beautiful

- **Have You Had Your Biodiversity Today?**
  various ways biodiversity affects our everyday lives—see background on pages 32–35 for examples

- **Biodiversity on the Cutting Edge**
  biodiversity-related research, including new discoveries about how ecological processes work, the latest information on numbers of species, new products derived from animals and plants, and so on

- **Over the Meadows and Through the Woods**
  a look at ecosystem variety around the globe

- **News of the Wild**
  strange-but-true stories from nature about weird animals or plants, bizarre behaviors, and so on

- **Future World**
  possibilities, both positive and negative, of what the future might bring for biodiversity and humans

Have several students prepare the bulletin board for the daily addition of articles and other items (see step c) by creating the title and adding artwork or photos to highlight the theme.

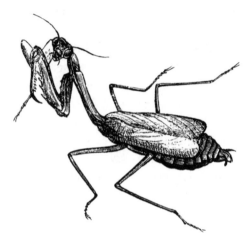

b) **Add to the bulletin board daily.**
   Have the students add one or more items to the bulletin board each day. One way to do this is to assign each student or group of students a day that they'll be responsible for bringing in a newspaper article, picture, Internet story, short essay, or other information based on the bulletin board's theme. Have students explain their item and how it relates to biodiversity before adding it to the bulletin board.

c) **Change the bulletin board periodically. (optional)**
   If your biodiversity or environment unit is ongoing, you might want to change the specific theme of the bulletin board every week or two to expose the students to different aspects of biodiversity.

WRAPPING IT UP

Assessment

Use this activity as either an introductory or a summary activity. No separate assessment is needed, although the bulletin boards or brochures can be assessed.

Portfolio

Have the students initial their contributions to the bulletin board and include them in the portfolio. The *Planet Earth: Your Vacation Destination* brochure can also be placed in the portfolio along with the article from *You Read It Here First!*

Writing Idea

As a class, have the students "publish" *a Best of Biodiversity* book featuring facts about the most bizarre, unusual, beautiful, and interesting aspects of biodiversity. Each student could be in charge of researching and writing one or more articles.

Extension

Either individually or in groups, have students create biodiversity collages. The collages can be thematically broad, covering many aspects of biodiversity, or they can focus on a specific theme. In the latter case, you may want to have students choose one of the bulletin board themes listed on page 96.

Magazines

Audubon, Discover, International Wildlife, National Geographic, Ranger Rick

Web Sites

Visit the National Wildlife Federation Web site at <www.nwf.org>.

Visit the New Scientist Web site at <www.newscientist.com/lastword/plants.html>.

Visit the Wildlife Kid Web site at <www.kidinfo.com/Science/Wildlife.html>.

Visit the World Wildlife Fund Web site at <www.worldwildlife.org>.

Books

The Collins Guide to Rare Mammals of the World by John A. Burton (Greene Press, distributed by Viking Penguin, 1988).

Eco-Inquiry by Kathleen Hogan (Kendall/Hunt Publishing Company, 1994).

WOW!—A Biodiversity Primer (World Wildlife Fund, 1994).

5 | Ten-Minute Mysteries

SUBJECTS

language arts, science, social studies

SKILLS

analyzing (identifying components
and relationships among components,
identifying patterns)

FRAMEWORK LINKS

33

VOCABULARY

biodiversity, burr, dependence,
ecosystem, eucalyptus, **microorganism,**
organic, penicillin, sequin, vampire bat

TIME

one session

MATERIALS

copies of "Ten-Minute Mysteries"
(pages 17–22 in the Student Book)

CONNECTIONS

For a fun way to introduce your
students to biodiversity, try "What's
Your Biodiversity IQ?" (pages 80–89)
or "Something for Everyone" (pages
90–93). To learn more about amazing
biodiversity connections, try "Super
Sleuths" (pages 182-187), "'Connect
the Creatures' Scavenger Hunt" (pages
102-107), or "The Culture/Nature
Connection" (pages 214-225).

AT A GLANCE

*Solve "mysteries" as you discover some amazing
and little-known connections among people,
species, and ecosystems.*

OBJECTIVES

*Explain several ways biodiversity is
connected to human lives.*

There are countless connections between people and
biodiversity. Plants and animals are the sources and
pollinators of many of our foods, they provide the
chemical foundation for many of our medicines, and
they have inspired many of our important inventions. But how
much do we know—or think—about the behind-the-scenes benefits
that plants and animals provide? This activity helps your students
uncover some of these amazing (and surprising!) connections in a
lively, challenging mystery game.

Before You Begin

This activity can be organized in one of two ways. You may choose to lead the entire activity yourself, reading all the mysteries aloud and providing the clues as students try to solve the mysteries as a group. Or you may prefer to divide the class into teams of two or three students and then pair the teams, letting each side take turns posing mysteries to the other. If you choose the latter approach, make enough copies of the mysteries on pages 17–22 in the Student Book so that each *pair* of teams gets a complete set of the mysteries, and divide the mysteries so each team has half of the mysteries and answers.

What to Do

1. Introduce a practice mystery.

Tell your students that their mission is to try to solve some "ten-minute mysteries." (You don't need to explain that the mysteries relate to biodiversity; that will become clear later.) They can start by solving the practice mystery as a class.

Practice Mystery
A boy tells his mother that he has a sore throat and a fever. She turns to the boy and says, "Lucky for you, there's moldy bread in the world." **WHY?**

Answer: The boy has strep throat and will probably be treated with penicillin (a common antibiotic used to treat throat infections). Penicillin was originally discovered when a scientist noticed that mold, growing in his Petri dishes, seemed to produce a substance that killed bacteria. (Mold is a type of fungus that often grows on bread and other foods.)

2. Solve the practice mystery.

Read the practice mystery aloud. Explain to the students that the mystery may seem impossible to solve at first. But they can get more information by asking questions. The only rule is that the students must ask questions that can be answered with "yes," "no," or "I don't know." For example they could ask, "Is the boy really sick?" or "Will the doctor give the boy bread mold?" If they get stuck, help them out with the following clues (given one at a time):

- **Clue #1**–The boy has strep throat, an infection caused by bacteria.

- **Clue #2**–Bread mold contributed to an important medical discovery made in 1928.

- **Clue #3**–Strep throats are treated with penicillin, an antibiotic (or substance that kills bacteria).

Afterward, review the solution. You may want to explain that penicillin was discovered by the Scottish doctor Alexander Fleming. He discovered the fungus growing by accident in his lab and named it *Penicillium notatum*. Then he isolated the substance in the fungus that killed bacteria and called it penicillin. Penicillin was first used to treat patients in 1940. Today it is synthetically produced.

bread mold

3. Continue giving new mysteries.

Once your students have solved the practice mystery, challenge them to solve some of the other mysteries provided, either as an entire class or in teams. If you have the students work in teams, pair teams of two to three students so that each team can take turns posing mysteries to the other. Divide the nine mysteries so that each team has half of the mysteries and answers that they'll then read to the team they're paired with. Give each team copies of only the mysteries they'll read to the other team.

4. Discuss some of the connections as a group.

Discuss which connections the students found most amazing or surprising and why. Did any of the mysteries change the way the students feel about living things? Ask your students what important point is made by all of these mysteries. (Answers will vary, but students should observe that humans depend on plants, animals, and microorganisms in many important ways.) Discuss the term *biodiversity* with the students. (Biodiversity is the diversity of life on Earth, including the wealth of different genes, species, and ecosystems.) Be sure the students understand that our dependence on other living things is a dependence on biodiversity.

5. Write statements about the connections between people and biodiversity.

Divide your group into small teams. For each mystery it solved, ask each team to write a sentence or brief statement that describes the important connection to biodiversity the mystery revealed. For example, a statement related to the moldy bread mystery could be: "Biodiversity is a source of important medicines, such as penicillin." Or, "Some small organisms that we might not think are important can be very useful—and even lifesaving."

BIOFACT

Friendly Fires—Some species of pine tree depend on fire in order to spread their seeds. The seeds are sealed inside cones with a layer of sticky sap. When a forest fire heats the cones, they burst open and send the seeds flying in all directions. With luck, some may land where the fire has already passed. The rich, ash-covered soil is an ideal place for the seeds to sprout.

WRAPPING IT UP

Assessment

There is no direct assessment.

Portfolio

Statements created in step five can be used for the portfolio.

Writing Idea

Write an article (no more than 500 words) that explains some of the ways people rely on biodiversity, and submit it to one of your local newspapers. Think about whether the people who read the paper will know what biodiversity is. Try to make the article as lively and colorful as possible. (You can use information from the mysteries in your articles.)

Extension

Have students work individually or in small groups to write their own ten-minute mysteries. They can choose an idea from *WOW!—A Biodiversity Primer* or come up with an idea of their own. Encourage them to consider the variety of benefits that species can provide: medicines, foods, pollination services, inspiration for inventions, and so on. Read the mysteries aloud and have the students try to solve them. Then discuss what makes a good mystery. (Assess for clarity and whether the mystery is solvable.) You can also have them create mysteries that focus on interconnections among species, such as ants and acacia trees.

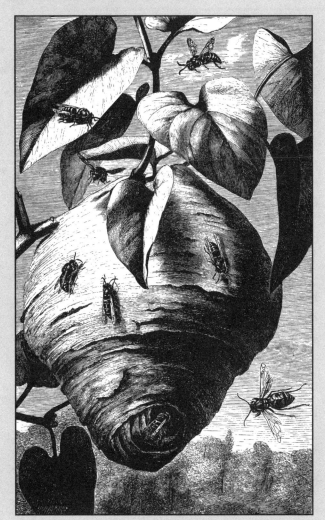

Resources

Amazing Biofacts: The Human Body, Animals, Plants by Susan Goodman (Peter Bedrick Books, 1993).

"Biodiversity—The Fragile Web." *National Geographic* 195, no. 2 (February 1999).

WOW!—A Biodiversity Primer (World Wildlife Fund, 1994).

"Connect the Creatures" Scavenger Hunt

SUBJECTS

science

SKILLS

gathering (collecting), organizing (manipulating materials), analyzing (identifying components and relationships among components), presenting (describing, illustrating)

FRAMEWORK LINKS

1, 3, 12

VOCABULARY

biodiversity, ecosystem, food chain, habitat, introduced species, invertebrate, native species

TIME

one session or stretched out as homework over a period of time

MATERIALS

copies of "Creature Connections" (page 23 in the Student Book), plus

- *for outdoor version: bags for collecting scavenger hunt items*

- *for indoor version: nature and science magazines, scissors, glue, and poster board or paper for each group of students*

CONNECTIONS

Before this activity, you can introduce the concept of biodiversity with "Something for Everyone" (pages 90–93) or "All the World's a Web" (pages 74–79). Try "Super Sleuths" (pages 182–187) to explore more connections among living things. For more on habitats and the species that use them, try "Backyard BioBlitz" (pages 134–143), "Mapping Biodiversity" (pages 252–265), or "Space for Species" (pages 286–301).

AT A GLANCE

Search for examples of connections among living things and their habitats by taking part in a scavenger hunt.

OBJECTIVES

Describe some of the ways organisms are connected both to their habitats and to other organisms, and give examples of these connections. Discuss specific ways that people are connected to the natural world.

It may sound like a cliché, but it's true: Every living and nonliving thing in the natural world, whether it's a rock, a river, a beetle, or a bird, is connected to many other things in complex ways. These connections are often so intricate that it's nearly impossible to separate the threads that weave them together. For example, nearly every leaf that falls on the floor of a forest in the Pacific Northwest is eaten by a certain kind of millipede *(Harpaphe haydeniana)*. As the leaves pass through the millipede's gut, they are broken down into a form that millions of other invertebrate species can use. These species break the leaves down even further, making the nutrients in them available to forest plants, which absorb the nutrients from the soil through their roots.

The more we learn about connections in the natural world, the more we realize just how much more there is to know. Here's an activity that reinforces the concept of connections—not only between and among animals and their habitats, but also between people and the natural world.

Before You Begin

Make copies of "Creature Connections" on page 23 of the Student Book (one per student if students will be working individually; otherwise, one per group).

Based on whether you'll be doing the hunt outdoors or indoors, complete one of the following sets of directions:

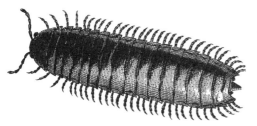

- **Outdoor Version:** Locate an outdoor area ahead of time and set boundaries within which the students can work. Gather one plastic collection bag for each student. Become familiar with the collection tips on page 104.

- **Indoor Version:** Collect nature and science magazines that the students can cut up. Gather scissors, glue, and poster board or paper for cutting and mounting the photos. (See the list of photo sources on page 104 for suggestions.)

What to Do

1. Introduce the activity and distribute copies of "Creature Connections."

Pass out copies of "Creature Connections" (either one per person or one per group). Explain to your students that they will be taking part in a scavenger hunt to look for connections in the natural world. Then follow the appropriate directions below:

Indoor version: Distribute the magazines, along with scissors, glue, and paper or poster board. Tell your students they can cut out or draw pictures that illustrate the items on the list.

Outdoor version: Distribute bags for collecting small plant samples and nonliving items such as acorns, pebbles, dead leaves, and so on. Whether you conduct the activity indoors or outdoors, explain to the students that they will have a certain amount of time to collect as many of the scavenger hunt items as they can. (You can do the hunt in one class period, or you can spread it out over the course of a couple of weeks and have the students conduct the hunt on their own time.) If the students will be working in groups, you might want to have each group divide up the items on the list.

2. Review the items on the list.

Go over the list with your students to be sure they understand each item. Explain that they need to be able to show the connections between what they find and the items on the list. (Remind the students using magazines to keep track of which picture goes with which scavenger hunt item.)

Photo Sources
(for Indoor Version)

There are a number of magazines that feature photos your students can use in this activity. Below are several examples:

Audubon	*International Wildlife*
National Geographic	*National Geographic World*
National Wildlife	*Natural History*
Ranger Rick	*Sierra*
Smithsonian	*Discover*

Collection Tips

If you decide to have your students collect real examples of scavenger hunt items, be sure to do each of the following before starting:

- Instruct students to collect dead leaves, nuts, or other parts that do not kill or harm the plant. Encourage the students to draw sketches of the plants they want to record or to take small samples from common plants.

- Tell students not to collect bird nests. In many areas, such as certain federal and state-owned lands, it is illegal to do so. But even if it's not illegal in your area, students should avoid collecting them. Sometimes it's hard to distinguish an active nest from an inactive one—and even a nest no longer being used by its original "owners" may serve as a home for other organisms.

- Encourage students to be gentle when turning over rocks and logs.* If they find insects, centipedes, salamanders, or other small animals, encourage students to observe closely, but leave the animals in their homes. (Collecting can be stressful for some animals. Plus, they can be very difficult to maintain in the classroom, even for a short period of time.)

- Have students return rocks, logs, and plant matter to their original location.

- Set boundaries and time limits so that students don't scatter or get lost.

- Check for poisonous plants before sending students into the area. If there are any poisonous plants, be sure students can recognize them and know they are not to be collected.

*Federal and state parks now request that rocks and logs be left unturned and untouched to avoid harming these small habitats and the animals that live under them.

3. Begin the hunt.

Give the students time to find as many items as they can. Tell them that if they can't find an item (or, in the case of the outdoor version, if an item they find isn't appropriate for collecting), they can draw it or fill in the answer based on what they know. An answer sheet is provided on page 105. Use it to stimulate ideas if the kids get stuck.

> Note: Remind students doing the outdoor hunt that they should not collect living animals and that they should take only small samples of common plants. Have them draw any living things they find that are not appropriate for collection. (See "Collection Tips" on the left.)

4. Discuss the results of the scavenger hunt.

At the end of the hunt, have the students share some of the connections they discovered. Explain that the creatures they found, and the connections between them, are just a small sample of what makes up biodiversity—the diversity of life on Earth. Not only is our planet covered with an amazing variety of life, but also, as your students discovered, all the different organisms are connected in complex ways. Then ask the following question: *Did this activity point out any connections you'd never thought of or known about before? Give examples.*

CREATURE CONNECTIONS–ANSWERS

1. a small animal that a bigger animal depends on in some way *(Insects are food for many animals; krill are eaten by certain whales and other sea life.)*

2. a big animal that a smaller animal depends on in some way *(Fleas and ticks depend on bigger animals such as squirrels and deer. Mosquitoes depend on humans and other warm-blooded animals. An oxpecker is a bird that gets its food by removing insects from the backs of rhinos.)*

3. a plant that grows on other plants *(bromeliads, Spanish moss [a member of the pineapple family], orchids, moss, lichens)*

4. an introduced species that has caused problems for native species *(kudzu, which grows aggressively and crowds out native plants; starlings, which take nesting sites that bluebirds and other native species would otherwise use; Norway rats, which eat eggs and nestlings of native birds in Hawaii. See "Aliens Are Among Us!" on page 49 of the background information for more about introduced species.)*

5. a wild animal that can thrive in or around people's homes *(German cockroaches, house flies, house mice, raccoons, gray squirrels, house finches)*

6. an animal that eats dead things *(wood beetles, springtails, termites, millipedes, crows, gulls, vultures)*

7. an animal home that is in or on a plant *(nests and dens in trees, cocoons attached to plant stems and leaves, insect galls)*

8. a) a plant that benefits humans in some way *(crops; trees [timber, shade trees, food, windbreaks]; medicinal plants such as certain herbs or plants from which pharmaceuticals are derived; plants that provide fibers)*
   b) a plant that harms humans in some way *(poison ivy, nightshade, stinging nettle)*

9. a) an animal that benefits humans because of the role it plays in its habitat *(bees, bats, butterflies, wasps, and moths pollinate crops; dragonflies and bats eat mosquitoes; earthworms aerate and mix the soil; and some snakes eat rodents that might otherwise eat crops or cause other damage)*
   b) an animal that harms humans in some way *(mosquitoes, fire ants, Norway rats)*

10. an animal that looks like a plant *(insects such as praying mantids or certain moths that mimic twigs, leaves, bark, and other plant parts)*

11. two species that are useful to each other in some way *(honey bees and the flowers they pollinate, clownfish and sea anemones, rhinos and oxpeckers, lichens [symbiotic relationship of fungi and algae])*

12. an animal that spends its life in two different habitats *(certain migratory birds; amphibians and insects such as dragonflies that spend part of their life in water and part on land; fish that travel back and forth between freshwater and saltwater habitats)*

13. an animal that eats seeds or fruits and then spreads the seeds by passing them as waste *(birds such as thrushes, tanagers, and warblers; fruit bats and other animals)*

14. something that turns into soil *(rotting log, crumbling rock)*

15. a plant that depends on animals in some way *(any plant that is pollinated by animals or that has its seeds dispersed by animals)*

16. a plant that is part of a food web *(any plant, because all plants provide food for other organisms)*

17. an animal that is part of a food web *(any animal—including humans—because all animals must eat other animals, plants, or both)*

WRAPPING IT UP

Assessment

Divide the class into pairs or small groups before conducting the hunt. Have the students complete the activity as a team. After the group discussion, have each team select four or five items on its list and label and describe each connection in one collage, poster, or drawing. Students should include a few sentences that explain why they selected those particular connections to display. For the outdoor hunt, the collage, poster, or drawing can show the connections within the habitat they observed. The display should show that some organisms are connected to more than one other organism or habitat component and that most organisms are indirectly connected to each other. Have each student sign the artwork with a statement of how he or she is connected to the natural world.

Unsatisfactory—The team is unable to produce a document that shows connections with the observed habitat.

Satisfactory—The team produces a document that illustrates and clearly labels at least four items and the relationships among the four. The connections selected should be clearly explained and depicted in the collage, poster, or drawing.

Excellent—The team shows creative insight into relationships and is able to reveal relationships that go beyond the obvious. The rationale for choice indicates careful consideration leading to the selection of the connections chosen, as does the individual statement of connectedness.

Portfolio

The artwork should be part of the student's portfolio. A special note in the portfolio could reflect student statements on how the students are connected to the natural world.

Writing Ideas

- Have students write a paragraph that describes a connection that they noticed for the first time when they conducted the scavenger hunt.

- Have students write a paragraph that describes one connection they observed during the scavenger hunt. Then have them write a second paragraph that speculates on what the effects might be if one of the "connectors" were to suddenly disappear.

- Have students write one or more paragraphs either supporting or refuting the following statement: *People are a part of nature.* Have them express their agreement or disagreement with the statement, along with examples that support their view.

Extensions

- Use one of the connections students discovered during the activity to create a web of life. To do this, write the name of one of the organisms on the board. Also write the name of the organisms or habitats to which it has a connection. Draw an arrow between the two words. Then have the students add other items from the scavenger hunt to the board, along with arrows connecting them to one another and, if possible, to the other connected words on the board. Continue until a "web" of words develops. (See "All the World's a Web" [page 76–78] for sample webs.) To conclude the activity, have each student create a web of life that depicts some of the ways he or she interacts with, is influenced by, or depends on the natural world.

- Have students choose a tree, a bird, an insect, or some other organism to observe. During their observations, they should take notes on how their organism is connected to other organisms and to its surroundings. (They can include connections they didn't specifically observe, such as how the organism benefits humans.) Have the students present their findings to the class, and encourage them to be creative in their presentations. For example, they could act out certain connections or present entertaining graphics.

Resources

The Diversity of Life by Edward O. Wilson (Harvard University Press, 1992).

Environmental Science: The Way the World Works by Bernard Nebel and Richard T. Wright (Prentice Hall, 1997).

*"One hundred trout are needed to support
one [person] for a year. The trout, in turn, must consume
90,000 frogs, that must consume 27 million
grasshoppers that live off of 1,000 tons of grass."*

—G. Tyler Miller, Jr., writer

7 | Sizing Up Species

SUBJECTS

science, mathematics

SKILLS

organizing (classifying, estimating, graphing), analyzing (calculating), interpreting (relating)

FRAMEWORK LINKS

3, 21

VOCABULARY

abdomen, antennae, appendages, arthropod, **bacteria**, cephalothorax, class, classification, **evolution**, family, **fungi**, genus, kingdom, order, organism, phylum, **species**, taxonomy

TIME

two sessions to complete all three parts

MATERIALS

Part I—copies of "Arthropod Pictures" and "Arthropod I.D. Chart" (pages 24–25 in the Student Book), scissors and glue (optional)
Part II—stack of 100 sheets of paper or textbook with 100 pages, rulers, calculator (optional), set of number cards, signs for organism groups, markers, tape or glue, copies of "Sizing Up Species" on pages 12–13 of WOW!—A Biodiversity Primer
Part III—graph paper, rulers, colored pencils, Biodiversity! Exploring the Web of Life video (optional)

CONNECTIONS

Use "Biodiversity—It's Evolving" (pages 168–179) to show how closely related species develop. To explore arthropod diversity in your area, try "Insect Madness" (pages 144–157). For more on graphing, try "Space for Species" (page 286–301).

AT A GLANCE

Classify organisms using a classification flow chart, play a team game to find out how many species may exist within different groups of organisms, and make a graph to illustrate the relative abundance of living things.

OBJECTIVES

Use a classification flow chart to classify organisms. Name the major groups of organisms and the relative number of species identified in each group. Construct bar graphs that compare the number of species in different groups of organisms.

Did you know that a single tree in a rain forest can be home to more than 1,000 different kinds of insects? Or that a coral reef can support as many as 3,000 varieties of fish and other organisms? Or that the deep ocean floor may be home to more than 10,000 species of living things? The sheer number of organisms living on Earth is extraordinary. So far, scientists have identified about 1.7 million species, but there are actually many more. Estimates range from 3 million to more than 100 million.

This activity will help your students understand how scientists classify organisms and how many species have been identified within various groups. They'll discover, for example, that there are a whopping 950,000 different species of insects compared to about 4,000 mammal species. And there are still vast numbers of insects that are waiting to be identified—even though about 7,000 new insect species are described every year!

Part I introduces students to the biological classification system by guiding them through the identification of selected orders within the phylum Arthropoda. In Part II, students work in teams to estimate the total number of different species and the number of species in various organism groups. In Part III, students learn if their estimations were correct and then create graphs that illustrate which organism groups contain the most species that have been identified to date.

Make a copy of the "Arthropod Pictures" (page 24 in the Student Book) and the "Arthropod I.D. Chart" (page 25 in the Student Book) for each student. Depending on your students' choices, you may also need scissors and glue.

What to Do • Part I

1. Introduce classification.

Begin by explaining to your students that scientists classify living things into various groups. The system they use classifies organisms into ever more closely related groups and gives scientists from all over the world a common way to refer to particular organisms. To give the students a sense of how this classification system works, use the following information to compare the classification of a house cat with a dog. (See "Classification Chart" on page 111.) The students should notice that the cat and the dog share many classification groupings. Cats and

dogs are in the same kingdom, phylum, class, and order, but they belong to different families. You might also ask the students to name other species that would be in the same family as a house cat (lynx, bobcat, lion, tiger, puma, and other cats) as well as other species that would be in the same family as a dog (wolf, fox, coyote, jackal, and so on). You can also have the students name non-mammal chordates which are animals with backbones (fish, amphibians, reptiles, and birds), or non-cat and non-dog carnivores (bears, raccoons, weasels, mongooses, and so on).

Sorting Out Taxonomy

Naming Things

The work of classifying organisms is done by scientists called *taxonomists*. Taxonomists divide organisms into a hierarchical series of more and more specific groupings. The most general division of life is into five kingdoms: Monera, Protista, Fungi, Animalia, and Plantae. (See page 119 for a description of each kingdom.) Within each kingdom, there are groups of increasing specificity, each one containing fewer species of increasingly close evolutionary relationships to each other. These groups are phylum, class, order, family, genus, and species (see Figure 1 on page 111). This hierarchy enables taxonomists to group organisms based on their characteristics and evolutionary relationships. Species in any given order are more closely related to each other than to species in any other order; species in any given family are more closely related to each other than to species in any other family; and so on.

What's in a Name?

Most organisms have more than one common name. For example, what some people call a puma might be called a cougar or a mountain lion by other people. And a plant might be called cloudberry, salmonberry, or baked-apple berry depending on who is talking about it. This can be very confusing! Taxonomists use Latin words to give scientific names to organisms. Not only does this clear up the confusion over common names in any one language, but it also allows scientists who speak different languages to clearly identify any particular organism or group of organisms.

When scientists refer to a particular organism by its scientific name, they are using a combination of the genus (plural: genera) and species to which the organism belongs. For example, a coyote is referred to as *Canis latrans* (*Canis* is the genus name and *latrans* is the species name). Gray wolves, a closely related species, are *Canis lupus*. The genus and species names are always italicized. The genus name is capitalized but the species name is not.

Keeping Relationships Straight

Figuring out just where an organism belongs—how it should be classified—is not always easy. Scientists look for structural and genetic similarities among organisms that they classify together. But differences and similarities among living things are not always clear cut. Taxonomists sometimes disagree about where organisms should be classified, how genera should be arranged within families, and so on. As new information becomes available, taxonomists often revise where an organism is placed within the classification system. For example, giant pandas, which share some characteristics with raccoons and some with bears, have long been classified, along with red pandas, in their own group. However, recent genetic analysis has confirmed that giant pandas are actually true bears, and taxonomists are revising the species' classification based on those findings.

Defining a Species

A species is a population of organisms that interbreeds and produces fertile offspring in nature. For example, white-tailed deer and mule deer are different species because they coexist in many areas but they do not interbreed.

Taxonomy organizes organisms in increasing levels of specificity. A gray squirrel, for example, would be classified like this:

Kingdom: Animalia (animals)
Phylum: Chordata (animals with backbones)
Class: Mammalia (mammals)
Order: Rodentia (rodents)
Family: Sciuridae (squirrels and chipmunks)
Genus: *Sciurus* (squirrels)
Species: *carolinensis*

To refer to a gray squirrel, scientists call the animal by its scientific name: *Sciurus carolinensis*.

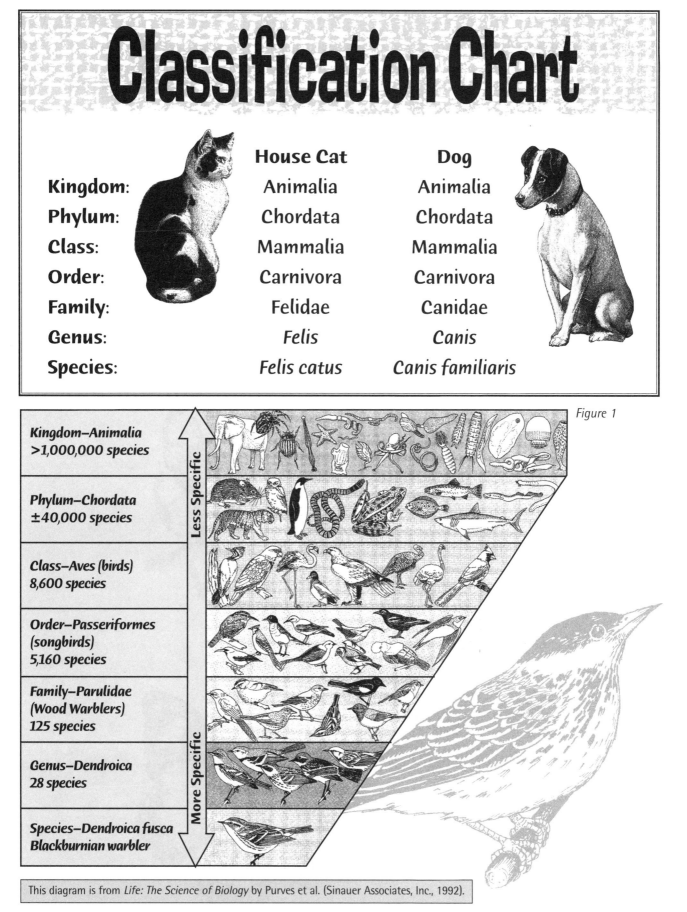

Classification Chart

	House Cat	**Dog**
Kingdom:	Animalia	Animalia
Phylum:	Chordata	Chordata
Class:	Mammalia	Mammalia
Order:	Carnivora	Carnivora
Family:	Felidae	Canidae
Genus:	*Felis*	*Canis*
Species:	*Felis catus*	*Canis familiaris*

Figure 1

Less Specific

Kingdom–Animalia
>1,000,000 species

Phylum–Chordata
±40,000 species

Class–Aves (birds)
8,600 species

Order–Passeriformes
(songbirds)
5,160 species

Family–Parulidae
(Wood Warblers)
125 species

More Specific

Genus–Dendroica
28 species

Species–Dendroica fusca
Blackburnian warbler

This diagram is from *Life: The Science of Biology* by Purves et al. (Sinauer Associates, Inc., 1992).

2. Use the identification chart to introduce the process of classifying organisms.

Hand out a copy of the "Arthropod Pictures" (page 24 in the Student Book). (Please note that this is a simplified chart and that all classes and orders in the phylum Arthropoda are not represented.) Start by writing the names of the five kingdoms on the board or overhead (see page 119). Ask your students if they can describe some of the characteristics of organisms that would be classified in each kingdom. Ask them if they can name which kingdom includes humans. Then ask them which kingdom they think the organisms included in the "Arthropod Pictures" belong to (Animalia). Ask them to look carefully at the drawings and try to determine what characteristics all these organisms share. Students may notice that all the organisms pictured have jointed legs, antennae, and other appendages. They may also know that all the organisms have a hard outer shell (exoskeleton). These common characteristics help organize them into a group, or phylum, within the kingdom Animalia called Arthropoda. Arthropoda means "jointed feet."

Hand out "Arthropod I.D. Chart" (page 25 in the Student Book) and explain that the organisms in the phylum Arthropoda can then be divided into different subgroups including subphyla and classes, which share more specific characteristics. For example, arthropods that are members of the class Insecta have six legs and bodies that are divided into three major parts—head, thorax, and abdomen. Scientists use keys such as this to identify unknown organisms and relate them to other more familiar species.

Write the following vocabulary definitions on the board to help your students go through the chart:

Appendage
any body part that extends outward from the main body, or trunk, of an animal, such as a leg, a claw, or an antenna

Antennae
sensory appendages located on the head or cephalothorax of some arthropods

Arthropod Pictures—Answers

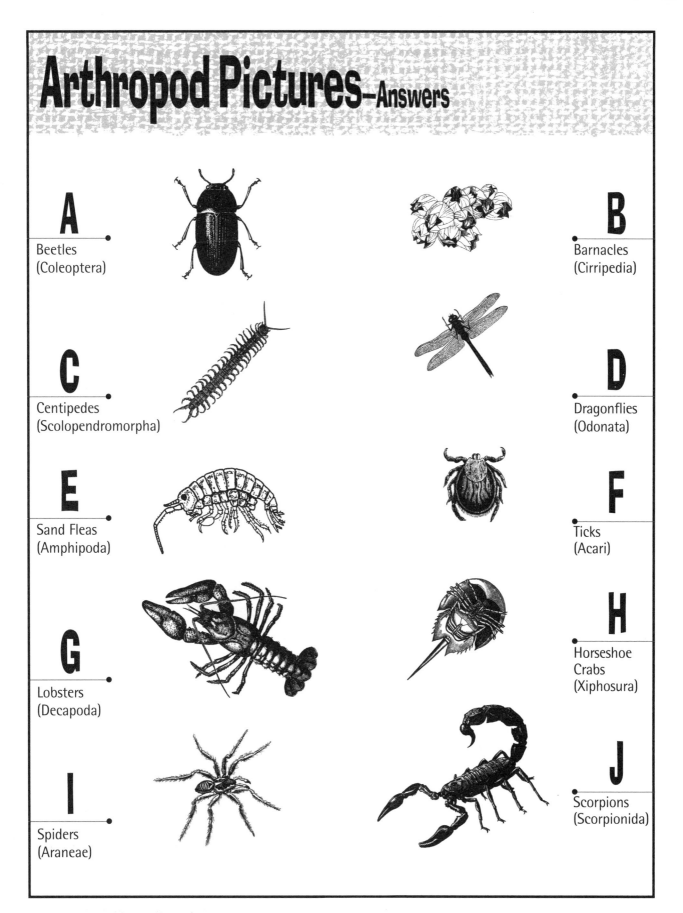

A Beetles
(Coleoptera)

B Barnacles
(Cirripedia)

C Centipedes
(Scolopendromorpha)

D Dragonflies
(Odonata)

E Sand Fleas
(Amphipoda)

F Ticks
(Acari)

G Lobsters
(Decapoda)

H Horseshoe
Crabs
(Xiphosura)

I Spiders
(Araneae)

J Scorpions
(Scorpionida)

3. Use the identification chart to classify scorpions, spiders, centipedes, and other arthropods.

Explain to your students that they will be using the identification chart to identify each of the organisms on the "Arthropod Pictures" page. Remind them that all of the organisms are in the kingdom Animalia and the phylum Arthropoda. Their job is to find out which class and order each organism belongs to.

Go over one example with the whole class to familiarize your students with the key. For example, hold up a picture of organism A, then ask the students to read the descriptions for each of the two subphyla on the I.D. chart and decide which subphylum organism A belongs to (Mandibulata).

Next have the students read the descriptions of each of the classes beneath Mandibulata and decide which class it belongs to (Insecta). Continue to order (Coleoptera [beetles]). Once your students figure out an animal's place on the chart, have them write the letter that corresponds to it under the order it belongs to.

Now have your students work individually or in pairs to classify each of the organisms on the "Arthropod Pictures" page. (The answers are provided on page 113.) Remind the students that they need to start at the top of the chart each time. If they'd like, they can cut out the organisms and glue the pictures to the proper places on the bottom of the chart.

Before You Begin • Part II

Your students will be working in groups of three to five. For the math problem, each group will need a ruler and a textbook with at least 100 pages in it (or a stack of 100 sheets of paper). Calculators are optional.

You'll also need to make a set of number cards for each group using a thick marker. To make each set of cards, you'll need six small pieces of paper (about 4 in. x 6 in.). Write the following numbers on separate cards: 4,000; 9,000; 19,000; 72,000; 270,000; 950,000. (Make the numbers large enough so they can be seen from a distance.) If you like, you can color-code each card set by using different colors of paper or markers.

Next make six signs using a flip chart or butcher paper. Write one of the following words or phrases on each sheet of paper: insects; trees, shrubs, and herbs; mammals; birds; fungi; and fish. Then hang the signs on the walls in your classroom. Have small pieces of tape or a glue stick handy for attaching the scraps of paper to each flip chart.

Finally, make a copy of the species-scape on page 120 for each student (if students are working alone) or for each group (if students are working in teams). (You can also make copies of the species-scape on pages 12–13 of *WOW!—A Biodiversity Primer*, although the numbers reflect an older estimate of the identified species in each group.)

What to Do • Part II

1. Discuss how many organisims there are on Earth.

Ask your students to estimate how many different *kinds* of organisms (species) they think there are in the biosphere. You may first need to explain that a species is an interbreeding population of organisms that can produce fertile, healthy offspring.

Discourage students from simply guessing a total number of species. Instead encourage them to reflect on prior knowledge and observations. Allow students to discuss their reasoning. Have each student make an estimate and explain how he or she arrived at that number. If estimates are low, ask students if they considered organisms of all sizes, including microscopic organisms. Finally reveal to the

students that so far scientists have identified approximately 1.7* million different organisms in the biosphere. But they predict that there may be an additional 2 to 100 million species that haven't been identified yet.

Help students gain an appreciation for how many 1.7 million is. Divide the students into small teams (three to five students per team). Provide them with a ruler and a stack of 100 sheets of paper or their textbook. Ask them to work together to solve this problem: If you were to write the name of every known living species (1.7 million) on a different sheet of paper and then stack up all the sheets, how tall would the stack be?

*Current estimates are 1.7 million. A few years ago the accepted number was 1.4 million, which is why the numbers in *WOW!—A Biodiversity Primer* say 1.4 million.

Arthropod I.D. Chart—Answers

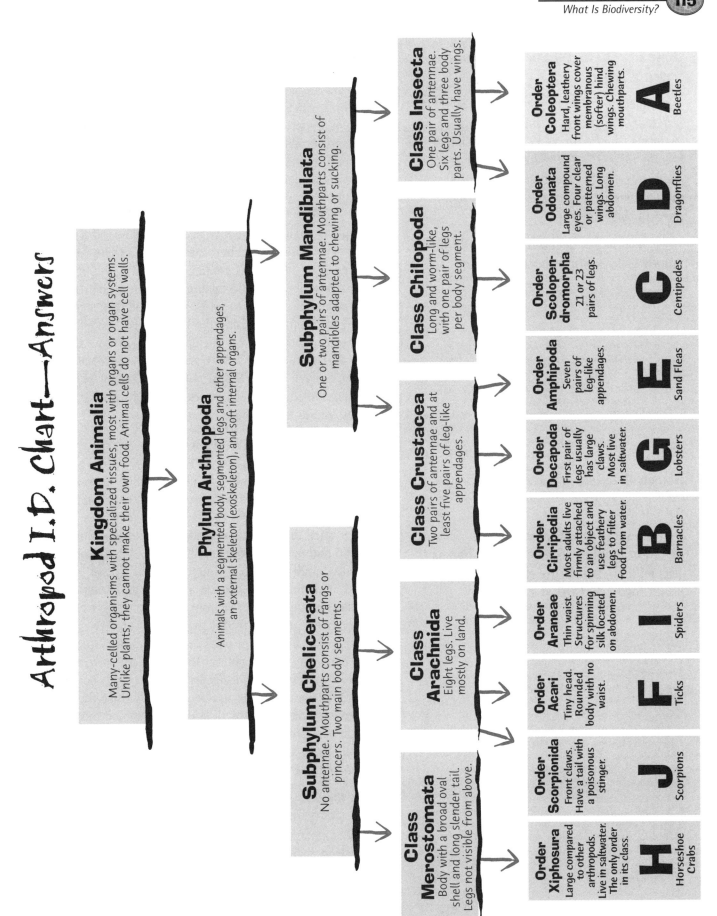

Kingdom Animalia
Many-celled organisms with specialized tissues, most with organs or organ systems. Unlike plants, they cannot make their own food. Animal cells do not have cell walls.

Phylum Arthropoda
Animals with a segmented body, segmented legs and other appendages, an external skeleton (exoskeleton), and soft internal organs.

Subphylum Chelicerata
No antennae. Mouthparts consist of fangs or pincers. Two main body segments.

Subphylum Mandibulata
One or two pairs of antennae. Mouthparts consist of mandibles adapted to chewing or sucking.

Class Merostomata
Body with a broad oval shell and long slender tail. Legs not visible from above.

Class Arachnida
Eight legs. Live mostly on land.

Class Crustacea
Two pairs of antennae and at least five pairs of leg-like appendages.

Class Chilopoda
Long and worm-like, with one pair of legs per body segment.

Class Insecta
One pair of antennae. Six legs and three body parts. Usually have wings.

Order Xiphosura
Large compared to other arthropods. Live in saltwater. The only order in its class.
H
Horseshoe Crabs

Order Scorpionida
Front claws. Have a tail with a poisonous stinger.
J
Scorpions

Order Acari
Tiny head. Rounded body with no waist.
F
Ticks

Order Araneae
Thin waist. Structures for spinning silk located on abdomen.
I
Spiders

Order Cirripedia
Most adults live firmly attached to an object and use feathery legs to filter food from water.
B
Barnacles

Order Decapoda
First pair of legs usually has large claws. Most live in saltwater.
G
Lobsters

Order Amphipoda
Seven pairs of leg-like appendages.
E
Sand Fleas

Order Scolopendromorpha
21 or 23 pairs of legs.
C
Centipedes

Order Odonata
Large compound eyes. Four clear or patterned wings. Long abdomen.
D
Dragonflies

Order Coleoptera
Hard, leathery front wings cover membranous (softer) hind wings. Chewing mouthparts.
A
Beetles

Number Crunching

Having trouble with the math? Follow these steps to find the height of your tower of paper.

First, measure the height in inches of a stack of 100 sheets of paper. In this example, the height is ¼ inch. (We'll use the decimal .25.) If the height of your 100 page stack is different, substitute the measurement of your stack for the .25 used in this example.

Then, use the following ratio to find the height of 1.7 million sheets of paper.

1. $\dfrac{.25}{100} = \dfrac{x}{1,700,000}$

2. $100x = .25 \,(1,700,000)$

3. $\dfrac{100x}{100} = \dfrac{425,000}{100}$

4. $x = 4,250 \text{ inches}$

(To calculate feet, divide by 12 inches.)

$\dfrac{4,250 \text{ in}}{12 \text{ in}} = 354 \text{ feet}$

If you want to find the height of a different number of pages (1.6 million or 100 million), substitute that number of pages for 1,700,000.

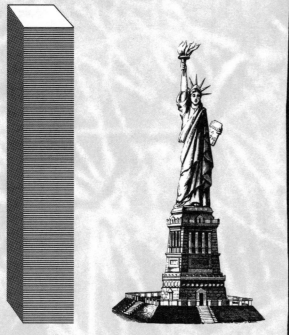

World Wildlife Fund

A number of different approaches may be used to solve the problem. One possible solution is to measure the height, in inches, of 100 pages from a textbook, and use this measurement to calculate the height of 1,700,000 pages. Answers will vary depending on the thickness of the paper. (See the box on the left for sample calculations.)

Compare your answer to a football field, which measures 300 feet long; the Statue of Liberty, which measures 302 feet high; and the Empire State Building (without radio antennas), which measures 1,250 feet high.

Now that students have a better feel for 1.7 million, challenge them to determine how tall the stack of paper would be if they added a sheet for each species that scientists predict exists but hasn't yet been discovered—1.6 to 100 million. Have the students use the same procedure to arrive at a range representing 1.6 to 100 million. Using the same type of paper as in the box to the left, we calculated a height of 333 feet to 20,833 feet (that's more than 1 to 65 football fields tall, or over 1 to 65 Statues of Liberty tall, or more than one-fourth to five times the height of the Empire State Building).

2. Decide how many species are in each group of organisms.

Hand out a set of number cards (see "Before You Begin") to each team, and explain that each card represents the number of species scientists have identified in a particular group of organisms. Hold up a number card (for example, 19,000) and explain that 19,000 refers to the number of bird, plant, mammal, insect, fish, or fungi species that scientists have identified. (Remind your students this is not the number of individuals but the number of species—there may be millions or billions of individuals.) Now explain that each team has to work together to decide which group of organisms listed on the signs posted around the room this number refers to. Once their decisions have been made, the teams should tape their number cards on or below the appropriate signs on the wall. (Teams should write down their choices so they remember them.)

3. Discuss the students' decision-making process.

Ask your students to share the methods they used for making decisions. Did they guess or reason? Many teams may start with what they believe are the groups with the highest and lowest number of species. Some may start with the number they are most certain about and then use a process of elimination. Other teams may base their guesses on experience and observation.

4. Reveal the actual numbers.

Go to each sign and tell your students the correct answers. Then have your students discuss their reactions. Did any of the answers surprise them?

Insects	950,000
Plants	270,000
Fungi	72,000
Fish	19,000
Birds	9,000
Mammals	4,000

5. Hand out copies of "Sizing Up Species" from *WOW!— A Biodiversity Primer.*

Ask your students to read the article and look at the species-scape to learn about the groups of organisms that weren't part of the game. Make sure they understand why the giraffe is much smaller than the dragonfly, spider, mushroom, and scallop in the drawing. And remind them that the numbers shown represent only the number of species that are already identified—not those that have yet to be named. Also have them look at page 120 to see how the numbers have changed. For example, new fish, bacteria, and fungi species have been discovered in the past few years. But the number of larger organisms, such as birds and mammals, has stayed the same. Ask the students if they can explain why. Also discuss their reactions to the species-scape in general. Then explain that they'll be making a graph to see how the numbers compare visually. (You can use either the newer or older set of numbers. The relative proportion is about the same and illustrates the point.)

Before You Begin • Part III

Each student or group will need four sheets of graph paper, rulers, and colored pencils. For the optional activity in this part, you'll need a copy of the *Biodiversity! Exploring the Web of Life* video.

What to Do • Part III

1. Discuss how the data can be presented in a graph. (See graph on page 120.)

Ask your students how they would present the species data given in "Sizing Up Species" (in *WOW!—A Biodiversity Primer*) on a graph. Which format is most appropriate? (A bar graph is probably the easiest to draw, but a pie graph most dramatically illustrates the contrast in numbers of species.) Review how to set up a bar graph on the chalkboard or overhead. Explain that the vertical axis represents the number of different species in each group of organisms. The vertical axis should increase in increments of 5,000 or 10,000. They might also have each number on this axis represent 1,000 different species, so that the number 250 on the graph would actually represent 250,000 species. The horizontal axis represents the different organism groups. You can have your students graph the groups of organisms used in Part II or all the groups used in the species-scape.

2. Create bar graphs to illustrate the number of species in different groups of organisms.

Hand out graph paper, rulers, and colored pencils, and have your students create their own graphs of the groups of organisms used in Part II. They'll each need about four sheets of graph paper to make room for the bar of 950,000 insect species. You can also have them make graphs using a computer. See the graph on page 120 for a sample.

3. Optional: Show the Terry Erwin video clip to summarize key points.

To wrap things up you might want to show a short video clip from *Biodiversity! Exploring the Web of Life.* Fast forward the videotape to the section that illustrates Terry Erwin's work with insects in tropical forests. The clip vividly illustrates the vast diversity of insect species as well as the enormous number of species that have yet to be identified. It can also be a discussion starter to talk about the pros and cons of Terry Erwin's methods of finding new insect species.

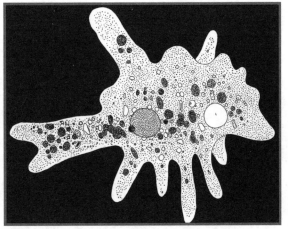

amoeba

KINGDOMS

Monera

The monerans are the Earth's bacteria.* They are single-celled organisms that are organized into two divisions—those that obtain energy by making their own food (autotrophs) and those that eat other organisms to obtain food (heterotrophs). Unlike the cells of other organisms, a moneran's cell has no nucleus, which is the control center in the cells of other organisms. In fact, monerans do not have many of the structures found in the cells of other living things.

Monerans are one of the oldest life forms on Earth. Scientists estimate that the Earth is about 4 billion years old and that the monerans have been around for 3.5 billion years.

Protista (or Protocista)

The kingdom Protista consists of single-celled organisms. Protista have a nucleus as well as other cell structures that perform specific jobs. Protists include certain types of algae, slime molds, amoebas, and diatoms.

*The number of kingdoms is often under debate, depending on how scientists interpret current research. For example, some scientists separate the monerans into two kingdoms: eubacteria (bacteria that get their nourishment from other living things), and archaebacteria (recently discovered bacteria that make their own food and live in extremely harsh conditions such as hot springs and hydrothermal vents).

Fungi

Most fungi are made of many cells. Mushrooms, molds, yeasts, and mildews are examples of fungi. Until recently, fungi were classified as plants. Scientists now place fungi in their own kingdom because, unlike plants, they are not able to make their own food from sunlight, carbon dioxide, and water. Instead, they get their food energy by digesting the organisms on which they grow (usually plants).

Plantae

As you might guess, this is the kingdom of plants. Most plants produce their own food energy through photosynthesis—a chemical reaction involving sunlight, carbon dioxide, and water. Flowering plants, mosses, ferns, and certain types of algae are members of this kingdom.

Animalia

Most animals are multicellular organisms that have specialized tissues, organs, and organ systems. Unlike plants, animals cannot make their own food, and their cells don't have cell walls. Fish, amphibians, reptiles, birds, mammals, and insects and other invertebrates are all part of the kingdom Animalia.

a species-scape (see page 120 for details)

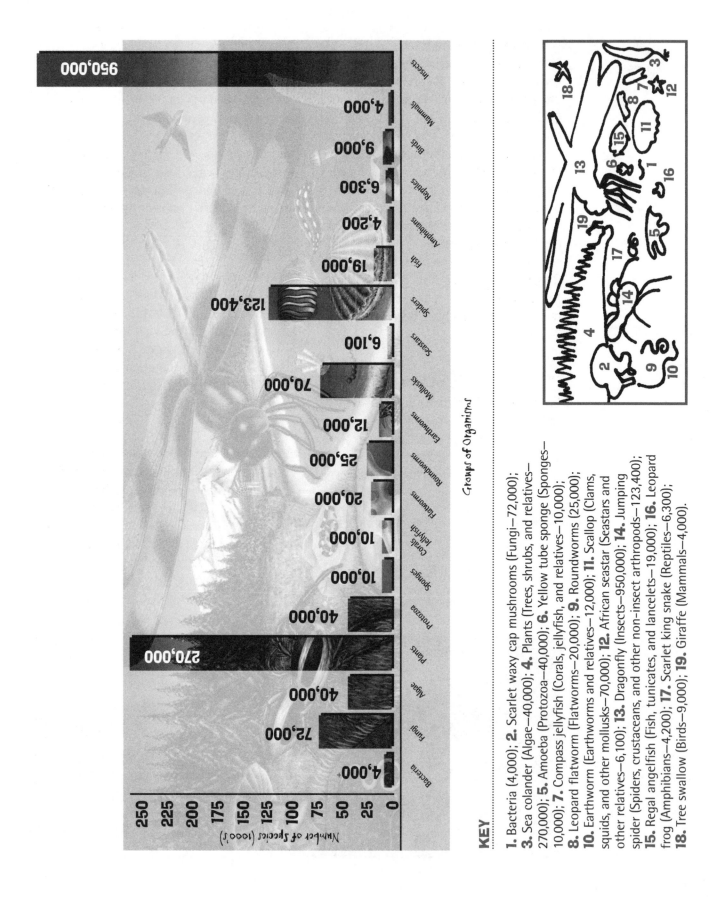

Groups of Organisms

Number of Species (1000's)

Group	Number
Bacteria	4,000+
Fungi	72,000
Algae	40,000
Plants	270,000
Protozoa	40,000
Sponges	10,000
Corals Jellyfish	10,000
Flatworms	20,000
Roundworms	25,000
Earthworms	12,000
Mollusks	70,000
Seastars	6,100
Spiders	123,400
Fish	19,000
Amphibians	4,200
Reptiles	6,300
Birds	9,000
Mammals	4,000
Insects	950,000

KEY

1. Bacteria (4,000); **2.** Scarlet waxy cap mushrooms (Fungi–72,000); **3.** Sea colander (Algae–40,000); **4.** Plants (Trees, shrubs, and relatives– 270,000); **5.** Amoeba (Protozoa–40,000); **6.** Yellow tube sponge (Sponges– 10,000); **7.** Compass jellyfish (Corals, jellyfish, and relatives–10,000); **8.** Leopard flatworm (Flatworms–20,000); **9.** Roundworms (25,000); **10.** Earthworm (Earthworms and relatives–12,000); **11.** Scallop (Clams, squids, and other mollusks–70,000); **12.** African seastar (Seastars and other relatives–6,100); **13.** Dragonfly (Insects–950,000); **14.** Jumping spider (Spiders, crustaceans, and other non-insect arthropods–123,400); **15.** Regal angelfish (Fish, tunicates, and lancelets–19,000); **16.** Leopard frog (Amphibians–4,200); **17.** Scarlet king snake (Reptiles–6,300); **18.** Tree swallow (Birds–9,000); **19.** Giraffe (Mammals–4,000).

WRAPPING IT UP

Assessment

Use both the classification activity and the graphing activity as bases for assessment. At the bottom of the I.D. Chart, have the students write an explanation of how the chart works. On the graph, have the students write the educated guesses the class discussed and how the data on the graph either do or do not support each guess.

> **Unsatisfactory—**The student does not complete any part of the assessment requirements.

> **Satisfactory—**The student provides an adequate explanation of classification and revealing relationships between educated guesses and data.

> **Excellent—**The student is able to clearly explain the classification and how data either support or do not support the class' educated guesses.

Portfolio

Notated graphs can be part of the portfolio.

Writing Ideas

- Write a short magazine article that discusses the amazing diversity of life on Earth. Include illustrations and captions.
- Have the students use the following as a journal starter: "What surprised me about this activity was . . . "

Extensions

- Survey the species diversity of your school grounds or a nearby park or reserve. Your students don't need to identify the species by name; they just need to be able to tell that one species is different from another. Afterward, find out if the ratios of species in different organism groups are similar to the ratios illustrated on the graphs your students made.

- Visit the collections at a local natural history museum or nature center, and explore the taxonomy of the organisms displayed.

- Have students research a class (or order) of organisms within the Phylum Chordata. Tell them to find out what characteristics the animals within the class share, examples of species within the class, and the approximate number of species that have been identified to date. Have each student write a paragraph to summarize that information. Then pool the data for the group and have each student create a bar graph (on graph paper or using a computer program) that illustrates the relative numbers of species in each group. (**Note:** Separate the classes and orders when creating the bar graphs.)

  Here are some suggested classes: Chondrichthyes (sharks, skates, rays), Osteichthyes (bony fish), Amphibia, Reptilia, Aves (birds), Mammalia.

  Here are some suggested orders: within Class Reptilia: Testudines (turtles, tortoises), Squamata (snakes, lizards); within Class Mammalia: Marsupialia (kangaroos, koalas, opossums), Insectivora (moles, shrews), Rodentia (rats, mice), Carnivora (cats, dogs, weasels, raccoons), Artiodactyla (even-toed ungulates such as deer, camels, hippos), Perissodactyla (odd-toed ungulates such as horses, rhinoceroses), Cetacea (whales), Pinnipedia (seals), Chiroptera (bats), Lagomorpha (rabbits), and Primates (apes, monkeys, humans).

Resources

Biodiversity by Edward O. Wilson (National Academy Press, 1988).

"How Many Species Inhabit the Earth?" by Robert M. May, *Scientific American*, October 1992.

"The Little Things That Run the World" by Edward O. Wilson, *Conservation Biology*, December 1987.

WOW!—A Biodiversity Primer (World Wildlife Fund, 1994).

SUBJECTS

science, social studies

SKILLS

gathering (reading comprehension), organizing (arranging, sequencing), analyzing (comparing and contrasting, questioning), citizenship (working in a group)

FRAMEWORK LINKS

53

VOCABULARY

hypothesis, *objective*, **scientific inquiry**, *subjective*

TIME

three sessions to complete all three parts, plus additional time depending on the options selected

MATERIALS

Part I—*copies of "Amazing Discoveries in Science" (pages 26–28 in the Student Book), flip chart paper or poster board, markers, scissors, and glue*
Part II—*OPTIONAL—flip chart paper or poster board and markers*

CONNECTIONS

To reinforce the role of observation in scientific inquiry, try "Backyard BioBlitz" (pages 134–143) after this activity. You might also want to do this activity before you try "'Connect the Creatures' Scavenger Hunt" (pages 102–107), "Space for Species" (pages 286–301), or any of the other outdoor activities in this module because this activity can help your students focus on how to make good observations.

 AT A GLANCE

Focus on scientific inquiry by asking questions, reading about scientific discoveries, making observations, and designing an investigation.

OBJECTIVES

Describe the process of scientific inquiry. Explain how good questions and accurate observations can lead to new discoveries in science. Discuss the difference between objective and subjective observations. Develop a hypothesis and an investigation to test it.

Somewhere within each of us there's an "inner scientist"—even if we've never looked into a single microscope or poured a drop of solution into a test tube. This built-in scientist is naturally curious about the universe and how it works. It's the part of us that wonders about all sorts of things, from how a hummingbird is able to hover in front of a flower to whether there's life on other planets.

This activity will help your students get in touch with their own inner scientists. It's designed to help them understand the important role that good questions, good observations, and good thinking play in the process of scientific inquiry. It also encourages students to look around at our diverse planet, ask questions about it and the life it harbors, and make their own discoveries.

Before You Begin • Part I

Students will work in groups of six or seven. Make one copy of "Amazing Discoveries in Science" (pages 26–28 in the Student Book) for each student. Also gather scissors, markers, glue, and at least one sheet of flip chart paper or poster board for each group. Each group may also want some extra blank paper, poster board, or other materials to make props for their presentations.

What to Do • Part I

How a Question Can Lead to Discovery

1. Solicit student questions.

Start by asking the following (or similar) questions: *Have you ever wondered why the sky is blue? Or how many stars there are in the night sky? Or how birds know when it's time to fly south for the winter?* Point out that it seems to be human nature to wonder about things. Then have the students share some of the questions they've always had about animals, plants, or other natural phenomena they've seen or heard about. Remind them that questions about people and human behavior can fall into this category, too, because people are part of nature and are greatly influenced by (and have a great influence on) the natural world around them.

2. List student questions.

Make a list of the students' questions where everyone can see them. You might want to include a few of the following to get them started: *Do fish sleep? Where do butterflies go when it rains? Why are rainbows made up of certain colors? How do cats purr? Do animals dream? Why is the ocean salty? Why did the dinosaurs become extinct?*

3. Introduce the concept of scientific inquiry.

Point out that most scientific discoveries, from the mundane to the sensational, are rooted in the fact that somebody wondered about something and posed a question about it. But getting from the initial question to the discovery itself often requires undergoing a careful process, one that can involve many steps and many people.

Next have the students brainstorm to come up with a list of words and phrases that describes the general process they think scientists might use to answer some of the questions listed in step 2. Write their responses where everyone can see them. Then explain that scientists answer questions by using a process called *scientific inquiry*. Point out that scientific inquiry can lead to some incredible discoveries. In many cases these discoveries dramatically affect people's understanding of how the world (or even the universe) works. In other cases the discoveries may seem obscure and pertinent only to scientists interested in a very specialized field. But even the most mundane discoveries can lead to other, seemingly more "earthshaking" ones.

"The mere formulation of a problem is often more essential than its solution. To raise new questions, new possibilities, to regard old problems from a new angle, requires creative imagination and marks real advances in science."

–Albert Einstein

4. Create teams to investigate "Amazing Discoveries in Science."

Tell the students that they'll be working in teams to take a look at some of the questions scientists have asked that have led to important or interesting discoveries. Then divide the class into teams of four or five students and give each person a copy of "Amazing Discoveries in Science" (pages 26–28 in the Student Book). Next, have each team select a leader who will help lead a team discussion, a spokesperson who will report on the team's discussion, and a recorder who will take notes during the discussion.

5. Discuss discoveries within the teams.

Have the students in each team "count off" so everyone has a number. Then have each student silently read the discovery (question and answer) that matches his or her number. Make sure that all the stories are assigned. Depending on the size of the team, some students may have to read two stories. When the students have finished reading, explain that each team should discuss what each of the scientists did to make his or her discovery. Have the team leaders call on each team member to describe, for the rest of the team, the scientific discovery he or she read about. Then have the student use the following questions to lead a discussion about the discovery. (During the team discussion, the recorder's job is to try to capture as many responses as possible.)

- What were the different observations that led to each of the discoveries?

- Besides observing, what are some other approaches or steps that the scientists used to answer their questions or learn something new?

- Thomas Edison once said, "Invention is 1 percent inspiration and 99 percent perspiration." What does this statement mean? How might it apply to scientific discovery?

As the students are discussing their stories, go around to each team to make sure the students are identifying elements of the scientific inquiry process in each story. Each team should make a list of elements that were the same in each story and elements that were different. Some of the stories started with a question, for example, while others started with an observation, and yet all the stories included observation of some kind.

6. Prepare team presentations.

Give the teams time to work together to create a short presentation about their discussion of how the discoveries were made. As part of their presentation, have them create a diagram or other visual aid that shows the steps the scientists took when trying to answer a question or solve a problem. (Provide flip chart paper, markers, and other materials, as needed.) Two examples of such diagrams are shown below.

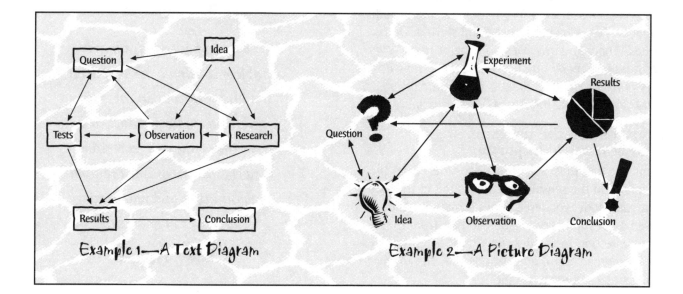

Example 1—A Text Diagram

Example 2—A Picture Diagram

Remind the students that not every scientist will follow the same path. Tell them to look for the steps that each scientist took to make his or her discovery and to be able to talk about how these steps are the same as or different from those steps taken by the other scientists they read about.

7. Use team presentations to identify different aspects of the scientific inquiry process.

Have each team present what it came up with and leave its visual aid about the process of scientific inquiry at the front of the class. Afterward have the students look over each team's information. While they're looking, launch a class discussion by asking the following questions:

- In terms of the steps that scientists took to make their discoveries, how are the scientists' processes the same?

- How are the scientists' processes different?

As the students discuss ways the scientists' processes were the same and different, they should begin to consolidate their ideas about how the scientists made the discoveries. On blank pieces of paper, write short descriptions (a few words) of each of the steps the students identify as a part of the scientific inquiry process. (Put only one step on each piece of paper. For example, one paper might say "making observations" and another might be "forming a hypothesis.") Tape each of the steps up on the wall as the students come up with them, but don't put them in any order yet.

8. Arrange student ideas into steps reflecting one or more ways scientists make discoveries.

Use the information under "The Road to Discovery" (pages 126–127) to help with the discussion of the process of scientific inquiry. If the students don't identify any of the steps described in the information provided, prompt them with questions that will lead them to identify aspects of the process that they haven't yet considered. If students come up with ideas that aren't part of the information provided on pages 126–127, feel free to add these ideas to the list on the wall. When you have taped all of their ideas to the wall, ask the students how the ideas could be arranged.

You might want to start by asking for a volunteer to identify a logical first step in the process of scientific inquiry. Have the student come up to the wall and place the paper where you would like your diagram to begin. Then ask, "Does anyone have a different idea for the first step? If so, what are some other options?" Next ask for a volunteer to identify a possible next step. Have this student draw an arrow from the first step and place the second step at the end of the arrow. Again, ask for students' opinions on the step. Continue this process until all the steps have been placed in an order. Then allow the students to draw in more arrows to show other options about how the steps could be arranged.

Emphasize that not all scientists use all steps of scientific inquiry every time they try to answer a scientific question. A scientist's particular process depends on what he or she is trying to find out, what materials are available, and many other issues. Regardless of the specifics, all scientists rely on observation, critical thinking, creativity, and analysis to solve problems and answer questions. Science also involves a lot of trial and error and learning from the work of others.

"Human knowledge will be erased from the world's archives before we possess the last word that a gnat has to say to us."

–Jean-Henri Fabre, naturalist

The Road to Discovery

Whether they're looking at plants, people, or parks, most scientists follow a similar approach in their efforts to answer a scientific question or solve a problem. This approach is known as scientific inquiry. Scientists adjust the process to fit the subject they're exploring, the resources they have, and the type of research they're conducting.

We've outlined some of the key components of scientific inquiry below. You can use this guide and your own experience to help your students better understand the importance of questioning, observing, building on the work of others, and thinking creatively and critically.

Making Observations

Many scientific discoveries begin with an observation; that is when someone notices something and wonders about it. Observation isn't limited to sight. It can involve our other senses as well.

Careful observation is a cornerstone of science. It's crucial that a scientist's observations be accurate and as objective as possible. If they are not, his or her results will be biased and unreliable. (See "Option #3: Focus on Observation" on page 131 for more information about objective versus subjective observations.)

To illustrate, tell the students to imagine that they've walked into a room and flipped a light switch on the wall to turn on a lamp, but the lamp doesn't come on. At this initial stage of "scientific investigation," what observations might they make? (Obviously, they'd observe that no light came on when they flipped the switch. They might also look to see if the lamp was still in the room, if it was upright, if they had the right switch, and if there was a light bulb in it.)

Point out to the students that making observations can involve more than simply being observant. An important part of this aspect of scientific inquiry is having some sense of the implications of your observation. For example, when Alexander Fleming, a bacteriologist,

noticed that no bacteria were growing around the mold in his Petri dish, he understood that the absence of bacteria was a potentially important phenomenon—he had a "prepared mind." Other observers, less thoughtful, knowledgeable, or imaginative, might not have given the phenomenon a second thought—even though they had been observant enough to notice it in the first place. The Petri dish might have been tossed into the trash, and penicillin may have gone undiscovered.

Sometimes observation forms the bulk of the process that a scientist goes through in making a discovery. For example, a biologist may spend a lot of time observing animal behavior in the field. These observations may eventually lead to a question or hypothesis, but the observations themselves may remain the focus of the research. Later the observations might be used to come up with new questions or ways of looking at the subject.

Asking a Question

Observations are often followed by (or even preceded by) a question or questions. The best questions, as far as scientific inquiry goes, are those that are clear and specific. Such questions are easier to answer than vague or multi-faceted ones.

In our light switch example, a specific question might be, "Why doesn't the lamp come on when I flip the switch?" It's worth pointing out again to the students that, as with observation, asking a good question can involve imagination and creativity.

Gathering Information

Once a scientist asks a question, the next step is to gather as much information about the topic as possible. That means researching the scientific literature to see what others have written about it. Once a scientist has found out what has already been discovered, he or she can decide how to proceed. On one hand, a scientist may discover that no one has addressed the subject before and that little or nothing is currently known about it. If this is the case, whatever the scientist discovers will be new knowledge. On the other hand, the scientist's research may reveal that a lot of information has already been gathered on the subject, and perhaps even that his or her question has already been answered. In this case the scientist may decide to ask another question that addresses a lesser-known aspect of the subject.

Gathering information also involves collecting data on the subject being studied. A scientist studying the courtship rituals of whooping cranes, for example, might write down everything he or she observes about the cranes and their behavior during courtship episodes. The scientist might also record seemingly extraneous information such as the air temperature, amount of cloud cover, and time of day.

As far as science is concerned, the importance of keeping careful records can't be overstated. It's also important that the data collection be as accurate as possible. As much as a scientist may want data to support a particular hypothesis, the results of his or her work won't be considered scientifically sound if data collection is skewed to favor a particular viewpoint.

At this stage in our light switch example you probably wouldn't gather information by researching the mechanics of lamps and electricity. But you might think about the condition of the cord, the plug, and the light bulb. Or you might wonder if the electricity in your building or neighborhood is working.

Forming a Hypothesis

Often scientists come up with one or more hypotheses–possible answers to their questions or an explanation of a problem. Each hypothesis must be tested to see if it's correct.

In the light switch example, you might come up with the following hypotheses: (1) the lamp isn't properly plugged in, (2) the cord is broken,

(3) the light bulb is burned out, or (4) the electricity is out in the neighborhood. If, based on testing, none of the hypotheses are correct, you have to come up with another hypothesis.

Testing Your Hypothesis

There are many ways to test a hypothesis. The method a scientist chooses depends first on the hypothesis itself and second on the resources available (particularly time, money, and equipment). Scientists spend a lot of time trying to come up with the best ways to test their hypotheses. They need a plan that will give them the greatest amount of accurate information to determine if their hypothesis is true or false. Surveys, experiments, and field observations are all ways to collect data to test a hypothesis.

To increase the chances for unbiased results, scientists often design tests that involve control groups. This practice is particularly true in laboratory studies. To test whether a certain drug works to lower cholesterol, for example, scientists might try to find a group of similar individuals (same age, medical background, and so on) and divide them into different groups. They might give one group a low dosage of the medication, another group a higher dosage, and a third group–the "control"–a dosage of a substance that has no medical value (a placebo). Because you can expect changes in the control group to be limited, this group will serve as a basis for comparison. By carefully designing experiments to limit factors that could affect the results, scientists can gather more accurate data.

In our example with the light switch, you might do several things to determine which of your hypotheses is true. First you could check to see if the cord is properly plugged in. You might unplug it and plug it in again and then test your prediction by flipping the switch again to see if the light comes on. If it doesn't, you might examine the cord to see if it's broken or damaged. If the lamp still doesn't come on, next you could try putting in a new light bulb and flipping the switch again to see if the lamp comes on.

The Road to Discovery (Cont'd.)

Interpreting Your Results

At this stage, scientists try to make sense of the data they've collected. They may create graphs or charts or use mathematical tests to help them figure out what the information they've collected can tell them.

Our light switch example is so straight-forward that interpretation isn't necessary. The hypotheses were simple and testing them didn't require collecting a large amount of data. If a scientist was trying to determine the range of a certain species or the patterns of resource use in an ecoregion for example, he or she might need to create a series of maps and overlays that help explain the data.

Drawing Conclusions

After they've gathered and sorted all their information, scientists use the results to draw conclusions. They may find that one of their hypotheses is correct, or they may find that none of them seems to be correct. There are many reasons a hypothesis might not be correct. Perhaps a scientist didn't do enough research, or his or her plan for testing the hypothesis didn't work. Or it could simply be that the data do not support the hypothesis. In any case, whether a hypothesis is supported or not, a scientist can choose to ask new questions at this point and look deeper into the research he or she has begun. The research results themselves might suggest another hypothesis to test.

It's important for students to understand that experiments often fail to confirm a hypothesis, and that such failure is part of the scientific process. Scientists can learn as much from negative results as they can from positive ones, and they can use their results to spur new thinking and testing. It's also important for students to understand that all results need to be tested and retested to ensure that they are accurate. Everything in science is open to criticism, review, and reflection—and this constant critiquing can help to push our thinking ahead.

Consider the light switch example again. Let's say that the light turns on after we put a new light bulb into the lamp. This result lends support to our third hypothesis that the light bulb is burned out or damaged.

Sharing the Results

If scientists didn't share what they learned, science wouldn't work. An important part of research is reviewing what questions have already been asked and what answers have been found. If no one shared knowledge, no one would be able to benefit from what's been learned, and lots of work would be repeated. Students need to realize that scientific knowledge builds over time and that what we know about how the world "works" is based on centuries of research. It's also important they realize the value of criticism from peers and the importance of getting ideas from others to shape their own thinking.

Many scientists share their work in peer-reviewed journals. Most journals focus on one area of science, which makes it easy for people working in different fields to find information on subjects they might be working on. There are also many generalists and interdisciplinary scientists who try to increase our understanding of the world by bringing together the work of many disciplines.

After sharing results and getting feedback from peers, a scientist might find that he or she wants to re-ask a question, try a new approach, or retest a hypothesis. The scientist might also spark new thinking in a colleague who may then be able to take the research to the next step.

Before You Begin • Part II

You may want to put the chart that compares subjective and objective observations (page 131) on the board or flip chart for students to fill in.

What to Do • Part II

Making Your Own Discoveries

Now that your students are familiar with the process of scientific inquiry, you might want to have them give it a try. Or you might want to emphasize one or two aspects of the process, such as observation or questioning. Here are several options you can explore.

Option #1: Write a hypothesis and conduct an investigation.

Conducting an investigation as a class (or as teams) will help reinforce in students how the process of scientific inquiry works. You can start by having your students make observations—preferably outside—that lead to questions. (For more about observation, see Option #3 on page 131.) Have the students use the questions they come up with to develop hypotheses that they can test. Or you can present your group with the following writing prompts to focus their thinking. Have each student or team choose one of the writing prompts.

1. Write a testable hypothesis that focuses on the food an organism eats. The hypothesis can deal with aspects of food choices such as color, smell, energy content, and special feeding adaptations, to name a few possibilities. (Examples of questions leading to such a hypothesis include the following: Do house cats prefer one brand of cat food over another? Do chickadees prefer one kind of bird seed over another? Will frogs eat prey that is not moving?)

2. Write a testable hypothesis to determine how organisms use different areas within their habitat. (Examples of questions leading to such a hypothesis include the following: Do gray squirrels gather food from the ground or from trees? Does the location of food in a habitat determine whether an animal will eat it? Where do worms go when it rains?)

3. Write a testable hypothesis to determine how an organism's special adaptations help it survive. (Examples of questions leading to such a hypothesis include the following: Which body parts do insects use to escape from predators? Does the shape of a bird's foot help in feeding? Can animals that live in the ground sense sunlight?)

There are many resources that can help your students design and conduct an experiment. For example, if your students are interested in investigating the feeding habits of animals, we suggest that you use *Eco-Inquiry* (see page 461 in the Resources), which includes a unit on feeding habit investigations. Or, if your students are interested in biodiversity issues, you might want to look at *The Diversity of Life* (see page 456 in the Resources).

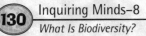

Exploding the Myth of the Wild-Haired Guys in White Coats

You might want to take a few minutes to explore the stereotypes that people often have about scientists. Begin by asking the students for their opinions on how the typical scientist looks and behaves. You could also have each person draw a picture. (Stereotypical images include the view that scientists are men who wear white lab coats, that they tend to be boring, that they work in labs and spend much of their time staring into a microscope, and that they're very serious. There's also the image of the scientist as a "mad professor" type who is brilliant but scatterbrained and has wild hair that reflects a lack of attention to personal appearance.)

Point out to the students that, as with all stereotypes, these ideas about what a scientist is and does are inaccurate. For example, there are many female scientists as well as males. Scientists can be young or old, and they come from many different ethnic and cultural backgrounds. Many don't work in a lab at all but spend much of their time working in the field—in places as far flung as tropical jungles, polar ice floes, rural villages, or the depths of the oceans. To reinforce a more realistic view of scientists, you might want to have your students look for articles and pictures of non-stereotypical scientists. (For an extra challenge, have them focus on scientists doing work in the field of biodiversity.) Students can present their findings in a "Scientists in the News" current events session, or you can have them create a "Who's a Scientist?" bulletin board.

Option #2: Research a scientist.

Another way to apply the process of scientific inquiry is to have your students read about the work of a real scientist. One way to do this is to have each person pick a scientist to explore in depth. Students can work individually or in teams to find out how each scientist used the process of scientific inquiry to make a particular discovery. Then the students can present what they learned to the rest of the class. Examples of scientists to research include the following:

Archie Carr
(conservation biologist who studied sea turtles in Costa Rica)

Rachel Carson
(zoologist who studied the effects of pesticides)

George Washington Carver
(biologist who studied agriculture)

Eugenie Clark
(marine ecologist who studies sharks)

Jason Clay
(anthropologist who links economics, culture, and other disciplines to address conservation problems)

Theo Colborn
(zoologist who studies the effects of toxic chemicals on living things)

Charles Darwin
(scientist who helped develop the theory of evolution)

Jared Diamond
(physiologist who researches life on islands and other conservation issues)

Paul Ehrlich
(population biologist who studies connections among human population, resource exploitation, and the environment)

Thomas Eisner
(biologist who studies insect defenses)

Terry Erwin
(entomologist who studies rain forest insects)

Dian Fossey
(zoologist who studied gorillas in Africa)

Jane Goodall
(scientist who studies chimpanzees in Africa)

Kai Lee
(physicist and political scientist who studies environmental and natural resource policy)

Barbara McClintock
(geneticist who studied gene transposition in corn)

Gregor J. Mendel
(a monk who "discovered" genetics)

Mario Molina
(chemist who studies atmospheric pollution)

Gordon Orians
(conservation biologist who studies behavioral ecology)

Edward O. Wilson
(entomologist who studies biodiversity)

Option #3: Focus on observation.

Tell the students that one of the most important skills a scientist uses is the skill of observation. That's because much of the work a scientist does is based on what he or she can observe. Explain that, in general, scientists try to be as objective as possible in their observations. It's important to remind students, however, that no one is completely objective, although scientists *do* try to be as unbiased as possible.

Using the following chart as a guide, ask the students for their input on what constitutes an objective, as opposed to a subjective, observation. (You might want to draw a similar chart on the board and have the students fill it in.)

Objective Observations	Subjective Observations
include facts only	may include information that's not necessarily factual
don't include personal feelings	may include personal feelings
don't include value judgments	may include value judgments
don't make assumptions	may make assumptions
often include specific measurements and details	may compare sizes to other objects rather than to specific measurements
are as unbiased as possible	are often biased

To help your students understand the difference between objective and subjective observations, read the following statements one at a time. Stop after each one to discuss whether the statement is objective or subjective, and why. Have students come up with objective alternatives to the subjective statements.

- *The mighty oak's dazzling fall colors sparkled and danced in the cheerful breeze.* (subjective because it includes value judgments and information that's not merely factual—e.g., "mighty oak," "dazzling colors," "cheerful breeze")

- *The dog barked.* (objective because it's a simple statement of fact)

- *I watched sadly as the hopeless mountain lion left her dead mate's body.* (subjective because it includes personal feelings and makes assumptions about the feelings of the animal being observed)

- *The snake was as long as a bus.* (subjective because the snake's length is compared to another object rather than being given a specific measurement)

- *At 9:02 AM the female robin returned to her nest and fed her chicks.* (objective because it's a factual statement that provides specific information)

- *My favorite elephant played in the dust.* (subjective because saying "favorite" indicates bias; also subjective because the phrase "played in the dust" does not give specific information about the elephant's activities and makes an assumption that the animal was playing, as opposed to dust bathing or some other activity)

After discussing the statements, ask the students why it's important that scientific observations be based more on objectivity than subjectivity. (Science, by its nature, deals with factual information. Subjective observations aren't solely, or even at all, factual, and they can therefore vary widely from person to person. Science relies on information that's specific and consistent from one observer to the next; without such information, scientists wouldn't be able to arrive at factual conclusions, and other scientists wouldn't be able to use those conclusions to support other conclusions.) Point out that there's nothing wrong with subjective observations per se. In fact, they are a valuable part of the work that artists, poets, writers, and many others do. But scientific knowledge could not exist without objectivity and facts supported by evidence.

After discussing objectivity and subjectivity, you also might want to conduct one or more brief observation activities. Before getting started, brainstorm about some of the characteristics of good observers. (They focus on their surroundings, take in details, remember what they see, look in more than one direction, look for change over time, can describe to others what they see, and so on.) Then try one or more of the following ideas with the students:

- Observe a burning candle, fish in an aquarium, a beehive, or some other object for five minutes. Write down everything you notice.

- Observe a volunteer from your class or another class. (The volunteer should sit quietly at the front of the class for several minutes.)

- Look at a photo or drawing of an animal or natural scene for 30 seconds. Then write down everything you remember about it.

- For 30 seconds, carefully look at a table full of objects. Then share what you remember about what was on the table.

- Observe one or more natural objects outside, such as trees, flowers, or other stationary objects. Write down what you notice about the object or objects.

- Look at a picture of someone doing something. Then make a list of observations and assumptions about the person. (You can use this type of activity to talk about stereotypes, making assumptions without facts to back them up, and subjective versus objective descriptions.)

- Go outside for 10 minutes and make 25 observations. Highlight any observations that surprised you. Share the results to see how many people noticed the same things.

- Select something outside to observe over time. Record your observations in a notebook and make sketches to illustrate what you see. Describe any changes you notice over time.

Afterward have volunteers share their observations. Ask if others noticed anything else. Also ask if the observations were subjective in any way.

Option #4: Make a question-and-answer bulletin board.

Use the examples in "Amazing Discoveries" to inspire your students to come up with their own questions and answers. Then have them design a bulletin board that highlights exciting discoveries in biodiversity research. They can focus on scientists who are looking at specific species or ecosystems, genetic diversity, ecological connections among species and ecosystems, or broader issues that affect biodiversity, such as global climate change and toxins. Or they can look at social scientists who study population dynamics, cultural diversity, economics, and other social issues.

WRAPPING IT UP

Assessment

Have each student choose one of the questions from step 1 of Part I and turn the question into a hypothesis. Then have each student develop a plan to test his or her hypothesis with one or more experiments. Finally have each student identify elements of the scientific inquiry process in his or her plan.

Unsatisfactory—Student does not complete the plan or identify elements of the scientific inquiry process.

Satisfactory—Student completes the plan and is able to label appropriate parts of the scientific inquiry process.

Excellent—Student develops a thorough study of the question and includes all components of scientific inquiry in his or her plan.

Portfolio

The diagram developed by the teams in Part I can be used in the individual's portfolio. From Option #1 of Part II, the notes on a hypothesis and investigation can become part of the portfolio.

Writing Idea

Have students imagine that they are journalists in the future who have been given the assignment of writing a news-breaking story dealing with a scientific discovery. Ideas include the discovery of a new species of animal or a lost line of human ancestors formerly thought to have died out, a breakthrough in genetic research that would eliminate a debilitating disease or medical condition, an inexpensive and easy-to-use energy source that would end human dependence on fossil fuels, and research leading to the discovery of life on another planet. Encourage the students to include information about how the scientists involved in the discovery used the process of scientific inquiry.

Extensions

- **Green Science**
  Have the students create time lines highlighting major scientific discoveries relating to biodiversity and the environment.

- **Paradigm Shifts**
  Remind students that there are many ideas that are now widely accepted in the scientific community that were once thought of as unlikely, impossible, or ridiculous—often by scientists and non-scientists alike. Examples include the idea of the sun (rather than the Earth) as the center of the solar system, the theory of plate tectonics, and the fact that the Earth is round. Have the students brainstorm a list of modern ideas that may one day be proven to be true or highly likely, despite current skepticism.

- **Science in Action**
  Visit a lab or field station at a local university or business to observe a working scientist.

Resources

Blacks in Science: Astrophysicist to Zoologist by Hattie Carwell (Exposition Press of Florida, 1977).

Eco-Inquiry by Kathleen Hogan (Kendall/Hunt Publishing, 1994).

Everyday Wonders by Barry Evans (Contemporary Books, 1993).

"In the Nose of Jaws," by Mark Wheeler, *Discovery*, March 1998.

Stepping into the Future: Hispanics in Science and Engineering by Estrella M. Triana, Anne Abbruzzese, and Marsha Lakes Matyas (American Association for the Advancement of Science, 1992).

Women Life Scientists: Past, Present, and Future by Marsha Lakes Matyas (American Physiological Society, 1997).

SUBJECT

science

SKILLS

gathering (collecting, observing, researching), citizenship (working in a group)

FRAMEWORK LINKS

3, 12, 21, 54

VOCABULARY

ecoregion, *ground-truthing*, *insect galls*, **migration**, **native species**, *noxious*, *precipitation*, **rapid assessment**, *sampling*

TIME

Part I—*long-term project*
Part II—*two sessions*

MATERIALS

Part I—*copies of "Ecoregional Survey" (pages 29–30 in the Student Book), field guides, regional almanacs, and other research materials*
Part II—*copies of "BioBlitz Survey" (pages 31–32 in the Student Book)*
OPTIONAL—*plastic bags and plastic containers with lids to collect specimens; thermometers; magnifying glasses; and field guides*

CONNECTIONS

Follow up with "Insect Madness" (pages 144–157) to learn more in-depth techniques to measure diversity, or "Inquiring Minds" (pages 122–133) to explore the scientific inquiry process. To learn more about global classification and surveys, try "Sizing Up Species" (pages 108–121) or "Mapping Biodiversity" (pages 252–265).

AT A GLANCE

Answer an ecoregional survey, then take a firsthand look at biodiversity in your community.

OBJECTIVES

Name several native plants and animals and describe your local environment. Design and carry out a biological inventory of a natural area.

You don't have to travel to the rain forests of the Amazon or the coral reefs of Australia to discover biodiversity. Just walk out the door and you'll find an amazing diversity of life in backyards, vacant lots, streams and ponds, fields, gardens, roadsides, and other natural and developed areas. In this activity, your students will have a chance to explore the diversity of life in their community. They'll also get an introduction to how scientists size up the biodiversity of an area—and why it's so hard to count the species that live there.

Before You Begin • Part I

You will need to gather field guides and other resources about your area. Your state's department of natural resources or fish and game department may have brochures and other information about trees, other plants, and animals in your state. You can use the "Ecoregional Survey" (pages 29–30 in the Student Book) as it is written, or you can adapt it more specifically to your area and situation. (For example, if you live in an urban area, you may wish to delete the word "native" from questions two, three, five, and six.) Either way, you'll need a copy for each student, plus one copy for each team of four to five students. You'll also want to take the survey yourself in order to generate possible answers. Check with a local naturalist or other knowledgeable person in your area for answers if you are unsure.

What to Do • Part I

An Ecoregional Survey

In this part of the activity, your students will get a chance to complete an "ecoregional survey." It is designed to get them thinking about their local area, the plants and animals that live there, and some of the factors that may affect where and how plants and animals live in your region.

Because some of the questions can require a good amount of research, Part I can be turned into a long-term project. Student groups can be assigned a particular set of questions or the entire survey to answer in a certain amount of time. Most of the answers can be obtained by calling a local nature center, but encourage the students to conduct other forms of research. The library and Internet can be great resources.

1. Take the ecoregional survey.

Pass out a copy of the "Ecoregional Survey" to each student, and review any unfamiliar terms such as *native* (see glossary). (You can have the students omit the second part of question two if you think it will be too difficult or take too much time.) Then give them about 10 minutes to complete the survey. Afterward ask the students how they think they did. (Don't share possible answers at this point.) Collect the completed sheets as a pretest of the students' knowledge.

2. Divide the group into teams to complete the survey.

Divide your group into teams of about four students each. Give each team a clean copy of the ecoregional survey. Tell the students that the members of each team should work together to complete the survey as accurately as possible (or assign each team a different set of questions from the survey). Explain that the students can use whatever resources they can find to answer the questions, including the resources you gathered, the library, the Internet, community elders, or a local naturalist. Stress that they should find the most accurate information they can, and encourage them to collect drawings or pictures of the animals and plants they list. You may want to provide the students with phone numbers of local nature centers, museums, and libraries.

3. Set a time limit on research.

Give the students at least two days to find answers to the questions. If you plan on doing the entire BioBlitz activity, this is a good place to stop and skip ahead to Part II. Research for the "Ecoregional Survey" should be done as homework on the days you spend on Part II, the "BioBlitz Survey." By the third day, Part II should be completed. You can go over the "Ecoregional Survey" results from their research as a wrap-up for this activity.

4. Go over the survey results.

Once the students have finished the survey, have them share the information they found and compare their answers to the pretest. Did students find different answers to some of the questions? (For example, how extensive was the group's list of native plants?) What sources proved to be the most helpful? How did the students find these sources? Were they surprised by any of the information they found? The quiz discussion can also be used as the wrap-up to Part II and as a way to discuss things the students observed during the "blitz".

Before You Begin • Part II

You will need to find a nearby natural area where the students can conduct their "BioBlitz Survey." School grounds, a nearby park, or the grounds around a neighborhood nature center can all work. Just be sure that your area is safe for your students (no broken glass or other hazards) and that you have the permission of the owners if needed. For example, if you're using your own school grounds, you probably don't need permission, but if you're using a nearby city park, you should check with the city

parks department first. You will also need to sketch a quick "site map" for the students. This map should show the boundaries of the study area and a rough delineation of different plant types. For example, areas with shrubs would look different from grassy areas (see samples below). Be sure to have a copy of the "BioBlitz Survey" for each student (plus optional plastic bags and plastic containers with lids to collect specimens, thermometers, magnifying glasses, and field guides).

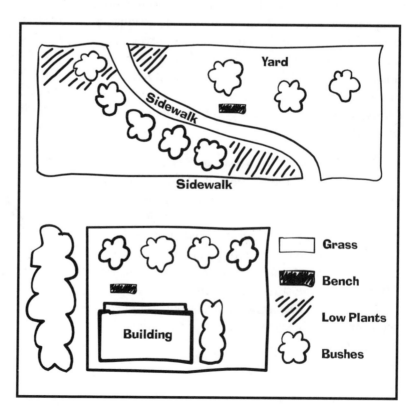

Getting to Know Biodiversity

When scientists want to know what lives in a particular area or region, they rely on a number of tools and techniques. Here's a quick look at some of them.

Bird's-Eye View

Aerial photographs and satellite images can give conservation biologists, wildlife managers, and others a lot of information about a region. (See photo on page 254.) For example, different cover types such as coniferous forests, deciduous forests, or grassy areas show up in these pictures as different colors or patterns. Scientists can use the photographs to delineate features on the ground before they ever visit an area. Knowing what grows on the ground can be useful for predicting where animals might be, whether areas could sustain certain animal populations, and for planning strategies to manage these populations. Scientists can also compare pictures taken at different times to look for changes in such things as forest cover and land-use patterns.

See It for Themselves

No matter how much aerial photographs or satellite images tell us about a study area, scientists like to visit the area to see for themselves if the information they gleaned from the photos is accurate. The process of going to an area to verify information is called ground-truthing. Ground-truthing gives scientists a firsthand look at the areas they're interested in and can help guide further studies.

Sampling

Scientists rarely have the time to identify every single plant and animal that lives in a particular area. And even if they did have the time, it would be extremely expensive for them to do so. For these reasons, scientists rely on statistics to get an idea of species diversity. The scientists look closely at only small portions—or samples—of the total area they're interested in. Then they use mathematics to extrapolate their findings to the larger whole. For example, as shown in the video **Biodiversity! Exploring the Web of Life**, entomologist Terry Erwin sampled some individual trees in the rain forest to get an idea of species diversity throughout it. Scientists frequently use aerial photographs or satellite images to decide where to do their sampling. If an area they want to study is covered by both woodlands and grasslands, for example, scientists will take samples in both.

Fast Fact-Finding

In the race to save the world's biodiversity, scientists have developed methods to find out as much information about a particular habitat as quickly as they can. In such rapid assessments, teams of scientists work together. Each member of the team has a specialty, such as botany (plants), entomology (insects), or ornithology (birds). The team members travel to the study area together, collect as much information as they can in the short time allowed, including carefully collecting specimens of individual organisms, and then return to their laboratories or offices to sort out and identify what they found. The video **Biodiversity! Exploring the Web of Life** features just such a rapid assessment in Papua New Guinea. Rapid assessments can be particularly effective in assessing the biodiversity of remote areas where it would be too expensive to employ researchers for more extended periods of time. However, rapid assessments provide only snapshots of what's found in particular areas and usually can't cover extensive geographic areas.

What to Do • Part II

A Look at Biodiversity

In this part of the activity, your students will have a chance to go outside and take a firsthand look at biodiversity in their own backyard. Observation is very important in science. This activity is a great opportunity for students to develop their observation skills. (For more on observation skills, see "Inquiring Minds," page 131.)

1. Set the stage.

Ask your students to imagine that the school board is planning to add another building to their school. One factor that's important in the board's decision to build is how biodiversity might be affected by the development. The board is planning to meet in just two days to decide whether they should add the building to the school, and it has asked the students for a list, or inventory, of all the species found on the site. (Rapid assessments are usually conducted because a decision about land use must be made quickly, and the species living on the land in question are being factored into the decision. If you're in a nonformal setting or if you can't use your schoolyard for the BioBlitz, adjust the school board scenario accordingly.)

What kinds of things would your students need to consider as they inventory the biodiversity of this area? List their ideas on the chalkboard. If the students don't suggest anything along these lines, ask them if there might be differences depending on the time of year. Would they expect to find the same species in areas covered by grass as in areas where trees grow? Do they think the relative numbers of individuals, or the population sizes, of each species might be important? Stress that knowing what lives in an area, knowing where different things live within the area, and having an idea of the size of the populations of different living things are all important pieces of information that wildlife

managers and conservation biologists try to find out when they investigate the biodiversity of different land areas. Save all the questions the students generated for the wrap-up (step 7).

Ask your students how they think scientists find out answers to questions like the ones they've generated. *(Scientists may use aerial photographs, satellite photos, and special maps; they may interview knowledgeable people and consult historical records; and they usually go to the areas of interest and look at the plants and animals firsthand.)* See "Getting to Know Biodiversity" on page 137.

2. Explain the task.

Explain to the students where their study site is located and pass out copies of the "site map" you sketched earlier. Also pass out copies of the "BioBlitz Survey." Explain each of the different categories listed on the survey sheet and give some examples of each.

Divide the group into teams of four to five students, and explain that the team members have to work together to design a way to fill out their sheets as completely as possible in a relatively short time. Where are they going to look? What are they going to look for? How will they record what they find? Are they going to draw sketches of different species, collect specimens, or take very detailed notes? How are they going to divide up the work?

Tell them they will have only 30 minutes to work at the site, and let them know whether they'll be able to bring samples back to identify. (Remind the groups that correct identification of different species is not a necessary goal of this activity. "Green needle bush" and "shiny black bug" are as correct as "juniper" and "patent leather beetle." However, depending on your group and the time you have available, you can teach your students to use field guides and incorporate accurate species identification into the survey.)

Review the range of animal signs the students should look for (see "Animal Signs to Look For" on page 140). Also review the "Do's and Don'ts of Field Work" (page 141), adding any additional points needed for your particular area.

Now give the students time to work in their teams to come up with their study plans.

3. Review the study plans.

Once the students have designed their study plans, meet with each group independently and have the group explain its design. Make sure that each group has evenly divided the amount of work to be done among the group members, will be getting to all areas of the study site, and has accounted for inventorying the full range of species types listed on its survey sheet.

4. Conduct the BioBlitz.

Take the students to the study area and give them approximately 30 minutes to conduct their surveys. Although identification is not the ultimate goal of this activity, you might want to have field guides available for students to use to help identify what they are seeing.

Students can collect specimens to take back to the classroom. Pass out plastic bags and small containers for use in specimen collection. Some things should not be collected: animals, delicate or rare flowers, dangerous plants (poison ivy and poison oak), and endangered plants. Have the students draw sketches of items that should not be collected or are hard to describe. (Again, refer to the "Do's and Don'ts of Field Work.")

5. Finalize findings.

Give the teams time to review and identify what they found, and consolidate information. Have them make notes on the sketch of the area to indicate where certain things were found or where animals or plants were concentrated. You may even have the students prepare a presentation around any specimens they collected to share with the class.

6. Share results.

Have the groups report on their findings and discuss the processes they used. How many different living things did they find? Where did they find different things? Did they find any native species? Non-native species? Were species evenly distributed across the site or did the students find greater variety in particular areas? If there were distribution differences, where did they find the greatest diversity? Do they think that as a group they found everything out there? What factors might have affected the number of species they found? For example, would they have expected to find the same number and types of species if they'd done their BioBlitz at a different time of year? Or with magnifying glasses? Did one team have a way to complete the investigation that worked particularly well? What was the hardest thing about conducting their BioBlitz? Were they surprised by anything they found or didn't find?

7. Wrap up.

Have the students look back at the questions they generated in step 1 of Part II. Based on their recent field experience, is there any other information they need to know about the land in order to make a complete inventory of its biodiversity? What kinds of organisms have they probably missed? Do they think these kinds of rapid assessments are useful? *(It's often difficult to find all the species in an area in a short amount of time. Because animals tend to come and go from different areas, they can be missed if the amount of time spent looking for them is too short. Very small or microscopic organisms can be hard to find and identify. Also, there are often seasonal changes in the organisms in an area, so an inventory conducted at one time of year might be very different from an inventory of the same area at a different time of year. But despite their problems, rapid assessments are often very useful because they are a way to quickly get a good idea of the diversity of species in an area. When time is short, a BioBlitz may be the only way to go.)*

Animal Signs to Look For

In addition to looking for animals, keep your eyes open for animal signs. These signs include the following:

- burrows
- nests
- digging and scratching marks
- tracks
- bones
- feathers
- insect galls
- cocoons
- spider webs
- nibbled leaves and branches
- droppings
- feeding holes in dead trees and logs
- runways and trails

Also, don't forget to look everywhere, including:

- on the ground
- on tree trunks
- in tree branches
- in leaf litter
- on plant stems and leaves
- under and around logs
- under rocks

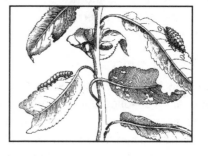

Do's and Don'ts of Field Work

Do's

✓ Do be sure that you have all the materials you need before you head to the study site.

✓ Do be a careful observer.

✓ Do take careful notes about what you find, including information about the locations and characteristics of plants and animals.

✓ Do handle animals with care—and handle them as little as possible.

✓ Do return animals you find to the places where you found them.

✓ Do replace logs and rocks to the position you found them.

✓ Do stay within the boundaries of your study area.

✓ Do try to identify unknown species while you're in the field.

✓ Do look for animal signs as well as actual animals.

✓ Do wash your hands carefully as soon as you return to the classroom.

Don'ts

✓ Don't damage trees or other plants by digging them up, ripping off leaves, or tearing at the bark. Be careful when collecting specimens.

✓ Don't put anything you find—such as berries, leaves, mushrooms, and bark—in your mouth. Also, don't put your fingers in your mouth until after you have returned to the classroom and washed your hands thoroughly.

✓ Don't chase after, yell at, or throw things at animals you see.

✓ Don't touch or collect animal droppings, dead animals, mushrooms, or human refuse such as bandages, broken glass, rusty cans, or needles.

✓ Don't reach under logs or rocks, crevices, or other spaces if you can't see into them.

WRAPPING IT UP

Assessment

Have each student write a mystery movie review of the "BioBlitz" that the class conducted. The critique should identify which members of the teams played what roles in the blitz, weaknesses in the "plot" or in their study plans, how the blitz was organized (directed), and so on. Encourage the students to use the movie metaphor to look for strengths and weaknesses in the BioBlitz as an assessment of diversity of an area.

Unsatisfactory—The student does not use the movie metaphor, critique the BioBlitz, or complete the activity. The student does not complete any part of the assessment requirements.

Satisfactory—The student uses the movie review to identify both what happened in the BioBlitz and the different roles of individuals in conducting the diversity measurement.

Excellent—The student uses the movie review to identify strengths and weaknesses in the BioBlitz as it was conducted and the use of a BioBlitz in general for measuring diversity.

Portfolio

The pretest and posttest can be used for the portfolio.

Writing Idea

Have your students write an article explaining the process they used to collect their data, including any conclusions they may have drawn during the activity. (Use the list of questions they generated in step 1 of Part II.)

Extension

If you use a natural area for this activity, you can have students keep track of changes in it from season to season and year to year by comparing their data with that collected by other groups in the past. You can also do an urban blitz to identify the plants and animals that live in a city block.

Resources

All the Birds of North America by Jack L. Griggs (Harper Collins, 1997).

The Audubon Society Field Guide to North American Mammals by John Whitaker (Knopf, 1980).

Eco-Inquiry by Kathleen Hogan (Kendall/Hunt Publishing Company, 1994).

Global Biodiversity: Status of the Earth's Living Resources edited by Brian Groombridge (Chapman and Hall, 1992).

Yuck! A Big Book of Little Horrors by Robert Snedden (Simon and Schuster, 1996).

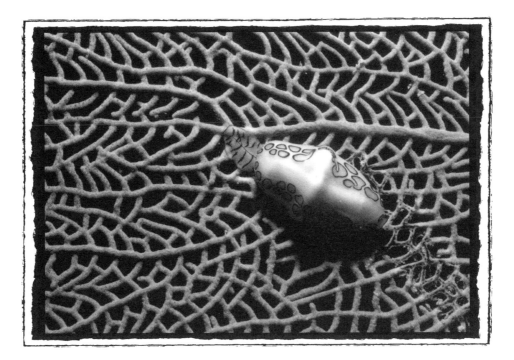

"One result of formal education is that
students graduate without knowing how to think
in whole systems, how to find connections,
how to ask big questions, and how to separate the
trivial from the important. Now more than ever—we need
people who think broadly and who understand systems,
connections, patterns, and root causes."

**–David Orr,
writer, professor**

SUBJECTS

science, mathematics

SKILLS

gathering (collecting), organizing (manipulating materials, classifying), analyzing (calculating), interpreting (drawing conclusions)

FRAMEWORK LINKS

3, 12, 21, 54

VOCABULARY

abdomen, class, ecosystem, habitat, indicator species, invertebrate, order, species evenness, species richness, thorax

TIME

Time will vary according to the level of students. Most classes will need at least three sessions to complete the activity.

MATERIALS

For the class: *insect field guides, model insects, scissors*
For each group of three students: *sweep net (each group can make its own—see "How to Make a Simple Sweep Net" on page 150 for list of materials); large, clear, sealable plastic bag; small paper bag; magnifying glasses; copies of "Insect I.D." (page 33 in the Student Book), "Putting Them in Order" (pages 34–36 in the Student Book), "Incredible Insects" (pages 37–38 in the Student Book), and "Insect Madness Data Sheet" (page 39 in the Student Book)*

CONNECTIONS

For a good introductory outdoor activity, try "Backyard BioBlitz" (pages 134–143). Use "Sizing Up Species" (pages 108–121) before this activity to give students background on the numbers and types of species that live on the planet as well as an introduction to the taxonomic system, and how insects fit into it.

AT A GLANCE

Create a diversity index for insects in your schoolyard.

OBJECTIVES

Use a simple chart to identify insects from different orders. Measure insect diversity of a site using the Sequential Comparison Index. Discuss how diversity indices are used by scientists.

Biodiversity is the diversity of life in all its forms, from tiny organisms to ecosystems that cover thousands of square miles, from a single human life to cultures and languages that connect people around the world. If biodiversity is made up of so many interacting individuals and systems, how can we measure it? How do we know that the rain forests are diverse ecosystems? And how do we know that biodiversity is in decline?

It isn't easy, but one tool scientists use to measure biodiversity is a diversity index. A diversity index measures the diversity of organisms in an ecosystem. Scientists might create an index for frogs in a pond or birds in a forest. In this activity, you'll create a diversity index for insects in your schoolyard.

Scientists use diversity indices in many ways. The indices are often used to help gauge the health of different ecosystems. The diversity of species in aquatic ecosystems, for example, can be used to determine the quality of stream or river water. And diversity indices can also be used to monitor levels of diversity over time. If scientists wanted to know how fires affect plant diversity, for example, they could use a diversity index to measure species diversity before and after a fire.

Computing a diversity index is a fun way to begin investigating local habitats. In this activity, your students will use the Sequential Comparison Index (SCI), one of many diversity indices, to investigate insect diversity. They'll collect the insects using a sweep net, a common tool of entomologists, and we'll show you how you can make your own nets that the students can keep.

Before You Begin

Find an open, grassy area where your group can work. You might use the school grounds, a nearby park, or some other open space.

Since the SCI diversity index does not work well with small numbers of insects, each team of students will need to find at least 10 individual insects (not necessarily 10 different species). Because insects are cold-blooded, they are most active on warm days. Be sure to do this activity on a warm day at a time of year when insects are numerous in your area. If some groups do not find at least 10 insects, several groups can pool their results so that they have enough data to calculate the index.

If possible, collect field guides, posters, model insects, and other references that will help your students become familiar with and identify insects in your area. For each team of students, make one copy of "Incredible Insects" (pages 37–38 in the Student Book), "Insect I.D." (page 33 in the Student Book), and "Putting Them in Order" (page 34–36 in the Student Book), and at least two copies of the "Insect Madness Data Sheet" (page 39 in the Student Book).

You will need enough nets, plastic bags, magnifying glasses, scissors, and small paper bags for each team of students to have at least one of each. Nets can be purchased from most biological supply companies and range in price from $5 to $50. If your budget won't allow for purchasing nets, or if you think your students would like to make and keep their own nets, you have several options. There are many ways to make sweep nets, and you can do some research to find the best type of net to suit the time and money available. We've included instructions for one easy way to make sweep nets (see page 150). The nets can be made in class, or you can have the students make them at home with the help of an adult.

"The bulk of what we call biodiversity isn't lions and tigers and bears. It's the little animals that support all these larger animals."

**–George Benz,
biologist**

The Biodiversity Index:
Species Richness and Species Evenness

A biodiversity index is a helpful tool for measuring diversity because it looks at two important aspects: the number of different species in an area (species richness) and the number of individuals of each species (species evenness). If we used only the number of species in an area to tell us about the area's biodiversity, then two areas (such as wetlands) that each contained three different kinds of plant species would have an equal level of biodiversity. But what if in one of those wetlands almost every plant was of one species, and there were only a few plants of the other two species? And what if the other wetland had about the same number of plants of each species? Which would be more diverse? (See example below.)

Wetland 1

Wetland 2

The two wetlands above each contain three types of plants: arum, marsh mallow, and cattail. Since both wetlands contain the same species, their richness is the same. However, since each wetland contains a different number of those three species, the species evenness is different. Wetland #1 is more even than wetland #2 because it contains about the same number of each species. Wetland #2 is dominated by one species, making it less even.

World Wildlife Fund

What to Do

1. Introduce insect types and characteristics.

Explain to your students that they're going to survey the diversity of insects on a site near their school. Tell them they'll be identifying different species of insects, so they'll need to be familiar with insects' major features and some of the major groups, or orders. They won't have to know the name of each species they find, but they'll need to be able to tell species apart. Use the handouts "Insect I.D." and "Putting Them in Order" (pages 34–36 in the Student Book) to go over the different types of insects. If you want, show the students some models of insects from different orders. They should feel comfortable identifying insect characteristics and be able to tell one species from another. Using the information on pages 152–153, discuss insect life cycles with your students to give them a clearer understanding of how to recognize and classify the insects they find.

2. Introduce diversity indices.

Explain that a diversity index is a way to measure the biodiversity of a group of animals in an area. The students will be using one to look at the diversity of insects in their schoolyard or nearby area. Discuss the concepts of species richness and species evenness. Focus on how these two terms are different and how they relate to biodiversity. You might ask your students why they think that a diversity index could be a useful tool for scientists. (Have them think about comparing the levels of diversity in different areas and in the same area over time.)

3. Explain the Sequential Comparison Index (SCI)*.

Explain how the SCI works. Draw a large square on the chalkboard and tell the students to imagine that it's somebody's yard. Suppose they went into that yard and found 9 birds in 15 minutes. (Make 9 marks within your square to represent the birds.) To calculate the SCI, which is one measure of diversity, you begin by writing down the name of each individual bird you saw. Then you put those names in a paper bag, pick them out at random, and write them down in the order in which you drew them. See the example below:

$$\left[\begin{smallmatrix} robin, robin \\ 1 \end{smallmatrix}\right]\left[\begin{smallmatrix} blue\ jay \\ 2 \end{smallmatrix}\right]\left[\begin{smallmatrix} robin \\ 3 \end{smallmatrix}\right]\left[\begin{smallmatrix} cardinal, cardinal \\ 4 \end{smallmatrix}\right]\left[\begin{smallmatrix} sparrow \\ 5 \end{smallmatrix}\right]\left[\begin{smallmatrix} robin \\ 6 \end{smallmatrix}\right]\left[\begin{smallmatrix} sparrow \\ 7 \end{smallmatrix}\right]$$

Now you figure out how many "runs" of different birds you have. Runs are groupings, or strings, of the same species. In the case of the bird count above, the first 2 robins represent 1 run, the blue jay a second, the next robin a third, and so on. There are seven runs in this sequence. Tell students to think about how the number of runs is related to the species richness and evenness in the area. (The more different kinds of species there are, the more runs you are likely to have.)

To calculate the SCI, you simply divide the number of runs by the number of individuals:

$$SCI = \frac{Number\ of\ Runs}{Number\ of\ Individuals} \quad or \quad \frac{7\ runs}{9\ birds} = 0.77$$

The SCI can range from 0 to 1, with a value of 1 representing the greatest diversity. In other words, the closer the SCI is to 1, the greater the variety of species is in a given area.

*It is important to note that the Sequential Comparison Index is just one of many indices used to determine diversity. This particular index is a very simple one and is not used widely in scientific research.

Averaging the Results

Because the SCI is based to a certain extent on chance (the order in which you draw organisms), the same data can give different results. In the bird illustration for example, there is a maximum of nine possible runs, as shown in the following computations:

$$\left[\begin{array}{c}robin\\1\end{array}\right]\left[\begin{array}{c}blue\ jay\\2\end{array}\right]\left[\begin{array}{c}robin\\3\end{array}\right]\left[\begin{array}{c}cardinal\\4\end{array}\right]\left[\begin{array}{c}robin\\5\end{array}\right]\left[\begin{array}{c}cardinal\\6\end{array}\right]\left[\begin{array}{c}sparrow\\7\end{array}\right]\left[\begin{array}{c}robin\\8\end{array}\right]\left[\begin{array}{c}sparrow\\9\end{array}\right]$$

$$\mathbf{SCI} = \frac{9\ Runs}{9\ Birds} = 1$$

There is a minimum of four possible runs:

$$\left[\begin{array}{c}robin,\ robin,\ robin,\ robin\\1\end{array}\right]\left[\begin{array}{c}blue\ jay\\2\end{array}\right]\left[\begin{array}{c}cardinal,\ cardinal\\3\end{array}\right]\left[\begin{array}{c}sparrow,\ sparrow\\4\end{array}\right]$$

$$\mathbf{SCI} = \frac{4\ Runs}{9\ Birds} = .44$$

Your students should not feel that because the same data can give different results that this is not a valuable tool. The number of runs, although it can vary, is related to the diversity of species: The more individuals of the same species there are (which means biodiversity is low), the lower the number of runs will be. All the students need to do to correct for possible variations is to repeat the exercise three times. They should put the names back into the bag and pick them out again. After getting three values, they should average the results for a more reliable SCI. In this example, the average—the final SCI value—is 0.7. Leave this example on the board so students can refer to it later during their own computations.

4. Practice determining the SCI.

Give your students a little extra practice figuring out the SCI before they go outside. Divide them into groups of three and give each group a paper bag and a copy of the handout "Incredible Insects" (pages 37–38 in the Student Book). Pass out scissors and ask the students to cut out the insect squares. Then ask them to put the squares into the paper bag and draw them randomly. Explain that they should line up the insects as they draw them and then calculate the SCI. Have them repeat this two times and average their three values.

5. Prepare for field collection.

Now that your students are familiar with both insects and the SCI, explain that they'll be going on an expedition to discover insects in their schoolyard or nearby open area and calculating an insect SCI for the area.

Keep students in teams of three. Each team should have at least two copies of the "Insect Madness Data Sheet" (page 39 in the Student Book) and at least one net and one plastic bag for identifying and collecting insects. Explain that each team will use the sweep net to collect insects. Then they should shake all of the insects from the net into its plastic bag where the insects will be easy to identify. (All insects should be returned to the wild after the activity.)

6. Demonstrate how to sweep for insects.

Sweep the net as if you are sweeping the grass with a broom. Get the net down into the grass so that it is nearly touching the ground as you make a sweep. Tell the students that in order to collect the most insects, they will need to make sure their nets get down into the grass where insects are abundant.

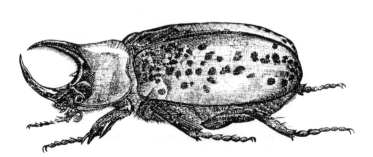

After you sweep through the grass for about 30 seconds, quickly swing the net through the air to force the insects to the bottom of the net. Then grab about one-third of the way up the net to prevent insects from escaping. Have a student hold the clear, sealable plastic bag. Turn the net inside out into the bag and shake the insects into it. Remind students that any large, winged insects should be recorded and set free before putting the rest of the insects in the plastic bag. Once all the insects are in the bag, tell students they should seal the bag and identify the insects. The students should record the insects on the "Insect Madness Data Sheet," and then set the insects free before doing their next sweep.

Since each student should get to sweep the net at least once, you might want to shorten the amount of time each student sweeps his or her net if insects are abundant in your area. While you demonstrate the sweeping technique, note about how many insects you collect in 30 seconds. If you collect more than 20 insects, you might suggest that each student sweep the net for only 10 or 15 seconds at a time.

7. Collect insects.

Have each team pick one small area of the yard in which to make its sweeps. The teams should be spread throughout the schoolyard. Have students take turns sweeping the net until they have collected at least 10 insects. Tell students not to collect more than 30 insects so that they don't get overwhelmed when they calculate the SCI. The number of sweeps you will need to get the right amount of insects depends on your location and the time of year. In some areas, you could collect more than you could handle in just a few sweeps. But, if conditions aren't right, you might have to sweep for hours to get enough insects. (Caution students not to collect bees, wasps, and other stinging insects. They can just write down what they catch and carefully let them fly out of the net. See "Be Careful!" on page 154.)

8. Identify the insects.

After each sweep, the insects in the net should be shaken into the plastic bag. Have the students identify the insects to the best of their ability. They can begin by dividing them into orders. If any insects don't fit into the orders we've listed, have the students record the characteristics that set them apart from other orders.

Next, the students should distinguish between each species. If they can't identify the species by using field guides, tell them to look at each insect's size, color, and shape. Then ask them to devise a descriptive name for each species. For example, they might call a spotted insect "mottled black beetle" and a flat, green one "leaf-like grasshopper." Have the students write the species name for each insect in a box on the "Insect Madness Data Sheet". They need to fill in one box for each individual insect. So a group that found two large orange butterflies would fill in two boxes with the words "large orange butterfly."

Once the students have recorded each insect in the plastic bag, have them let the insects out before they make their next sweep.

Sample Data Sheet

shiny beetle	black cricket	click beetle	grasshopper
honey bee	cockroach	millipede	black cricket
praying mantis	yellow butterfly	earwig	yellow butterfly
aphid	honey bee	aphid	earwig
walking stick	shiny beetle	millipede	shiny beetle
dragonfly	dragonfly	ant	aphid

How to Make a Simple Sweep Net

With a few minutes and a few dollars, you can make a sweep net using the following supplies:

- **one wire coat hanger**
- **one five-gallon nylon paint strainer (available at most paint stores for about $1.20 each)**
- **duct tape**
- **one ⅝-in. by 36-in. wooden dowel (available at most hardware stores for less than $1.00)**
- **a stapler or sewing kit**

First, bend the coat hanger into a square as shown below. Then attach the paint strainer to the coat hanger. For a more durable net, sew the mesh fabric of the paint strainer to the hanger. A faster method is to use duct tape to attach the net to the hanger and then to staple over the duct tape to reinforce the job. The hanger and strainer can be connected to the handle (the wooden dowel) using duct tape. Your finished product should look like the one below.

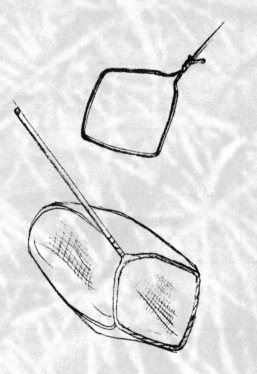

9. Compute your area's insect SCI.

Back in the classroom, have each team cut up the squares its members filled in with insect names and put the squares into a paper bag. Then have the students pull the squares out at random and write down the name of each insect as they pull it from the bag, just as they did in step 4. Have the students count their runs and calculate their SCI. Then have them repeat this two more times and average their SCI values. (If your students found very few insects, have groups combine results so that there are at least 10 insects total per combined group before they calculate an SCI.)

After each group has come up with its SCI value, average the values of the entire class to come up with an average diversity index of your schoolyard or nearby open area.

10. Compare and discuss results.

Because the SCI gives students only a decimal value that represents insect diversity, the number doesn't mean much on its own. Explain that diversity indices are useful because they allow us to make comparisons of diversity between areas or in the same area at different times. Your students could do this exercise again at a different time of year to find out how insect diversity can change throughout the year. They can also make comparisons between different areas of the schoolyard.

Because they collected insects in different areas, different teams could have encountered different insect habitats. Discuss how these different habitats support different types and numbers of species and how different habitats can lead to different SCI values around the yard. Can students explain why some teams had especially high or low SCI values? Were teams with similar SCIs near each other in the yard? Can the students identify any things that areas with high or low values have in common? *(Areas with high diversity may have been near water or a food source like dung or decomposing plants or animals. Areas with low diversity may have been low in plant life.)*

11. Discuss problems with the SCI.

Discuss with your students the method they used to sample the insects. Were some students better at collecting insects than others? How might that affect their results? *(Students who didn't do a good job of collecting insects didn't get a good sample. If, for example, a student swept only the top of the grass, he or she wouldn't collect any of the insects found lower in the grass and would have a lower SCI than students who swept lower and collected more kinds of insects.)*

Do the students think the insects they collected represent all of the insects in their schoolyard? *(Because they used sweep nets, students only collected samples of insects above the ground. Students may not have swept through the entire area. Some of the insects might have escaped the nets.)* There are many ways to collect samples of insects that live on and under the ground. See the second extension for ideas about how you and your students could collect different types of insects.

What are other problems with the SCI? *(There is a degree of randomness and the possibility of getting a high diversity index without really having much diversity in the survey area. For example, if every other species were the same, you'd get a high diversity index when there's really only a total of two species in the area.)*

12. Discuss species diversity.

Do your students think a diversity index is a good measure of a site's ecological value? Explain that although biodiversity is usually important for most ecosystems to be healthy, diversity isn't always an indication of ecological importance. Imagine going to a site and finding ten bald eagles and no other species. That area would get a low score on the SCI, but it would be extremely important for conservation. See the section entitled "Is More Better?" (page 154) for a discussion of this topic.

A high or low SCI does not tell us about the biodiversity of every type of organism in an ecosystem. The diversity of different types of animals, such as insects, birds, and mammals, can be very different on the same site. There may be a wide variety of plants but very few animals. Does that mean that we shouldn't pay attention to a diversity index? Definitely not. Indices of some species can be very important. Certain species, called indicator species, tell us a lot about the health of an ecosystem. Some scientists think that frogs are important indicator species, and the scientists are measuring the diversity of these animals to investigate the effects of air and water pollution around the world.

Growing Up an Insect

Most insects hatch from eggs and change form as they become adults. This process of change is called metamorphosis. Some insects go through three life stages: egg to nymph to adult. This is called simple or incomplete metamorphosis. More advanced insects go through four life stages: egg to larva to pupa to adult. This is called complete metamorphosis.

As your students are out collecting, they might find nymphs, larvae, or pupae. Because the insects may not be in their adult forms, sometimes it's hard to tell which taxonomic order they belong to. Encourage your kids to use field guides, which often have pictures of these growing stages. You can also tell them that nymphs usually look like miniature copies of their parents without wings (like young grasshoppers and cockroaches), but larvae and pupae look very different from an adult (such as the caterpillar and cocoon stages of a butterfly). And nymphs that develop in water, which are often called naiads, also look very different from the adults (such as dragonfly and mayfly naiads).

Insects that go through incomplete metamorphosis include springtails, dragonflies, mayflies, termites, stoneflies, crickets, grasshoppers, praying mantids, earwigs, cockroaches, true bugs, aphids, cicadas, and thrips. Insects that go through complete metamorphosis include lacewings, ant lions, dobsonflies, beetles, butterflies, moths, flies, fleas, ants, bees, and wasps.

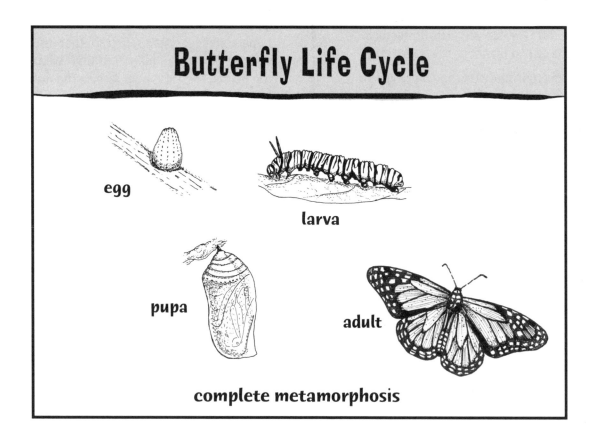

Butterfly Life Cycle

egg

larva

pupa

adult

complete metamorphosis

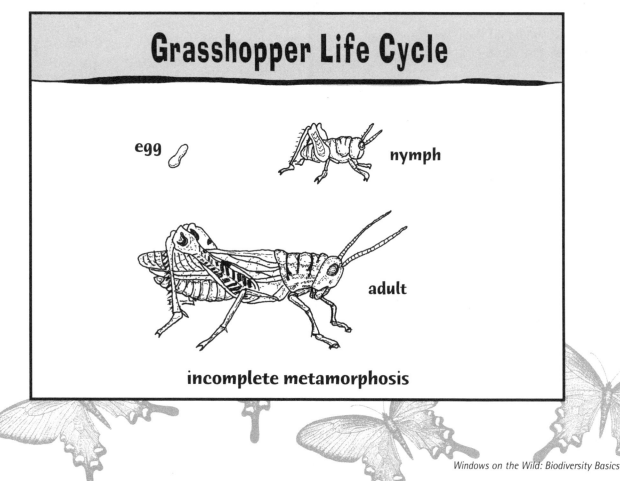

Grasshopper Life Cycle

egg

nymph

adult

incomplete metamorphosis

Be Careful!

Students will enjoy this activity because they can get outside and explore the diversity of life around them, but you need to make sure your students and the insects are safe. Remember the following when planning this activity:

- Although your students shouldn't need to handle any insects, there's always a chance that someone could be stung. Find out if any of your students have an allergy to any type of insect sting, and have the necessary first aid on hand or have students with severe allergies stay inside and work on a related project or extension while the others are outside.

- Some insects can be harmed by handling and should not be collected. These include large-winged insects such as dragonflies and butterflies, whose wings are easily damaged. Students who sweep up any of these insects should jot down a description of the insects and let them go.

- Remind students to limit the time they spend identifying insects in the plastic bags because most insects can't survive in the bags for long. Emphasize that one objective of this activity is to gain a new respect for the diversity of insects, so the students should do their best to preserve insect diversity in their schoolyard.

Conclude the discussion by explaining why insect diversity is important. Have students think about all the services that insects provide for humans, plants, animals, and ecosystems. Explain that insects are very important in the pollination of both wild plants and crop plants. They also help recycle the nutrients present in dead plants and animals and return the nutrients to the soil. We eat the honey made by bees and wear the silk made by caterpillars. In many countries, people also eat insects. Can your students think of other ways that insects are ecologically or economically important, and other reasons it's important to protect insect diversity?

IS MORE BETTER?

Does more diversity mean a healthier ecosystem? Not always. For example, when a patch of Central American forest is cleared, the number and variety of butterflies in the area may rise, but the overall productivity of the area is reduced. A cultivated garden may have a wider variety of plant species than a nearby field full of native plants, but the native plants play an important role in the local ecosystem that the garden does not. Although disturbed areas can host a larger number of species, many of which are likely to be introduced, they usually are not as ecologically important as areas filled with native species. So when scientists use a diversity index to measure environmental quality, they compare the results from the study site with the results from a pristine ecosystem to form their conclusions.

WRAPPING IT UP

Assessment

Explain to your students the following hypothetical situation:

Students at the Roger Tory Peterson School conducted an insect survey on their school grounds. When they did their first survey of species to calculate their SCI, this is how the insects lined up:

red grasshopper, long-horned beetle, long-horned beetle, long-horned beetle, blue tip butterfly

Students at the nearby Martin Luther King School conducted an insect survey on their school grounds on the same day. This is how their insects lined up:

green bug, green bug, long-horned beetle, blue tip butterfly, gold bug, red grasshopper, blue tip butterfly, red grasshopper, green bug, sparkly beetle

Ask your students to determine the SCI for both sites and to briefly compare the two diversity scores. They should develop a list of possible reasons the scores differ and a list of questions they would want answered before they could determine the actual reasons for the different diversity scores. What information about the insects would help them know if they had conducted an accurate SCI?

> **Unsatisfactory**—Student is unable to compute the SCI and fails to list reasonable explanations for the differences or to request more information.

> **Satisfactory**—Student is able to compute the SCI and lists one or two reasonable explanations for the differences, can identify one or two logical requests for information, and can identify one or two physical characteristics of insects that would help ensure the proper identification.

> **Excellent**—Student is able to compute the SCI and list several reasonable and creative explanations for the differences, can identify several requests for information, and can identify several physical characteristics of insects that would help ensure the proper identification.

Portfolio

The data sheet can be used as documentation for this lesson.

Writing Ideas

- Have students choose one of the insects they found and write a description of it. Tell them not to use the name of the insect, even if they know it. Instead, tell them to write down all the other ways they can describe the insect to others.

- Have students imagine themselves as insects for a day. They can write about their experiences from their new perspective.

- Have students keep a record of the insects they found, including the insects' distinguishing characteristics and location. They can put all these together to create a lab book with their drawings and observations.

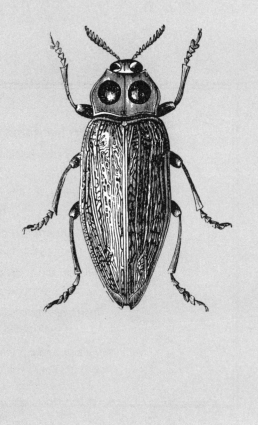

WRAPPING IT UP (Cont'd.)

Extensions

- Have students compare the SCI of different habitats such as a forest and a field.

- Have students investigate ways they can sample the diversity of insects on and under the ground. There are many fun and easy ways to find and study these insects. Pitfall traps, aspirators, and Berlese funnels are just a few. See the Resources for good sources of information on more ways to investigate insects.

- Encourage students to investigate aquatic insects. After collecting insects, students can use the SCI to compare the insect diversity of different streams or rivers. (For more information contact the Global Rivers Environmental Education Network at 721 Huron Avenue, Ann Arbor, MI 48104. <www.igc.org/green>.)

- Make a poster that illustrates the diversity of insects in your area.

- Let students choose an insect species as a mascot for the class and research its life history, its ecological significance, and its value to humans.

- Visit a nearby museum or insect zoo to observe examples of the worldwide diversity of insects.

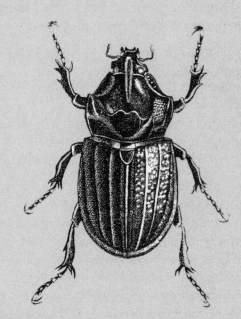

Resources

Adventures with Insects by Richard Headstrom (Dover Publishing, 1982).

Carolina Biological Supply (supplier of sweep nets) 2700 York Road, Burlington, NC 27215 <www.carolina.com>

The Common Insects of North America by Lester A. Swan and Charles S. Papp (Harper and Row, 1972).

Entomology: Adventures, Projects, and Ideas for Exploring the World of Insects by Ellen Doris (Thames and Hudson, 1993).

A Field Guide to the Insects: America North of Mexico (Peterson Field Guides) by Donald J. Borror and Richard E. White (Houghton Mifflin, 1998).

Incredible Insects: The Ultimate Guide to the World of Insects by Rick Imes (Barnes and Noble Books, 1997).

An Introduction to the Study of Insects by Donald J. Borror, Dwight M. Delong, and Charles A. Triplehorn (Holt, Rinehart, and Winston, 1997).

Skip Moody/Dembinsky Photo Associates

*"Beetles are the most diverse and among the oldest
of all land animals. The more than 290,000 species known worldwide
are only a small fraction of those thought to exist. Local diversity is
extreme: More than a thousand different kinds, ranging from carnivores
to wood borers and leaf eaters, have been found on single trees
in the forests of Central and South America."*

–Edward O. Wilson, biologist

SUBJECTS

science

SKILLS

gathering (simulating), analyzing (identifying patterns), interpreting (identifying cause and effect, inferring)

FRAMEWORK LINKS

4, 19, 20, 22, 28

VOCABULARY

chromosome, **evolution, gene, genetic diversity,** *herd, inherit, nucleus, population,* **species, trait**

TIME

three sessions

MATERIALS

Part I—*copy of the "Human Genetic Wheel" (page 41) and "Checking Out Your Genetic Traits" (page 40)*
Part II—*15 to 20 index cards*
Part III—*scissors, copies of "All About Giraffes" (pages 42–43), "Giraffe Genetic Wheel" (page 45), "Giraffe Cards" (pages 46–50) on white paper and colored paper, "Spotting Giraffes" (page 44), "Event Cards" (page 52), and "Giraffe Calf Cards" (page 51) on white and colored paper (The copy pages for this activity are all included in the Student Book.)*

CONNECTIONS

For more on the importance of genetic diversity, follow up with "Diversity on Your Table" (pages 230–237) or "Biodiversity—It's Evolving" (pages 168–179). To investigate how the loss of genetic diversity affects species, try "The Case of the Florida Panther" (pages 246–251).

AT A GLANCE

Play several different games that introduce genetic diversity and highlight why it's important within populations.

OBJECTIVES

Identify and classify genetic traits using a genetic wheel. Explain why genetic diversity may be necessary for the long-term survival of a population of animals or plants. Explain that lack of genetic diversity is one of the reasons why small and fragmented populations are vulnerable to extinction.

From a scientific perspective, conserving biodiversity means more than just protecting the variety of different species on Earth. It also means preserving the natural variation that exists among the individuals of each species. Just as humans vary in their appearances and abilities, so, too, do individual fish, mushrooms, oak trees, and amoebae. Preserving variety within populations of species is essential for preserving the ability of that species to cope with environmental change.

An organism's ability to adapt to environmental change determines how well it will survive in the long run. The greater the diversity of genes in a population, the greater the chances that some individuals will possess the genes needed to survive under conditions of environmental stress. As wild populations of plants and animals become smaller and more fragmented, it becomes less likely that the remaining individuals will possess the genes needed to survive environmental changes. The individual—and the species—is subject to destruction.

This three-part activity will introduce your students to the concept of genetic diversity within a population. In Part I they will observe and compare human traits within their classroom population. This exercise should demonstrate that each individual has a variety of traits that make him or her unique and that create a diverse population within the classroom. In Part II they will discover through a quick, active demonstration that increased diversity contributes to greater survivability. And Part III will reinforce this idea as your students play a game in which they represent populations of giraffes coping with changes in the environment over time.

Before You Begin • Part I

For each student, make a copy of the "Human Genetic Wheel" (page 41 in the Student Book) and "Checking Out Your Genetic Traits" (page 40 in the Student Book).

What to Do • Part I

1. Introduce genes.

Your students may know that the physical characteristics of all creatures on Earth are determined by their genes. But what are genes and how do they work? Genes are sections of DNA that manifest themselves as visible traits, such as eye color and hair texture, and nonvisible traits, such as a susceptibility to a certain disease. Genes form visible bars on threadlike structures called *chromosomes*, which are inside the central part, or *nucleus*, of every plant and animal cell. Chromosomes contain the genetic material of each cell, made up mostly of DNA. Chromosomes become visible under a microsope when any animal or plant cell divides (see illustration below).

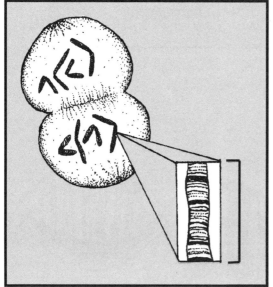
chromosomes and genes during cell division

In mammals, most healthy cells have two copies of each chromosome—one from each parent. Reproductive cells (eggs and sperm) have one copy of each chromosome. Different species have different numbers of chromosome pairs. In humans, for example, there are normally 23 pairs of chromosomes.

2. Discuss genetic diversity.

Explain that in a healthy population (a group of organisms of the same species living in a certain geographic area) there is a wide variety of genes that combine in many different ways to form a broad diversity of individuals. If the population is suddenly subjected to stress, such as disease or environmental change, the genetic variety makes it likely that at least some individuals will be adapted well enough to survive and continue the species.

Populations of some species have become so small or fragmented that they have lost much of their original genetic diversity. If these populations are suddenly subjected to a disease or other stress, there might not be any individuals with the genes that provide protection from the disease and enable the individuals to survive.

3. Determine the characteristics of the class population.

Give each student a copy of "Checking Out Your Genetic Traits." Go over the list of traits with your class. Have your students work in pairs to help each other determine their traits and check the traits off their worksheets. As you read the list, instruct your students to check the box that describes the trait they possess. They can also work in pairs to observe the traits in each other. For each trait, there are two possibilities:

1. Your ear lobes are either hanging loose or they are attached to the side of your head.

2. Your hair is either curly or straight.

3. You can either curl your tongue, or you cannot curl it. (This trait refers to whether you can or cannot roll the sides of your tongue to make it into a tube-like shape.)

4. You either have hair on your fingers, or you don't have it. (Look at the part of your finger between your knuckle and first joint.)

5. You either have light-colored eyes (blue or green), or you have dark eyes.

6. You either have a widow's peak, or you don't have one. (If your hairline comes to a point in the middle of your forehead, you have a widow's peak.)

7. Your little finger is either straight, or it is bent.

Point out to your students that their genes have determined each characteristic on the worksheet.

4. Use the "Human Genetic Wheel."

Pass out a copy of the "Human Genetic Wheel" to each student. Instruct each student to start at the inner band and find the appropriate letter code that describes his or her own ear lobe type (it will be either "L" for loose or "ll" for attached). Instruct them to continue moving outward on the wheel, finding their characteristics for each trait, until they have located their little finger type in band seven. Each person should then find the number next to his or her finger type and record this number on the worksheet.

5. Pool the results.

There are 128 possible combinations of the seven traits. To find out how many different combinations are present in the class population, go around the room and have each student give his or her Genetic Wheel number. Record the numbers on the board. If there is more than one student with the same number, place a check next to that number.

6. Discuss your findings.

Are there any two students in the class who have the same seven traits? Then ask the students if they can think of an eighth trait that would set these two people apart. Are there any numbers that have clusters of classmates? Why?

> *Every individual in any population is different from every other individual. Have students look at the variations among the people in their class as an example. But these variations don't make any individual a different species. Everyone in the class, regardless of his or her differences, is still a human being.*

DNA double helix

Before You Begin • Part II

You will need 15 to 20 index cards. On each card, write one characteristic that distinguishes one student from another. (See "Indexing Student Characteristics" on page 162.)

What to Do • Part II

1. Introduce the demonstration.

Divide the students into two teams and explain that they're going to do a demonstration that illustrates why genetic diversity is important. Show them your stack of index cards (see box on "Indexing Student Characteristics" on page 162), and explain that each one lists a characteristic that, for the purposes of the game, is going to represent a genetic trait. Tell them that once the game starts they are not allowed to change anything about themselves. Tell them that you're going to read several of these cards aloud and that if anyone on either team has the characteristic listed on that card, he or she will "die." Those students who are "dead" must sit down. The object of the game is to have at least one member of their team "alive" at the end.

2. Do the demonstration.

Have the students get into their teams and then stand facing you. Read one of the index cards you made earlier and ask all the students with the characteristic listed on the card to sit down. Repeat until you have gone through about three or four of the cards. (At least one of the teams should still have members standing.) Tell the students that if there's anyone still standing on their team, they can all regenerate and join back in. If both teams still have members standing, play another round, reading through three or four additional cards. Then go on to step 3.

3. Discuss the demonstration.

Ask the students what happened. Did any "characteristics" wipe out more people on their team than others? Did one team do better than the other? Why? (Answers will vary depending on what happens with your group. However, students should be figuring out that their team has a better chance of surviving when the characteristics of the team members are more diverse.)

4. Do the demonstration again.

Restore each team to its full number of "live" members. Then tell the teams that they're going to try the demonstration again, but that before you start they are allowed to make any adjustments they want on their teams. (Students should do things that give the group a wider range of traits. For example, some team members may untie their shoes while others may leave them tied, and some may add layers of clothing.) Shuffle the stack of cards and then read through several of them, having students with any of the characteristics "die" and sit down.

Sample Student Sheet

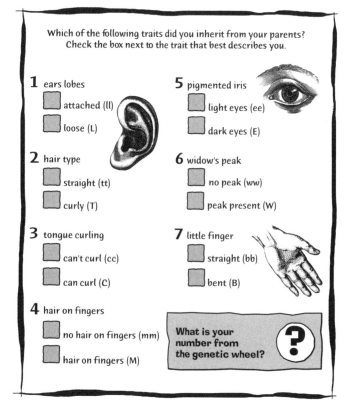

Which of the following traits did you inherit from your parents? Check the box next to the trait that best describes you.

1 ears lobes
☐ attached (ll)
☐ loose (L)

2 hair type
☐ straight (tt)
☐ curly (T)

3 tongue curling
☐ can't curl (cc)
☐ can curl (C)

4 hair on fingers
☐ no hair on fingers (mm)
☐ hair on fingers (M)

5 pigmented iris
☐ light eyes (ee)
☐ dark eyes (E)

6 widow's peak
☐ no peak (ww)
☐ peak present (W)

7 little finger
☐ straight (bb)
☐ bent (B)

What is your number from the genetic wheel? **?**

Indexing Student Characteristics

To do this demonstration you will need a stack of index cards, each of which has a "genetic" characteristic that can distinguish your students from one another. Because it may be difficult to come up with enough truly genetically based traits, you should feel free to use traits, such as clothing color or type of shoes, in the demonstration. Below are some possibilities for the cards. You will need to choose characteristics that will weed out your group—but not wipe out the entire class all at once. During the demonstration, each time you read one of these traits, every student who has the trait will "die out" for the rest of the round.

- **Light-colored eyes**
- **Bent little finger**
- **Not wearing glasses**
- **Shoes laced and tied**
- **Shoes without laces**
- **Not wearing red**
- **Attached ear lobes**
- **Not able to curl tongue**
- **Wearing earring(s)**
- **Wearing a sweater**
- **Wearing hair clips of any kind**
- **Wearing a watch**
- **A widow's peak**
- **Wearing a hat**

5. Wrap up.

Have the students describe what happened. Did their team last longer or shorter this time? What helped them or hurt them? What can they say about how genetic diversity might help wild populations of animals or plants survive? (Students should understand that the more diverse their team was, the greater the chance it had of having at least one member left at the end of several rounds. They should also be able to generalize that the more genetically diverse a wild population is, the greater its chances of surviving over time. However, if the students can't quite make this leap yet, don't worry. They'll get a chance to apply these ideas in Part III.)

Before You Begin • Part III

Make several copies of the "Giraffe Genetic Wheel" (page 45 in the Student Book) for each group. Also make two copies of the "Giraffe Cards" (pages 46–50 in the Student Book) for each group (one copy on white paper and one copy on colored paper), as well as one copy of "Spotting Giraffes" (page 44 in the Student Book) for each group. You'll need to make two copies of the "Giraffe Calf Cards" (page 51 in the Student Book) on white paper and two copies on colored paper, cut the cards apart, and put them in a container. Then make one copy of the "Event Cards" (page 52 in the Student Book), cut them apart, and put them in another container. Have scissors for each group. (If "All About Giraffes" [pages 42–43 in the Student Book] is used as a homework assignment, copy one for each student.)

What to Do • Part III

1. Introduce the giraffe game.

Tell students that they will play a game that illustrates why genetic diversity is important. The game focuses on the giraffe. You may want to read "All About Giraffes" (pages 42–43 in the Student Book) to the class as an introduction to the activity or give it to the students to read for homework the night before. Also give the students a copy of "Spotting Giraffes" (page 44 in the Student Book) to illustrate the characteristics discussed in "All About Giraffes."

2. Set up for the game.

Divide the class into five groups and give each group its two sheets of "Giraffe Cards" (one on white paper, one on colored paper). Have the students cut the cards apart.

Explain that each group of students is "watching over" a small population of giraffes, represented by the giraffe cards. Each card identifies the characteristics (genetic traits) that each giraffe will have during the game. The genetic traits used in the game are as follows: sex, migratory behavior, resistance to plague, spot pattern, and leg length. Colored cards

represent males and white ones represent females. The other genetic traits are written on each card.

3. Determine the genetic number of the giraffes.

Pass out several copies of the "Giraffe Genetic Wheel" (page 45 in the Student Book) to each group. Using the genetic traits provided on each giraffe card, tell the students to work together to determine the genetic number of each giraffe in their population. They should use the "Giraffe Genetic Wheel" to find the number of each giraffe in the same way they used the "Human Genetic Wheel" (Part I) to find their own numbers. Students should write the genetic number of each giraffe on each giraffe card.

4. Determine the genetic diversity of each group's population of giraffes.

Next ask the students to determine the genetic diversity of their group of giraffes. Ask the student groups to count how many different individual genetic numbers are exhibited by their 20 giraffes. This is the group's diversity number. Consider that a student group has a population of giraffes with the following genetic numbers:

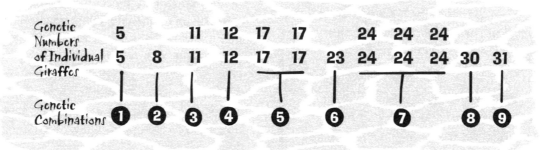

Rules and Strategies

Before students begin the game, share the following information:

- *If a giraffe dies, the students should turn the card that represents that giraffe face down.*

- *Only the dominant male giraffe can mate with the females. If the dominant male dies, a new dominant male must be designated. If a group loses all its males or females, it cannot reproduce.*

- *Events usually affect half of a population. If you have an odd number of giraffes that are affected by an event, round down to find the number of giraffes affected.*

- *Female calves cannot reproduce.*

- *During reproduction events, each qualifying female will receive a calf card. Students must choose traits for each calf based only on the traits of that female and the dominant male. See the following example:*

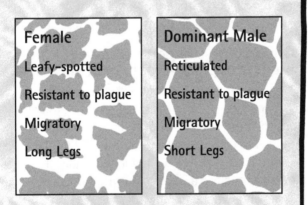

Female	Dominant Male
Leafy-spotted	Reticulated
Resistant to plague	Resistant to plague
Migratory	Migratory
Long Legs	Short Legs

*Then the calf can be either leafy-spotted or reticulated and have either long legs or short legs, but the calf **must** be resistant to plague and migratory (because both parents are).*

Every time a female has a calf, the students will have to assign traits in this manner. Circle the traits on the calf cards.

In this case, the student group would have a total of **nine** different genetic combinations represented by their giraffe group so the diversity number is nine. Write a tally on the board, recording each student group's number of giraffes and diversity number. The larger a group's diversity number, the more genetically diverse the population of giraffes.

> *Each student group should start with 20 giraffes. The diversity number of group one should be 4, group two should be 8, group three should be 12, group four should be 14, and group five should be 20. Some students may realize that they have an advantage–or disadvantage–at this point.*

5. Have each group select a dominant male.

Each group of students should select one male in its giraffe population to be the dominant male. Students should place a big letter "D" on the dominant male giraffe's card. This giraffe will be the only one that mates with the females in the population during the course of the game. If this male dies or joins another group of giraffes, the group will have to designate a new dominant male to take its place.

6. Have students choose cards from the "Event Cards" and read them to the class.

"Event Cards" depict scenarios of environmental change that the giraffe populations must confront. Bold text on the cards indicates the impact that the environmental change has on individuals in the population: (1) loss (death), (2) intermingling, and (3) reproduction. Remind your students that this exercise is a simulation of what *could* happen to a real giraffe population. While the events are not real, they do represent some of the many pressures exerted on populations by natural and human forces. Allow your students to take turns picking an event card at random and reading it aloud to the class. Tell your students to pay attention to the event being read and respond to that event based on the giraffes they have in their population. Every group follows the directions of each event card.

7. Record how many giraffes are left after the events have been read, and analyze the results.

After all the "Event Cards" have been read, record on the board the number of giraffes (adults and calves) surviving in each group's population. Compare different groups of giraffes and determine which ones were more successful. Did genetic diversity contribute to this success? How?

8. Discuss the results of the game.

After you finish the game, discuss genetic diversity using the following questions:

a. Why is genetic diversity important? *(Generally speaking, a more genetically diverse population is more likely to contain some individuals that have the traits necessary to survive and adapt to changes in the environment than populations that aren't as genetically diverse.)*

b. What is the relationship between the size of a population and its genetic diversity? *(As a population becomes smaller, some variation in traits is lost. Because there are fewer individuals in a smaller population, it is less likely that there will be individuals with the traits necessary to survive in times of environmental stress. This is one reason smaller populations are more vulnerable to extinction. Many species that once had large populations, such as the Florida panther, nene goose, and American bison, have lost a great deal of their genetic diversity in a short time because of habitat loss and overhunting.)*

c. What can be done to prevent the loss of genetic diversity? *(To preserve genetic diversity, it is important that wild populations of plants and animals do not become small or fragmented. This is becoming more and more challenging as human populations expand and increase their level of consumption as well as demand for space.)*

d. In the simulation, which events did humans cause? Which were caused by nature? Do you think humans can have both a positive and a negative role in influencing the survival of other animals? *(In the simulation, human actions had both positive consequences [the creation and protection of national parks] and negative consequences [poaching]. The better we are able to fulfill human needs while maintaining genetic diversity, the more a population will be able to withstand both natural and people-related pressures.)*

e. Did some traits seem to be favored over others? Were there any traits that were favored in one instance but selected against in another? How does this relate to the importance of genetic diversity? *(A trait that is advantageous under one set of environmental conditions may be detrimental under another. For example, migratory giraffes may do better in times of drought, but they will be more vulnerable to poaching because they're more likely to leave the safety of protected areas.)*

WRAPPING IT UP

Assessment

Have each student write a short response to the question: What does giraffe diversity (as represented in the "Giraffe Cards") have to do with the fate of a giraffe group (whether the group lives, dies, or successfully reproduces) in the game?

Unsatisfactory—The student is unable to make a connection between giraffe genetic diversity in a population and the fate of that group in the game.

Satisfactory—The student makes logical connections between giraffe population genetic diversity and the group's fate in the game.

Excellent—The student is able to incorporate the concepts of giraffe genetic diversity and the vulnerability of certain populations when confronted with environmental stresses ("Event Cards").

Portfolio

On the back of "Checking Out Your Genetic Traits," have students record their ideas about using a genetic wheel to compare human traits and their understanding of genetic diversity from the game.

Writing Ideas

- Students can compose creative stories that illustrate how genetic variety within species can help them survive over time. For example "Once upon a time, tigers were plain orange. Then a cub named Splasher was born. People called him Splasher because he was born with weird black stripes. Though everyone made fun of him, when he grew up, he found that he could stay hidden in the grass much longer than the ordinary, plain orange tigers. He caught more food and fathered many kittens. They had stripes, too."

- Pick a common animal or plant, and describe several distinct individuals, noting their physical traits. (Dogs and cats work especially well.) Students may illustrate their descriptions. How are the individuals different from one another? What sort of advantage or disadvantage might their characteristics provide?

Extension

Have students work individually or in groups to create displays focusing on how people have created genetic diversity in species to fill certain human needs and desires. They could highlight the dichotomy between the "original" species as they appear in the wild and the domesticated creatures they have become (e.g., wolves and chihuahuas, cougars and house cats.) Cattle and other livestock, crops, ornamental plants, and goldfish provide other dramatic examples. The students should also point out why the animal or plant was developed. After student presentations, ask how human manipulation of genes might help or hinder biodiversity.

Resources

The Cartoon Guide to Genetics by Larry Gonick and Mark Wheelis (Harper Perennial, 1991).

Giraffe by Caroline Arnold (William Morrow and Co., 1987).

The Giraffe, Its Biology, Behavior, and Ecology
by A. I. Dagg and J. B. Foster (VanNostrand Reinhold Co., 1976).

Giraffes, the Sentinels of the Savannas
by Helen Roney Sattler (Lothrop, Lee, and Sheppard Books, 1989).

Grizmek's Animal Life Encyclopedia by B. Grizmek (Van Nostrand Reinhold Co., 1972).

A Natural History of Giraffes by Dorcas MacClintock (Charles Scribner's Sons, 1973).

The genetic wheel approach was inspired by similar activities in *Losing Biodiversity* by Katherine Barrett, Global Systems Science, Lawrence Hall of Science, University of California at Berkeley (1996); and in *Biological Science: A Molecular Approach*, D. C. Heath and Co., (1985).

*"I had time after time watched the
progression across the plain of the giraffe,
in their queer, inimitable, vegetative
gracefulness, as if it were not a herd of
animals but a family of rare, long-stemmed,
speckled gigantic flowers slowly advancing."*

–Isak Dinesen, writer

Biodiversity—It's Evolving

SUBJECTS

science, language arts, social studies

SKILLS

gathering (simulating), organizing (charting), analyzing (identifying components and relationships among components), interpreting (identifying cause and effect), presenting (explaining)

FRAMEWORK LINKS

3, 4, 12, 19, 20, 21, 22, 23, 28

VOCABULARY

adaptation, **evolution, extinct, mutation, natural selection,** *population, selective pressures,* **speciation, species,** *trait*

TIME

Part I—*one to two sessions*
Part II—*one session*

MATERIALS

Part I—*plastic sandwich bags, plastic spoons, masking tape, hard candies preferably all the same size and shape*
Part II—*copies of "The Journals of Charlotte Durand, Explorer" (pages 53–56 in the Student Book), "Twitomite Table" (page 57 in the Student Book), "Twitomites: What Do You Think?" (page 58 in the Student Book), and "Twitomite Time Line" (page 59 in the Student Book)*

CONNECTIONS

For an introduction to genes and the importance of genetic diversity, try "The Gene Scene" (pages 158–167). For more on genetic diversity in food crops, try "Diversity on Your Table" (pages 230–237). Follow up with "Sizing Up Species" (pages 108–121) to give your group a sense of how many species inhabit the planet today.

AT A GLANCE

Take part in a short simulation about natural selection and analyze a set of fictional journals to see how natural selection can lead to new species.

OBJECTIVES

Define natural selection, evolution, and speciation. Describe how natural selection can lead to changes in a population over time and eventually to the evolution of new species. Describe the relationships between genetic diversity and natural selection, and between evolution and biodiversity.

Ever since the first single-celled organism appeared on Earth nearly four billion years ago, life has been in a constant state of change. New species have evolved and others have become extinct in response to ever-changing environmental conditions. Today, millions of species thrive on Earth in every habitat imaginable—from the acid in our stomachs to the world's vast oceans. And what is perhaps most remarkable about this biodiversity is that every species—every lion, sunflower, shark, butterfly, fungus, and dog—has evolved from single-celled ancestors!

The process of evolution is very complex. Although this activity provides a fun introduction to some of the most important concepts, your students will need at least a basic understanding of what genes and genetic traits are and how they are passed on. (See "The Gene Scene" for more on this topic.)

In Part I of this activity, your students will take part in a simulation showing how birds in a population undergo natural selection. In Part II, they'll explore how, over millions of years, the same process of natural selection that led to evolution in the birds from Part I also led to the formation of eight fictional lizard species.

Gather plastic sandwich bags (one per student), masking tape, individually wrapped hard candies preferably all of the same size and shape (about four or five per student), and plastic spoons (two per student). Clear a space so the students can play the game. (The space should be large enough for your students to move around comfortably.) Mark the boundaries of the game area with masking tape. Do not use walls for boundaries.

The Spooners and Grabbers

1. Introduce the simulation.

Begin by writing the question, "How do new species develop?" on the board. Ask the students if they have any ideas about how to answer it. After getting some responses, tell the students that they are going to participate in a simulation called "The Spooners and Grabbers" that will help them understand how species evolve over time.

Explain that for this simulation everyone in the class belongs to a population of birds. These birds, called Spooners, obtain their food by spooning up prey in their beaks. Gather the students in a circle at the front of the class and give everyone two plastic spoons and a plastic bag. Tell the students that they're going to use their plastic spoons as beaks. They'll need to spoon up as many food pieces, or candies, as they can and place the food in their plastic bags, which represent the birds' stomachs. They can't use their hands except to hold the spoons and the bag. (They might want to place the bag on the ground and put the candies in it as they spoon them up.)

Tell the students that the birds usually feed independently and that they should avoid any body contact with other birds during feeding.

2. Run through the first round of the simulation.

When students are ready, scatter the candies on the floor. Then let the game begin. After 30 seconds, have the students count their candies. Ask a few students to share their results to see the number of candies they collected.

3. Select two mutants.

Next ask two students to volunteer to be mutants. Explain that these two birds have been born with a genetic mutation that has changed the shape of their beak. Stress to your students that a mutation can be any change in the genetic information of a cell and that the change isn't necessarily bad or good—that status is determined by whether the mutation helps the animal or plant survive. In this case, the mutation causes the birds to be born with a "grabber" beak instead of a spoon beak. Leaving the thumb free, tape up the other four fingers on one hand of each mutant.

roseate spoonbill

The Nuts and Bolts of Evolution

This activity introduces many terms and concepts related to evolution. Here is a brief overview to help you sort them out and think about how they all connect to biodiversity.

What is evolution?

To start thinking about evolution, consider a **population** of pack rats, living their lives in some sun-soaked desert locale. What makes these pack rats a population? They're all the same **species**. They live in the same geographic area. And they are most likely to mate with other pack rats in their population. From an evolutionary perspective, it's also important to keep in mind that, while they are genetically similar, none of the pack rats are genetically identical.

Because individual pack rats within the population have different genes, they'll have different chances for survival and reproduction under particular environmental conditions. Suppose that some of the pack rats have genes that make them better able to store fat than the other rats. On the one hand, if food supplies become seasonally low for many years, the fat-storing individuals will likely live longer and reproduce more successfully than other pack rats. As these fat-storers pass on their fat-storing genes to their offspring, the fat-storing genes will become more prevalent in the population. On the other hand, if food supplies are consistently plentiful and if fat-storing pack rats can't run as fast as their thin counterparts, the effects may be different. Suppose a population of fast-moving predators moves into the area. If the predators are more likely to catch fat-storing packrats, the thinner pack rats may be the ones living to pass on their genes.

Both pack rat scenarios described above provide an example of **evolution**. Evolution is, quite simply, the change in the genetic makeup of populations of organisms over time. If the relative amount of a certain gene changes in a population, evolution has taken place.

How does natural selection fit in?

More specifically, the pack rat scenarios exemplify **evolution by natural selection**. Natural selection occurs because there are differences in genetic **traits** among individuals in a population, and because some sort of **selective pressure** causes certain individuals to be more successful than others. In the case of the pack rats, fat storage is one of many genetic differences among individuals, and the difference becomes significant in the presence of the selective pressures of food shortage or fast-moving predators. Natural selection is one of the two driving forces behind evolution. The other is **mutation**.

pack rat

Mutation is an important concept to understand because it is a central part of evolution and natural selection. A mutation is an abrupt change in the genetic information of a cell. In most cases, an individual with a disadvantageous mutation dies before it can reproduce and pass the mutation on. But in other cases, a mutation can give an individual an advantage, and the "new" gene gets passed on as a result of selective pressures and natural selection. Over time such mutations can accumulate in a population.

The process of evolution gives species a way to adapt and survive in a constantly changing environment. Depending on the type of organism, evolution can happen at different speeds. Species that grow and reproduce quickly usually evolve faster than species that take a long time to mature and reproduce. For example, viruses and bacteria can evolve significantly in a matter of days, while some insects can evolve in a matter of months. A pack rat population evolves more slowly, but much faster than larger mammals, which have a longer time between generations.

(Continued on page 172)

Explain that the taped hand will function as the bird's beak, which will be able to grab an individual candy and drop it into the bag. (You may need to emphasize that these two mutant animals are the same species as the others but that they have a unique genetic structure not inherited from their parents.) Explain that you'll be doing a series of exercises to see how the introduction of these two mutants affects the population as a whole.

Make two columns on the board with the headings "Spooners" and "Grabbers." Record the number of Spooners and Grabbers under each heading. (After the first round, the number of Grabbers should be two.)

4. Repeat the simulation.

Gather everyone's candies and scatter them on the floor again. Have the students pick up the candies and count their take. Find out how many candies each of the two Grabbers collected. Did any Spooners get that many? Who had the fewest number of candies?

Explain to the students that each round of the simulation represents a new generation of birds, and each student represents the next generation of his or her family. The two animals with the greatest number of candies will produce two offspring in the next generation because they have been very successful. The two animals with the smallest numbers of candies will produce no offspring because they didn't get enough food to survive and reproduce. The rest of the group will produce one offspring.

Most of the students will play the next round as their own offspring. In other words, in the first round, the student plays a bird that produces one offspring and eventually dies. In the second round, the same student represents that offspring. The birds with the two lowest takes, however, do not survive and have no offspring in the next generation. These students come back as the offspring of the two birds with the highest takes. If both Grabbers had the highest takes, then the two Spooners with the lowest takes will return as Grabbers. Settle ties for the lowest or highest positions by randomly selecting two students from those tied.

Before you begin the next round, record the number of Spooners and Grabbers on the board. How many Spooners do you now have? How many Grabbers?

5. Carry out the simulation for four more generations.

Run through the simulation for four more rounds or until you have a clear majority of one kind of bird. Keep track of the number of each type of bird with each round. When you finish, you will likely have a majority of Grabbers, if not all Grabbers.

6. Graph the results.

Ask for a few volunteers to make a graph of the results of this game. On the board, have the students run the numbers from 0 to x (x = number of students in your class) along the y axis and the number of rounds you played along the x axis. They should make one line showing the number of Spooners over time and another line showing the number of Grabbers.

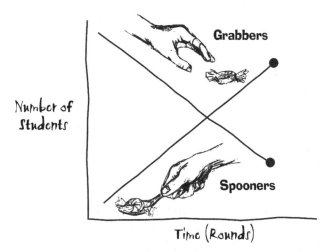

7. Review the simulation and its limitations.

Ask students to explain what they learned during the simulation. Why did the Grabbers, which numbered only two at the beginning of the game, end up dominating the population? *(It was easier—and more efficient—to pick up food with a taped hand than with the spoons; therefore, the Grabbers more consistently ended up with ample food supplies.)* Point out that when food supplies are plentiful, efficiency may not be as important as when food supplies are scarce. In times of limited food supply, the animals that can obtain food most efficiently

The Nuts and Bolts of Evolution (Cont'd.)

What does evolution have to do with biodiversity?

Suppose one pack rat population is separated into two distinct, isolated populations. This might happen after a major environmental event such as a flood or an earthquake. And suppose the population in one area experienced substantial seasonal food shortages over many years. The other population remained in an area with ample food supplies, but it was soon invaded by a population of fast-moving predators. Over time the first population would likely evolve to have more pack rats able to store fat in times of need. The second population would likely develop into a thinner and speedier bunch. Over subsequent centuries, mutation and natural selection would continue to lead the two populations down very different evolutionary paths. Through a process called **speciation**, *the populations could at some point become so different that they would actually be considered two different species.*

How do you know when speciation has occurred? That's something that scientists don't agree on. Some say speciation occurs when two populations can no longer reproduce even if they were put into contact with one another again. Others suggest that there are too many exceptions to this rule. In some cases, different species can reproduce to form hybrids. And although some other species may not seem very different genetically, they may not be able to reproduce. Regardless of the exact definition, scientists agree that evolution leads to speciation and is the force behind the remarkable diversity of organisms living on Earth today.

will likely survive and reproduce, while others simply will not get enough food to live. Changes will often happen fastest during these "crunch" periods when important resources are in short supply.

Explain to the group that this process by which organisms with particular genetic traits become more prevalent in the population is called *natural selection*. Natural selection leads to evolution—changes in the genetic makeup of populations. In the simulation, the Grabbers evolved to develop an ability to more easily pick up their food. Natural selection and evolution make it possible for species to adapt and survive in constantly changing environments.

Tell the students that natural selection can happen quickly or over long periods of time. Can they think of any reasons the time periods might vary? *(Species with long life cycles change more slowly than those with short life cycles, and major environmental change may more radically affect a species than less devastating trends. For example, a fatal virus might wipe out all animals except those with a gene making them immune, while a variable food supply might cause populations with particular traits to take turns dominating.)*

Ask the students if they think evolution in the simulation is occurring at a realistic speed. *(No, it's occurring too fast, although the tiniest organisms such as bacteria and viruses can evolve in a short period of time.)*

Ask the students to consider what might happen in subsequent rounds of the simulation. *(Certain traits would probably become more prevalent while others would become rare or non-existent.)* How could you change the simulation to make the outcome different?

Consider this: Elephants produce only 1 baby every 3 to 4 years, and their reproductive life usually is about 20 years. That means that the average female elephant has about 5 babies in her lifetime. Cockroaches, however, produce more than 1,000 offspring during their 4-year life span. Evolution and speciation occur much more quickly in animals that have a short time between generations and produce many offspring.

(*Expand the food supply to include more than one kind of food, or introduce a "disease" that affects only individuals with certain traits. In other words, change the selective pressure.*)

In the simulation, genetic variations were passed on to subsequent generations more quickly than they would have been in real life. Can anyone explain why? (*Grabbers and Spooners produced offspring with the exact trait they had—which means that Grabbers mated only with Grabbers, and Spooners only with other Spooners. In real life, however, Grabbers and Spooners probably would have reproduced with each other. The offspring could be either Grabbers, Spooners, or something in between, and it would take much longer for Grabbers to become the most common birds.*)

When thinking of the mutation that developed, was the grabber mutation a good one or a bad one? Can the students think of mutations that could have occurred that would have made the birds' food-gathering abilities worse instead of better? (*The grabber mutation was beneficial because it made picking up the pieces of food easier. If the mutation had caused the birds to have a skinnier and less effective beak than the original spoon beak, they might not have been able to pick up the hard candies at all.*)

Ask the students if they think that the Spooners and Grabbers had become two different species by the end of the simulation. Why or why not? (*No. The simulation demonstrated only a change in the relative amount of a certain gene. The number of birds with a spoon beak changed when compared with the number of birds with a grabber beak. Since the birds would still be able to interbreed, speciation didn't occur.*) What would have to happen for the two groups to become distinct species? (*Isolation is a key factor in speciation. It is important for two groups, or populations, to be separated, because the separation exposes them to different selective pressures caused by different environmental conditions. Isolation keeps the genes of each population from mixing, so the genetic changes can accumulate. If enough genetic changes accumulate to make the two populations unable to interbreed, then they have become two different species.*)

8. Review the concepts learned in the simulation.

Now ask the students to think over what they have learned in the simulation. Can they define evolution and natural selection and list the "ingredients" of this process?

Evolution is the change in the relative amount of a certain gene in a population overtime. Evolution can, over time, lead to the formation of new species. *Natural selection* is the process by which genetic traits that allow individuals to survive and reproduce are passed on to each successive generation, allowing species to become better adapted to their environment. Also known as "survival of the fittest," this is the driving force behind the process of evolution.

The following factors are necessary for natural selection and evolution to take place:

- *Selective pressures* influence whether an individual survives and reproduces successfully.

- *Genetic diversity* provides a variety of different genes in the population.

- *Time* allows genetic changes to be passed from one generation to the next.

Ask your students to define speciation and list the factors necessary for it.

Speciation is the process by which one or more populations of a species accumulate enough genetic differences to become a new species. This process often requires a long period of evolution to occur after populations have been isolated.

The following factors are necessary for speciation to occur:

- *Isolation* exposes the populations to different selective pressures caused by different environmental conditions. Isolation also keeps the genes of each population from mixing, so the genetic changes can accumulate.

- A *longer time* is required for speciation than for evolution through natural selection.

━━━━━━━━━━━━━ **Before You Begin • Part II** ━━━━━━━━━━━━━

For each student, make one copy of "The Journals of Charlotte Durand, Explorer" (pages 53–56 in the Student Book), the "Twitomite Table" (page 57 in the Student Book), the "Twitomites: What Do You Think?" questions (page 58 in the Student Book), and the optional "Twitomite Time Line" (page 59 in the Student Book).

━━━━━━━━━━━━━ **What to Do • Part II** ━━━━━━━━━━━━━

Twitomite Evolution

1. Discuss Charles Darwin and his research.

Begin by asking students what they know about Charles Darwin and his research. When students have shared as much as they know, explain that Charles Darwin, an English naturalist, sailed on a voyage of the HMS *Beagle* to survey the natural history of the west coast of South America and some Pacific Islands. In 1835 he surveyed the Galápagos Islands where, among other things, he observed and collected nine different types of finches, small birds found in many parts of the world. Darwin wondered why there were so many different kinds of finches on the islands. And in his attempt to explain why, he came up with one of the first theories of evolution by natural selection. As we now know, there are 13 different species of finches in the Galápagos, and all of them are believed to have descended from a common ancestral species that probably arrived in the Galápagos from South America some 100,000 years ago.

Charles Darwin

Explain that in this activity, the students are going to do some of the same kind of detective work that biologists such as Darwin have done to understand the evolutionary history of organisms. In other words, they'll look at biological and geological clues to see if they can figure out why and how an ancestral creature evolved into a number of distinct species.

2. Hand out "The Journals of Charlotte Durand, Explorer" and the "Twitomite Table."

Tell the students that you've given them the journals of a fictional 19th century explorer who describes a group of lizards she finds on an imaginary island chain off the west coast of Mexico. For homework, ask the students to read the journals, look at the accompanying sketches, and fill out the "Twitomite Table" (page 57 in the Student Book).

3. Discuss homework assignment.

In your next session, discuss any questions the students have about their homework assignment. Then divide them into pairs and ask them to compare their answers on the "Twitomite Table." After the students have come to a consensus on their answers, have them edit any parts of the table that they think they filled out incorrectly the first time.

4. Hand out "Twitomites: What Do You Think?" and the "Twitomite Time Line."

Working in their pairs, have the students use the journal, table, and time line to answer the "Twitomites: What Do You Think?" questions (page 58 in the Student Book).

5. Share results.

Discuss the answers to these questions as a group. The questions will encourage the students to think about and compare how the different twitomites might have evolved.

1. How many species of twitomites were there 20,000 years ago? Were the individual twitomites genetically identical? Explain. *(Twenty thousand years ago, there was just one species of twitomite. Individuals were not genetically identical because, even within a population, individuals have a lot of genetic variation.)*

2. When did the twitomites first become separated into three distinct populations? Do you think the twitomite populations on the three islands were fairly similar or radically different at first? *(The three populations would have formed 10,000 years ago when the peninsula became separated into three islands. At first the populations were probably quite similar because they'd so recently been part of the same large population.)*

3. Over time, what could have caused different species of twitomites to arise on each of the three islands? Be specific. *(Three distinct species could have arisen because the populations were isolated and because their selective pressures and habitats—shrubby desert, sandy beach, and forest—were different, making different traits more advantageous for individuals on each of the three islands.)*

4. What had happened to the islands by the time the explorers arrived? How did this affect the twitomites? *(The three islands separated into seven distinct islands. Once again the twitomites were divided into even more populations, and each population inhabited a different kind of habitat.)*

5. Explain how each of the eight twitomites is particularly well adapted to its habitat. *(Many of the twitomites' adaptations helped them find food. The different snouts gave certain species the ability to feed on the most abundant type of food in their habitat. The green twitomite was specially adapted to hide from predators through camouflage. The marine twitomite had webbed feet that helped it feed on marine creatures and get around more easily in its beach and marine habitat.)*

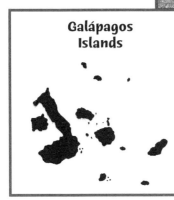

one species of
Galápagos finch

Galápagos
Islands

6. Some scientists think about the process of speciation in terms of this formula:

isolation + time + evolution = speciation.

Apply this formula to the twitomites to explain how their single ancestor evolved into eight different species. *(First, there was one common twitomite ancestral species 100,000 years ago, but then the sea level rose to divide the peninsula where the twitomites lived into smaller and smaller pieces of land. Because none of the twitomites was equipped to swim long distances, the separation of the islands by water prevented the populations on the different islands from intermixing. This element provided the "isolation" part of the formula. Also, because the islands were separated for 10,000 years, the twitomites were isolated for a long period of time. This element accounts for the "time" part of the formula. Finally, the twitomites in each habitat adapted to new conditions. Like the birds in Part I of the activity, the twitomites in different areas developed certain traits that helped them thrive in their habitat—beach, desert, or forest. Natural selection occurring over a long period of time on each of the islands led to evolution and speciation of each of the different groups of twitomites. Each of the species is adapted to survive in its habitat.)*

6. Review major concepts.

Once students have had a chance to discuss their answers, review the major concepts of the activity. Can anyone clearly answer the question posed at the beginning of Part I: How do new species develop? Ask the students to list again some of the key factors necessary for evolution (genetic variation among individuals in a population, selective pressures, time, and so on). Then see if they can apply their understanding to the more advanced questions posed below.

How does evolution affect biodiversity? *(Through the process of natural selection, an incredible diversity of species has evolved on Earth.)*

What effect will declining biodiversity have on the process of natural selection in the future? *(With many species' populations declining, there is much less genetic variety for natural selection to "work with." Fewer individuals means less genetic diversity in the population. Without a wide range of genetic traits in a population, there is less chance that enough individuals will have the traits necessary for survival. Rather than adapting to environmental changes, small populations of animals could die out, accelerating the loss of biodiversity.)*

What can we do to help protect and conserve biodiversity, and the processes of natural selection and evolution? *(We can protect biodiversity by maintaining large, diverse populations of species throughout the world by protecting species' habitats, keeping introduced species from out-competing native species, preventing pollution from harming living things, and responsibly managing wild populations of plants and animals we use for food, clothing, medicines, and other products.)*

TWITOMITE TABLE–ANSWERS

	Snout	Island/Habitat	Food Source	Outstanding Trait	How does this trait help it survive?
Shovel-Nosed Twitomite	shovel-shaped	Sandy Island / beach	insects and sea creatures	shovel-shaped snout	helps to easily dig through the sand and eat insects and sea creatures
Saw-Toothed Twitomite	long, with saw-shaped teeth	Cactus Island / desert	cactus fruit	saw-shaped teeth	helps to cut into the insides of cactus fruit
Long-Tongued Twitomite	long, with long, curled tongue	Flora Island / desert	nectar and insects from desert plants	long, curled tongue	helps to reach the nectar inside the long, tubular desert flowers
Vampire Twitomite	long and pointed with piercing tip	Bloodlost Island / desert forest	blood from forest animals	pointed, piercing snout	helps to feed on blood from mammals
Big-Jawed Twitomite	big jaw	James Island / moist forest near water	fruit	big jaw	helps to feed on the large fruits from the forest
Long-Toothed Twitomite	big jaw with long teeth	James Island / pine trees	insects	long teeth	helps to gnaw at the outside of trees and eat insects living underneath the bark
Green Twitomite	big jaw	Bird Island / moist forest near water	fruit	green coloration	helps to blend in with the forest canopy habitat, thereby avoiding predation
Marine Twitomite	small, less rounded than shovel-nosed	Rock Island / rocky beach and tidal pools	sea creatures	webbed feet	helps to swim and dive, allowing it to catch small sea creatures

WRAPPING IT UP

Assessment

Have each student "create" an insect or other creature that would live in his or her bedroom. Assume it is a species that rapidly adjusts to change through evolution. The students should describe how their rooms may change over time and how the insects or creatures would evolve to survive the changes. This assignment can be done as either a series of pictures (cartoon or storyboard) or a written story.

Unsatisfactory—The student does not demonstrate an understanding of evolutionary change.

Satisfactory—The student provides one example of how changing selective pressures affect evolution.

Excellent—The student uses creativity and logic in determining multiple evolutionary outcomes as the room changes.

Portfolio

The students can write comments on their reactions to the evolution simulation game (Part I). Collect the comments for portfolios. The "Twitomite Table" and answers to "Twitomites: What Do You Think?" can also be part of the portfolio.

Writing Idea

Have the students keep a journal describing the unique features of some of the plants, animals, or other living things in their own backyard or neighborhood. Tell the students to spend a few days looking closely at several species from one major group (mammals, insects, flowering plants, and so on) as they encounter the organisms around the yard or neighborhood. Ask the students to write about how they imagine the distinct species within these groups could have evolved from similar ancestors. They should think about what kinds of selective pressures influenced these species and look for adaptations these animals or plants have developed over time.

Extensions

- For a real-life example of how evolution and speciation occur, hand out copies of "The Case of the Peppered Moth " (page 60 in the Student Book). Tell the students to read the paragraph to themselves and answer the "Think It Over" questions. Then review their answers as a group. Be sure the students understand that the peppered moths always had a range of coloration (much as humans can be fair skinned or dark skinned or something in between) but that changing environmental conditions gave an advantage to darker-colored moths for many years.

  Emphasize that natural selection can operate on any genetic trait—including color, size, body shape, and behavior (such as ways of obtaining food)—that influences the survival of a species.

- Have your students research Charles Darwin and Alfred Wallace. How did they "discover" evolution? Which one is usually credited with the idea and why? How similar or different were their theories?

Resources

The Beak of the Finch: A Story of Evolution in Our Time by Jonathan Weiner (Vintage Books, 1995).

Life: Origins and Evolution by Alessandro Garassino and translated by Rocco Serini (Raintree, 1995).

The Origin of Species by Charles Darwin (Available from various publishers including Viking [1982], Mass Market Paperback [1991], Grammercy [1995], W. W. Norton and Co. [1975], and Random House [1993]; originally published in 1859).

The Song of the Dodo: Island Biogeography in an Age of Extinctions by David Quammen (Simon & Schuster, 1996).

Teaching About Evolution and the Nature of Science
by the Working Group on Teaching Evolution (National Academy of Sciences, 1998).

Where Worlds Collide: The Wallace Line by Penny Van Oosterzee (Reed, 1997).

*"Humanity co-evolved with the rest of life on this
particular planet; other worlds are not in our genes.
Because scientists have yet to put names on most kinds of
organisms, and because they entertain only a vague idea
of how ecosystems work, it is reckless to suppose that
biodiversity can be diminished indefinitely without
threatening humanity itself."*

–Edward O. Wilson, biologist

Why Is Biodivers

The activities in this section illustrate the complex and amazing ways biodiversity supports ecosystems, affects human health and societies, and helps sustain life on Earth. For background information, see pages 30-37.

World Wildlife Fund

ty Important?

"Biodiversity contains
the accumulated
wisdom of nature and the
key to its future."

–Donella H. Meadows,
environmental writer

13 | Super Sleuths

SUBJECTS

science, social studies, language arts

SKILLS

gathering (reading comprehension), interpreting (drawing conclusions, reasoning), presenting (designing)

FRAMEWORK LINKS

1, 2, 3, 4, 12

VOCABULARY

disperse, diurnal, **ecosystem**, **food chain**, food web, **genes**, **habitat**, nocturnal, **pollination**, seed dispersal, **species**

TIME

one session, plus overnight homework assignment

MATERIALS

copies of "Super Sleuths" handouts (pages 63–69 in the Student Book), and extra paper

CONNECTIONS

"Something for Everyone" (pages 90–93), "All the World's a Web" (pages 74–79), or "'Connect the Creatures' Scavenger Hunt" (pages 102–107) can provide good introductions to this activity. For more on the relationship between species and their habitats, try "Space for Species" (pages 286–301), "Mapping Biodiversity" (pages 252–265), or "Backyard BioBlitz" (pages 134–143).

 **AT A GLANCE**

Solve a series of logic problems, word games, and other brain teasers to learn about some of the fascinating relationships that exist among living things.

 OBJECTIVES

Give several examples of how species are connected to one another and their habitats. Give an example of a plant that has evolved in a way that ensures pollination and seed dispersal. Draw a food web. Describe the three levels of biodiversity and several adaptations of desert plants and animals.

I t can be easy to make the mistake of looking at living things in isolation from their habitat. For example, some conservation programs designed to save a particular wildlife species have not always been mindful of saving the habitat that the animal depends on. Or people have introduced exotic species to control a pest without giving much thought to how the new species will change local food webs. And some conservationists have mistakenly tried to protect ecosystems without addressing the needs of the people who live there. The more we learn about the connections among species, ecosystems, and people, the better we can devise conservation plans that protect species and ecosystems and meet human needs.

This activity is designed to help students explore some ecological connections through a series of fun but challenging brain teasers. The teasers introduce students to some of the fascinating relationships between plants and their pollinators, seed design and dispersal mechanisms, the complex relationships of a prairie food web, the levels of biodiversity, and the adaptations of some desert plants and animals.

After finishing the puzzles, the students can read about some remarkable examples of ecological relationships on the "Believe It or Not!" handout (page 69 in the Student Book), and they can come up with their own amazing connections.

Before You Begin

Make copies of the Super Sleuths handouts: "Pollination Puzzle," "Seeds on The Move" (Part I and Part II), "Draw a Prairie Food Web," "Levels of Diversity," "Life in the Desert," and "Believe It or Not!" (pages 63–69 in the Student Book) for each student.

What to Do

1. Introduce connections in the natural world.

Ask your students what ideas come to mind when they think about connections in the natural world. (You might suggest they think specifically of some of the ways species depend on one another, for example, or ask the class how species depend on their habitats for food, shelter, water, and living space.) When the students run out of ideas, tell them they're going to solve a series of puzzles that explore some of these fascinating relationships.

2. Pass out the "Super Sleuths" pages.

Tell the students to work individually to solve the "Super Sleuths" puzzles. If class time is limited, you may want to assign them as homework. Hand out extra paper for students to draw their prairie food webs.

3. Discuss answers.

Review the puzzles one by one. The following list provides solutions and some discussion questions to stimulate additional thinking on each of the topics.

Answers:

"Pollination Puzzle"

Ask for volunteers to share their answers and explain the process they used to find those answers. Ask if anyone used a different process for his or her answers. The correct combinations are Strelitzia (also known as bird-of-paradise) and ruffed lemur, trumpet creeper and ruby-throated hummingbird, saguaro cactus and Mexican long-nosed bat, and Angraecum orchid and hawk moth.*

Question: Do the students think there could be many other pollinators equipped to pollinate the orchid? *(The orchid and hawk moth evolved together over a long period of time, and now the hawk moth is the only pollinator that we know of that can reach the nectar inside that flower. There are many cases like this in the natural world, and, therefore, when the pollinator experiences a decline in numbers the plant usually will too, and vice versa.)*

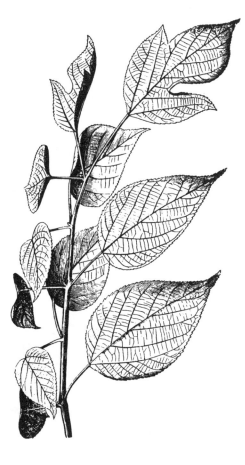

*The concept for the "Pollination Puzzle" was adapted with the permission of the National Wildlife Federation (NWF) from the "Rain Forests: Tropical Treasures" issue of *NatureScope*. For more information about NWF and its education programs, please call (800) 822-9919.

"Seeds on the Move"

If you can, find examples of the fruits mentioned in Part II and bring them into the classroom for students to examine. If you can't find the specific examples, try to bring in at least one example of each dispersal mechanism.

Part I

Answers may vary, but students might mention the following:

Wind: Seeds that are carried by wind are light and have "wings" or other ways of staying aloft.

Water: Seeds that float are durable, light, and often have "pockets" that trap air.

Animals: Seeds that are tasty are durable enough to pass through an animal's gut, and sticky seeds cling to animal fur.

Part II

red maple (wind), coconut palm (water), pin oak tree (animal), red mangrove (water), wild cherry (animal), dandelion (wind)

Questions: What might happen to coconut trees if people started collecting all the coconuts for themselves? *(There may be fewer adult trees in later years.)* How is it possible that seeds that are eaten by animals are able to germinate? *(Most animal-dispersed seeds are very durable and pass through the digestive tract intact after being eaten by the animal. And some seeds may even need to pass through an animal's gut to germinate.)* Squirrels like to eat acorns. Can you guess how they end up dispersing acorns? *(Squirrels bury acorns for winter food. Some of the acorns that they forget to dig up germinate and sprout into young oak trees.)*

badger

"Draw a Prairie Food Web"

There are two different food web examples on the worksheet. You should introduce your students to the concept of a food web before having them try this part of the activity. Be sure your students understand that arrows in a food web point in the direction energy is going. In other words, the arrows point *to* the thing that's doing the eating. After students have completed their webs, ask volunteers to draw the webs on the chalkboard. Although the students' drawings may differ, the arrows should connect in a way similar to the arrows in the example below.

Questions: What might happen if people killed a lot of predators (hawks, snakes, and so on) that eat mice and other rodents? *(In the case of this web, there might be more prairie dogs, mice, and pocket gophers.)* What might happen if a disease attacked the grasses and they did not produce many seeds? *(The animals that eat the seeds might die, decrease in numbers, or select an alternative food—all of which would have a ripple effect through the food web.)*

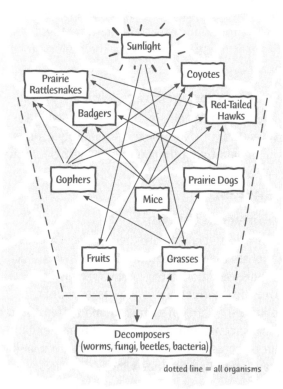

"Levels of Diversity"

Ask students to show their three boxes for the different levels of biodiversity in the Congo: ecosystems (mangrove swamps, tropical dry forests, grasslands, rain forests), species (African elephants, duikers, driver ants, termites), and genes (height, tail length, trunk size).*

Review several examples presented by students of genetic, species, and ecosystem biodiversity in your state.

"Life in the Desert"

You might want to start by describing how paragraphs are typically organized to provide information. Point out that they usually start out with a generalization, then give details, and finish with a conclusion. These paragraphs all follow that format. Following are the correct orders for each of the paragraphs listed: A. 4-2-1-3; B. 3-1-2-4; C. 2-4-3-1.

Questions: Ask the students if they can think of other desert plants and animals that have special adaptations. What might happen to the ocellated sand lizards if much of their habitat were paved over? *(They might not be able to survive.)* If these species have adapted so well to the desert, why might they not be able to adapt to human-caused changes? *(The changes that humans make usually take place over a much shorter period of time than other environmental changes, so many species don't have time to adapt.)*

4. Discuss the connections the students uncovered.

Ask the students to think about the connections they discovered in the different brain teasers. Were they surprised by any of the relationships? Were they aware of all the special adaptations of animals and plants living in harsh desert habitats? Did they know about the ways plants have adapted to disperse their seeds?

Have your students consider the complexity of all the levels of biodiversity and the ways that the levels are connected. Then ask the following question: "Given the complexity of interactions in the natural world, do you believe people should be concerned about the loss of biodiversity? Why or why not?" (See Chapters One and Two of the background information on pages 22–37 for points you can use as students discuss their ideas. In particular, see the sections entitled "Countless Connections" [page 28] and "Life-Support Systems" [page 32.])

5. Hand out "Believe It or Not!"

Read over the page with the students and discuss some of the amazing relationships described on it. You could also have the students create a "Believe It or Not!" bulletin board by illustrating examples of amazing connections in the natural world.

---

*The concept for the "Levels of Diversity" activity was adapted with permission from *Conservation Biology* by Robert B. Blair and Heidi L. Ballard (Kendall/Hunt Publishing Company, 1996).

WRAPPING IT UP

Assessment

Have each student choose one local animal and plant. Provide each student with a sheet of paper. Have the students diagram as many ways as they can think of in which their selected plant and animal are related to each other, and label each of the connections. Have students consider food and reproduction relationships. Note that they may have to go through several steps to find relationships between some plants and animals.

Unsatisfactory—The student cannot identify at least one relationship for food and one relationship for reproduction between the plant and the animal selected.

Satisfactory—The student identifies at least one relationship between the plant and the animal. The relationship is logical and includes several indirect relationships, and the direct and indirect relationships are clearly displayed.

Excellent—The student identifies at least four relationships between the plant and the animal. Several of the relationships are indirect and involve extended connections. Relationships are clearly explained.

Portfolio

The "Super Sleuths" puzzles can be maintained in the students' portfolios, along with their individual corrections.

Writing Idea

Have students write a dialogue between two species that illustrates the relationship between the species and how their lives are interconnected.

Extensions

- Have students work in teams to figure out some of the brain teasers. For example, by giving clues to each student in a team, the pollination puzzle can serve as a great cooperative learning activity. (Each student would get two to three clues.) Explain that the students should not show their clues to each other but that they can ask questions and read their clues aloud. Then, as a group, they can try to figure out the pollination puzzle.

- Have students research and develop their own biodiversity brain teasers to use with other students.

- Have students find seeds in their own neighborhood and try to guess how the seeds are dispersed. Or have them find flowers and try to guess what kind of animal pollinates them.

Resources

A Child's Place in the Environment series by Olga Clymire (California Department of Education, 1994).

Conservation Biology by Robert B. Blair and Heidi L. Ballard (Kendall/Hunt Publishing Company, 1996).

The Private Life of Plants by David Attenborough (Princeton University Press, 1995).

Wildlife Communities from the Tundra to the Tropics in North America by Clarence Hylander (Houghton Mifflin Company, 1966).

"The unsolved mysteries of the rain forest are formless and seductive.
They are like unnamed islands hidden in the blank spaces of old maps, like dark
shapes glimpsed descending the far wall of a reef into the abyss. They draw us forward
and stir strange apprehensions. The unknown and prodigious are drugs to the
scientific imagination, stirring insatiable hunger with a single taste. In our hearts,
we hope we will never discover everything. We pray there will always be a world like
this one at whose edge I sat in the darkness. The rain forest in its richness is
one of the last repositories on Earth of that timeless dream."

–Edward O. Wilson, biologist

SUBJECTS

science, social studies

SKILLS

gathering (reading comprehension, researching), analyzing (questioning), interpreting (identifying cause and effect, relating), applying (problem solving)

FRAMEWORK LINKS

12, 35, 52, 55

VOCABULARY

environmental health, fertility, **global warming***, hormones, indicator species, infertility*

TIME

two sessions, plus research time

MATERIALS

copies of "The F-Files" comic strip (pages 70–71 in the Student Book) and "Get on Your Case!" (pages 72–74 in the Student Book)

CONNECTIONS

The video "Biodiversity! Exploring the Web of Life" introduces the topic of environmental health. Try using "Easter's End" (pages 330–333) and "The Many Sides of Cotton" (pages 278–285) to further illustrate connections between environmental health and human health.

AT A GLANCE

Explore the connection between toxic chemicals and living things as you read a comic strip about an investigation of frogs in Minnesota.

OBJECTIVES

Explain some of the ways biodiversity can be affected by toxic chemicals and other environmental threats. Research environmental health topics using books, journals, interviews, and the Internet. Produce a summary and position statement on an environmental health-related research topic.

In August 1995, a Minnesota middle school class took a field trip to a farm in the south-central part of the state. While hiking, the students tried to catch the frogs that they saw on the trail. One of the students caught a frog with only one leg. Other students caught frogs with deformed legs, eyes, and other parts. In fact, half the frogs they caught had some kind of deformity.

Because of what the students found that day, people began investigating the frogs in that area. Since 1995 there have been high rates of physical deformity in frogs found not only in Minnesota but in many other states as well. In fact, there have been deformed amphibian findings in many parts of the world. No one knows exactly what's causing the deformities. But many scientists believe the frogs are being affected by contamination of the air, water, and/or soil. These researchers wonder: If frogs are being affected now, could humans be next?

The connection between environmental health and human health is a new, complex area of study called **ecotoxicology**. Your students may be familiar with some environmental health issues that have received a lot of media attention—for example, the link between air pollution and respiratory problems, or the links between ozone depletion, increased UV radiation, and skin cancer. But your students might not know about the wide variety of environmental concerns—

everything from global warming to pesticide use to household chemicals—that scientists are researching today to find out about their effects on human health. The frogs in Minnesota show that humans aren't the only ones whose health may be affected by changes in the environment. Some scientists think other species of animals and plants are even more vulnerable than humans. But what creates health problems for wildlife could very well create health problems for humans, and vice versa. In addition, since living things are connected in the web of life, if one species declines because of environmental contamination, the effects are likely to be felt by many other species in the ecosystem. The field of ecotoxicology presents more questions than answers. This activity is designed to guide your students on an open-ended investigation of environmental health topics. Introduce them to the "F-Files," then set them off to find out what's happening with the health of frogs, people, and lots of other creatures across the planet.

Before You Begin

Make copies of "The F-Files" comic strip (pages 70–71 in the Student Book), keeping the two parts on separate sheets of paper. Hand out only the first page at the beginning of the activity. Then hand out copies of the second page halfway through the activity. Make copies of "Get on Your Case" (pages 72–74 in the Student Book) for each team.

What to Do

1. Pass out the first page of "The F-Files" comic strip.

Tell the students that the comic strip you're passing out is based on a real unsolved mystery. Have them read it to themselves. Then ask one or two volunteers to summarize what Agents Croaky and Hopper are investigating.

2. Ask students the "Questions for the Class."

Be sure to let the students know that there is no single correct answer to any of the questions. Record their suggested answers on the board.

Questions for the Class

1. If you were Croaky and Hopper, what sorts of questions would you ask in order to begin your investigation? *(Answers will vary. They may include the following:*

When did the changes begin? What other changes related to the environment or human activity occurred at that time? Have changes occurred in other types of animals?)

2. If you were Croaky and Hopper, where would you begin looking for clues to solve this mystery? What would you ask a laboratory team at headquarters to analyze? *(Answers will vary. Might check contaminant levels in the air, water, and soil [samples checked against records of historical data may show changes in air and water quality and contamination in the soil]; might check for contaminant levels in the frogs themselves; might look for high densities of parasites in the water; and so on.)*

3. What do you think might have caused the problem with the frogs? *(Answers will vary.)*

4. How might the problems with the frogs affect other plants and animals? *(The problems affecting the frogs could be affecting both the animals that eat the frogs and whatever the frogs eat; the problems affecting the frogs could be affecting other species, too; and so on.)*

3. Pass out the second page of "The F-Files" comic strip.

Pass out the second page of the comic strip and have the students read it. After the students have finished, discuss the comic strip. Ask the students if the scientists at the Bureau of Scientific Investigation agreed on the cause of the frog deformities. What were some of the different explanations the scientists provided? Did all of the theories sound equally credible? Which explanations sounded most likely? Why?

Encourage the students to follow their hunches when choosing a theory that sounds the most plausible. Also point out that it's important to evaluate the sources of the scientists' information. *International Tabloid*, for example, doesn't sound like a credible journal. Explain to the students that whenever they hear conflicting information about a topic, they should find out the sources of the information and assess how well respected those sources are.

Minnesota Deformed Frog Discovery Team

New Country School, Henderson, MN 1995

4. Introduce the topic of environmental health.

Explain to the students that the investigation of frog deformities is one of many areas of research in which scientists are looking at possible connections between environmental changes and the health of wildlife, plants, and humans. Ask the students if any of them can name illnesses that are thought to be tied to environmental problems. If you need to jump-start students' thinking, suggest that they consider such conditions as asthma, cancer, and birth defects. Then ask students to think about how these illnesses may be connected to environmental quality. As students name connections, ask them if they know whether the links are proven or suspected. Where did they get their information?

5. Pass out copies of "Get on Your Case!"

Tell the students that there are many things scientists still don't know about the links between the state of the environment and the health of living things. But there are a lot of ideas out there—some more fully researched than others. Explain that you've passed out a list of topics that are currently being investigated by people all over the world. Using magazines, newspapers, books, and the Internet, the students' job is to try to collect as much information as they can on one of the topics. Remind them to look hard at the credibility of their sources. They'll need to be especially careful to scrutinize Web sites—even a site with an official-sounding name might be nothing more than one amateur's unsubstantiated ideas.

Organize the class into teams of two or three. Then allow each team to pick one of the topics. More than one team can pursue the same topic, but encourage the students to investigate as many of the alternatives as possible. Explain that they should do their best to answer the questions listed under their case descriptions.

Give the students several days in class or after school to conduct their research. They should use the school's library and computer lab (for Internet access) for information.

6. Have a group reporting session.

When the students have finished their investigations, have them share their results with the class. Each team should describe the topic it investigated, explain the research methods it used, and describe the results of its search.

7. Discuss connections.

Afterward see if any teams can name connections between their area of study and that of one or more of the other teams. As they cite connections, have the students keep a list of the relationships, or, if possible, record a complete web of interrelationships on the board. Finish the discussion by asking the students if they have any final reflections on their investigations. You may want to re-emphasize that there are many complex interrelationships among living things as well as between living things and the physical environment.

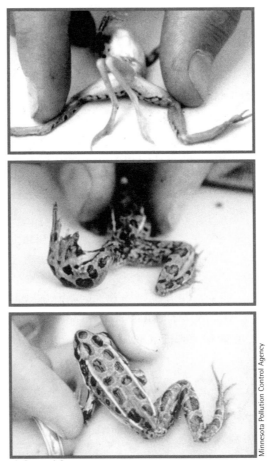

deformed frogs

Minnesota Pollution Control Agency

WRAPPING IT UP

Assessment

The students' reports should serve as one means of assessment. In addition interview the students (individually or in groups), asking the following questions about their reports. (You can also have them write answers to the questions.)

- How did you conduct your search for information?
- Which sources were the most valuable and why?
- What were indicators of bias in the materials you examined?

Unsatisfactory—The student cannot adequately answer the questions.

Satisfactory—The student is able to address each question on a satisfactory level.

Excellent—The student is able to fully address the questions and provide in-depth answers.

Portfolio

Use a copy of the team report in the portfolio.

Writing Idea

Have students write a report on their findings, including the following sections: (1) background information on the issue, (2) results of their research, (3) a discussion of the opposing views, and (4) a conclusion summarizing their position on the issue.

Extensions

- Have the students turn their own investigation into a comic strip that is in "The F-Files" style. They should feel free to turn themselves into fictional investigators and play up the dramatic elements of their research.

- Have your students conduct their own amphibian survey to monitor amphibian deformities in your area. They can visit the North American Amphibian Monitoring Program Web site at <www.im.nbs.gov/amphibs.html> or the North American Reporting Center for Amphibian Malformations at <www/npwrc.org/narcam> (or call 800-238-9801) for more about conducting and reporting the results of their survey.

- In many areas, poor communities are sometimes exposed to more environmental health risks than their more affluent neighbors. Have students research environmental justice issues and write a short case study that highlights one example of the link between poverty and environmental health factors.

Web Sites

Visit the Center for Global Environmental Education Web site, "A Thousand Friends of Frogs," at <www.cgee.hamline.edu/FROGS/index.htm>.

Visit the FrogWeb site at <www.frogweb.gov>.

Visit the Minnesota New Country School Frog Project Web site at <mncs.k12.mn.us/frog/>.

Visit the Minnesota Pollution Control Agency Web site at <www.pca.state.mn.us/hot/frogs.html>.

Visit the North American Reporting Center for Amphibian Malformations Web site at <www.npwrc.org/narcam>.

Books, Articles, and Videos

Biodiversity! Exploring the Web of Life Education Kit
by World Wildlife Fund, Earth Force, and WQED/Pittsburgh (1997).

"Conference Provides Update on Deformed Frog Research"
by Tom Meersman. *Star Tribune*, 5 December 1997.

"The Observatory: The Appearance of Deformed Frogs" by Tom Meersman. *Star Tribune*, 25 March 1998.

Tracking the Vanishing Frogs: An Ecological Mystery by Kathryn Phillips (St. Martin's Press, 1994).

"*We cannot solve the problems
that we have created with the
same thinking that created them.*"

–Albert Einstein

15 Biodiversity Performs!

SUBJECTS

science, art (drama), language arts

SKILLS

gathering (reading comprehension),
analyzing (identifying patterns),
applying (creating), presenting (acting)

FRAMEWORK LINKS

2, 8, 10, 11, 12, 25, 28, 33

VOCABULARY

atmosphere, **decomposers**, **ecosystem**,
ecosystem services, erosion, **photosynthesis**, **pollination**, population,
respiration, sediments, **wetland**

TIME

one session

MATERIALS

one set of "Biodiversity Performs!"
cards (pages 75–78 in the Student
Book)

CONNECTIONS

After "acting out" ecosystem services
in this activity, let your students
perform a series of simulations to find
out more in "Secret Services" (pages
226–229). To learn about other ways
that humans benefit from biodiversity,
try "Something for Everyone" (pages
90–93), "Ten-Minute Mysteries"
(pages 98–101), "The Culture/Nature
Connection" (pages 214–225), or
"Diversity on Your Table" (pages
230–237).

 **AT A GLANCE**

Play a charades-like game to learn about some of the
"free" services that ecosytems provide.

 OBJECTIVES

Work in teams to act out different ecosystem services.
Describe several "free" services that biodiversity
provides to humans and explain how these services
make life on Earth possible.

Whether you realize it or not, ecosystems and the species
within them are working around the clock to perform
many jobs that help make human life possible and
more livable. Much of the work ecosystems do is
difficult to watch or see, so it's easy to take these "secret services"
for granted. Just remember, while you're eating, sleeping, working,
driving, or watching TV, forests are helping to clean the air, oceans
are regulating the climate, and wetlands are helping to purify water,
minimize storm damage, and perform flood control. In this activity,
your students can learn about some of the secret services that
ecosystems and the species within them provide by creating
charade-like performances for their peers.

Before You Begin

Make a copy of the "Biodiversity Performs!" cards (pages 75–78 in the Student Book) and cut the cards apart.

What to Do

1. Introduce ecosystem services.

Begin the activity by asking students to give examples of ecosystems (such as deserts, temperate forests, wetlands, coral reefs, and so on). If your students aren't familiar with the word "ecosystem," you might want to do an introductory activity that focuses on what an ecosystem is (see Resources on page 196). You can also lead a discussion that helps them understand that an ecosystem is a community of plants, animals, and other organisms that interact with each other and the physical environment and that are linked by energy and nutrient flows. The soil beneath a fallen log, a salt marsh in an estuary, and the Brazilian rain forest are all examples of different-sized ecosystems.

Explain that ecosystems as a whole and the individual species that live within them perform many functions. Through their day-to-day activities aimed at their own survival, individual organisms end up performing jobs that help make life more livable for us. And the combined efforts of many species in an ecosystem can have effects that also help make our lives more livable. To help illustrate this idea, ask your students to explain how plants "help" the environment (produce oxygen, help absorb carbon dioxide, help reduce erosion, and so on). The students' answers will begin to underscore the important services that ecosystems and the species within them provide. Be sure to emphasize that species don't perform certain activities to make the planet a better place to live. Rather, the actions that organisms take in order to survive can have beneficial side effects for other living things.

2. Set the stage.

Divide the class into seven groups. Explain that each group will be given a card that describes one of the free services that ecosystems and species provide. They will have time to discuss and practice a performance that gets across the idea described on the card. The audience (members of the other teams) will need to try and figure out what the service is and how it works. Emphasize that the performers should act out the secret service rather than the words on the card. Some groups may need to divide their performance into several segments in order to get the full message across. Students may not use words during their performance, but if you would like to make the game less challenging, they can use sound effects and props. Group members can individually imitate an animal, plant, or other object, or the entire group can form an organism or object.

3. Set the students loose.

Give each group a "Biodiversity Performs!" card and allow the groups plenty of time to develop and practice their performances. When the students are ready, call one group at a time to perform. Make sure you know which card each group has. You will judge whether a student's guess is a correct answer. You may want to make a list of possible answers on the board. Let the performers finish their skit before the audience guesses. If the students can't guess correctly, you may want to provide additional clues as the performance is repeated. When you feel a correct answer has been given, choose a student to read the card to the class. You may also want to go over some of the services as a group to be sure everyone understands how each service works.

4. Summarize the activity.

Remind your students that whatever we call the important jobs that ecosystems and their species provide—"secret services," "ecosystem services," "nature's services," or "free services"—those services are happening around the clock and help make our lives more livable. Ask your students to explain any connections they see between the services and biodiversity. *(Biodiversity is the diversity of life on Earth—including genes, species, and ecosystems. It's because of Earth's natural diversity that we enjoy these benefits. The free services that ecosystems and species provide make life on Earth possible and are, some people feel, one of the most important reasons we need to protect biodiversity.)*

WRAPPING IT UP

Assessment

Upon completion of the game, have each student summarize the services presented in the various skits and how those services occur in the local community.

Unsatisfactory—The student does not participate with the group in the presentation. The student is unable to summarize services or make connections to the local community.

Satisfactory—The student participates in the presentation. The student is able to summarize most of the services presented and can make connections to the community.

Excellent—The student is centrally involved in the group presentation. The student summarizes all the services presented and clearly explains how the services occur in the local community.

Portfolio

There is no portfolio documentation for this activity.

Writing Ideas

- Based on the service they acted out for the class, have the students write a script with dialogue among at least three characters.

- Have the students select one of the ecosystem services portrayed in the skits and, using a metaphor or analogy, write a description of the service.

Extensions

- Challenge your students to work in groups and develop a list of additional services that ecosystems and species provide. Then, play charades again using the students' own examples.

- Have your students create an "ecosystem services" bulletin board for a local library or other community center.

Resources

Earthwatch: Earthcycles and Ecosystems by Beth Savan (Addison-Wesley, 1991).

Ecology for All Ages: Discovering Nature Through Activities for Children and Adults by Jorie Hunken (Globe Pequot Press, 1994).

Nature's Services edited by Gretchen D. Daily (Island Press, 1997).

The Random House Book of How Nature Works by Steve Parker (Random House, 1993).

Web Sites

Visit the National Museum of Natural History's Web site, "Welcome to Exploring Ecosystems Online," at <www.bsu.edu/teachers/academy/ecosytems>.

Visit the Bureau of Land Management's Web site, "Understanding Ecosystem Management," at <www.blm.gov/education/ecosystem/classroom.html>.

Michael Durham/ENP Images

*"The humblest of beetle species is still the product
of a considerably longer creative process than any painting, statue,
or piece of music . . . the natural habitats being destroyed today
and the species lost are crucial to our spiritual well-being,
in addition to being vital to the Earth's long-term health."*

—Andrew P. Dobson, professor, writer

16 The Nature of Poetry

SUBJECT

language arts

SKILLS

gathering (reading comprehension), analyzing (discussing), presenting (writing)

FRAMEWORK LINKS

42, 43, 58

VOCABULARY

See the vocabulary listed with the discussion questions provided for each poem (pages 199–202).

TIME

two or more sessions

MATERIALS

copies of poems (pages 79–83 in the Student Book) and "Types of Poems" (page 84 in the Student Book), paper, pencils

CONNECTIONS

This activity can be enhanced by introducing your students to any of the activities in the "What Is Biodiversity?" or "Why Is Biodiversity Important?" chapters.

AT A GLANCE

Read and discuss several poems related to biodiversity, then write original biodiversity poetry.

OBJECTIVES

Compare several poems that relate to the theme of biodiversity. Recognize the tone, meaning, rhythm, and use of specific language and imagery in poems. Write poetry to convey personal reflections on biodiversity.

Living things and wild places often elicit strong emotional reactions in people. Some people feel awe at the sight of graceful birds; others feel terror at the sound of a rattlesnake. Some people have felt threatened and fearful in a forest at night; others have felt peaceful on a secluded lake or in an urban park. For all these types of experiences, poetry can be highly effective for sharing our reflections with others. In this activity, your students will read poems that relate to biodiversity and compare the poems' different themes and tones. Then the students will try their hand at writing poetry of their own to see if they can capture an emotion or insight they've had about species or landscapes.

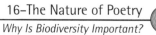
Before You Begin

Make copies of the poems (pages 79–83 in the Student Book) and "Types of Poems" (page 84 in the Student Book) for each of your students.

What to Do

1. Hand out copies of the poems.

Give one set of poems to each student. Give the students time to read the poetry in class, or assign it as homework. Tell the students to pay attention to the tone of each poem, the message of each poem, and the imagery the poets use to get their messages across. Because many insights about poems come in repeated readings, you may want to have a student read each poem aloud just before it is discussed. Also, you may want to provide a list of vocabulary words for the students to review either before or after reading the poems.

2. Discuss poems.

Encourage your students to discuss their reactions to each of the poems. You may want to start with general questions such as "Which poem did you like best?" "Why?" "Did any poem capture your feelings about nature and/or biodiversity?" You may also want to use the suggested questions (following each title) to organize your discussion. Please note that the italicized responses are possible interpretations, not necessarily "correct" answers. Encourage your students to speak freely, and give recognition to any answer that they can support thoughtfully.

"In Time of Silver Rain"
by Langston Hughes

1. During which time of year does this poem take place? *(The poem takes place in spring.)*

2. What is the tone of this poem? (In other words, how does the poet appear to feel as he's writing it?) *(The tone is jubilant, joyful, wondrous.)*

3. Which words and rhythms convey the tone most effectively? *(The tone is conveyed through "joy," "wonder," "rainbow," "sing," repetition of "to life!")*

4. What examples of different kinds of living things does the author give? *(Different living things include grasses, flowers, butterflies, boys and girls.)*

"The Panther"
by Ogden Nash

1. What is the tone of this poem? How is that tone conveyed? *(The tone is humorous; it is conveyed through playful use of words, jokes.)*

2. How much do you think the poet knows about panthers? Does that matter here? *(The poem does not convey detailed biological knowledge.)*

"Rain Forest"
by Joseph Richey

Vocabulary

los palos del sol and *cedars*: two types of trees

industrious: hard working

nourish: sustain with food, enrich

organic: containing carbon, i.e. living or once-living

biospheric: related to the land and atmosphere occupied by living things

1. What is the setting of this poem? *(The setting is a tropical rain forest.)*

2. Which images in the poem best helped you to see the rain forest? *(Answers will vary.)*

3. How do the images of this poem differ from the images of "In Time of Silver Rain"? *(The images in this poem convey more specific, detailed descriptions of the ecosystem; this poem has more of an ecological focus.)*

4. How similar is the tone to that of "In Time of Silver Rain"? *(The tone is not quite as exuberant, yet it still seems wondrous, delighted.)*

5. What colors does the poet mention? What effect does the use of these colors have? *(Colors include white, green, orange, blue, pink, red, black, crimson; they highlight the diversity, colorfulness, and richness of the rain forest.)*

6. What different plants and animals are mentioned? *(The poem mentions orchids, vines, cedars, butterflies, caterpillars, ants, and so on.)*

7. Name some of the different relationships within the rain forest that are described. *(The poem describes vines that live on trees, and ants that use leaves to grow fungus for food.)*

8. Langston Hughes mentions humans in his poem. Why do you think Joseph Richey does not? *(Answers will vary.)*

9. Which words does the poet use that are more commonly associated with human habitation than wild places? What effect does using them here have? *(Notice words such as "commuting," "plumbing," "disneyesque." Answers will vary, but you can point out that these words help make imagery come alive because the words are somewhat surprising. [You might want to contrast this with examples of clichés, which tend to make a poem flat.])*

10. Read the sentence that begins with "tiny crimson roots." What effect does repetition have in this sentence? *(Repetition conveys the sense of the roots twisting around; it is active and expressive.)*

"What Makes the Grizzlies Dance"
by Sandra Alcosser

Vocabulary
numinous: spiritual, awe-inspiring

lacquered: covered with a hard glossy protective coating

serrated: notched like a saw

penstemon: a kind of wildflower (lousewort)

1. What is the setting of this poem? *(The poem is set in June in the mountains.)*

2. List some of the living things the poet mentions and what they are doing. *(The poet mentions that "snowpeas sweeten," "ladybugs swarm," grizzly bears "swirl" and "bat" color against their mouths, and berries swell. The poet also mentions poppies and penstemons.)*

3. Do you notice anything interesting about the above list of verbs? What do you think the poet is doing? *(Most of the verbs begin with "sw." The poet is creating a repetitive sound. Students may have different ideas about what these sounds do for the tone of the poem.)*

4. What is the mood of the poem? Show how that mood is reflected in images or words. Have you ever been in this kind of mood? When? Which other poet describes a similar feeling? *(The mood is joyous, spring-feverish, euphoric; perhaps it is most like the mood in Langston Hughes' poem. Other answers will vary.)*

5. How does the relationship between Sandra Alcosser and the grizzly bear compare to the one between Ogden Nash and the panther? *(Ogden Nash, while humorous, plays upon the fears we have of large predators. Sandra Alcosser connects with the mood of the grizzly, almost trying to walk in its [hairy, leathered] feet. She also likens the way the bear eats ladybugs to the way people taste the "language of early summer"—or perhaps how the poet herself "tastes" the language of this poem.)*

6. How did you feel about grizzly bears before reading this poem? Did the poem support that feeling or change your attitude about grizzlies? *(Answers will vary.)*

"For a Coming Extinction"
by H. S. Merwin

(Note: Since this poem was written, gray whale populations have made a dramatic comeback. In 1994 gray whales were removed from the endangered species list. This poem is for advanced students.)

Vocabulary
fore-ordaining: already determined, predestined

1. In what ways does this poem feel similar to or different from those by Hughes, Nash, Alcosser, and Richey? *(This poem focuses on the natural world and, like Richey's, on species diversity; but the poem focuses on the author's sadness about species decline, not on exuberance about what's there.)*

2. How would you describe the mood of this poet? What is his attitude toward wildlife? *(The poet conveys a somber, solemn, somewhat angry mood; he has a respectful and caring attitude toward wildlife.)*

3. What is it that the poet finds so upsetting? *(The poet is lamenting the decline of the gray whale and the extinction of a host of other species, as well as people's attitudes about species.)*

4. Who is the "I" in the poem, and who is the "we"? How does the poet feel about the "we"? *("I" is the poet; "we" is humankind. The poet admits his membership in humankind but scorns its attitude toward other animals.)*

5. What do the phrases "we who follow you invented forgiveness," "we were made on another day," and "black garden" have in common? What purpose do these references serve in the poem? *(All the phrases have religious [Judeo-Christian] connotations: the theme of forgiveness, how God made humans on a different day than the rest of the animals, and the reference to the Garden of Eden [here a "black garden"]. These references remind us of one view of the Judeo-Christian attitude that believes humans are superior to other creatures, an attitude that the author clearly disagrees with.)*

6. Do you think the poet believes his last two sentences? Why or why not? *(No. The phrase reflects what most people seem to think, but the poet clearly questions this attitude.)*

"A Song for New-Ark"
by Nikki Giovanni

Vocabulary
disgorge: discharge

1. What is the setting of this poem? *(The setting is a city scene in Newark, New Jersey.)*

2. What effect does hyphenating New-ark have? Where else in the poem does the poet get playful with language? *("New-ark" makes the word a kind of pun—the old ark was Noah's ark, a vessel made to save all the animals from a flood; the new-ark is a place devoid of most wildlife. It also plays off the site, "Newark," New Jersey. Other examples of playful language are: "unyoung, unable . . . unto the unaccepting streets" and the "rat-tat-tat" sound of bullets aimed at rats.)*

3. Is this a poem about biodiversity? How so? *(It isn't necessarily what we think of as a traditional "nature poem," but it does recount the history of how biodiversity was lost in this region.)*

4. What does the poet mean by "obvious . . . family . . . ties"? *(Human beings and rats are related as mammals.)*

5. Compare the two phrases: "forest creatures and their predators" and "people and their predators." Are they referring to the same kinds of predators? What effect does this repetition have? *(Wild animals were the predators of forest creatures; other humans and the ills of urban life are the predators of people. Repetition highlights not only this difference, but also the parallel—that we, too, have our predators.)*

6. Which of the other poems does it remind you of, and why? *(Answers will vary, but students may mention that the tone is most like that of H. S. Merwin's poem. The city setting could be what Langston Hughes was writing about many years earlier.)*

7. Do you think it would be possible to write a poem about biodiversity in the city today that had the joyful tone of Hughes' poem or the amazed descriptions of Richey's? *(Answers will vary.)*

3. Have students write their own poetry.

To assist students in writing original poetry, it is important to help them focus their writing. You might begin by asking them to recall some of the different emotions or ideas conveyed in the poetry they have read (awe, delight, humor, fear of wildlife, and so on). Have they had similar or different reactions to the plants, animals, and places they have seen or read about? With the students working in small groups, encourage them to share specific emotions about significant experiences with nature that they have had. You might even want the students to share images that capture the feeling or idea they are remembering.

Alternatively, you might want to take students outside for a nature walk. Encourage them to look closely at one or a few specific elements of the natural world. They can draw pictures or jot down ideas before going back inside.

Pass out copies of "Types of Poems" (page 84 in the Student Book). Go over the descriptions of the different kinds of poems to make sure the students understand them. Explain to the students that they will be writing their own poems about biodiversity. They can choose any form of poetry, from one of the forms described on the sheet to rhyming verse. And they may write about any aspect of biodiversity they choose. They should write at least two poems each. Encourage them to try using two different styles.

Allow the students to begin writing their poetry. Again, they can write about any aspect of biodiversity—a specific place they have visited, a particular plant or animal, their general feelings or ideas about nature, and so on. Remind the students that images and rhythms can all be used to reflect their ideas and feelings. And stress again that poetry can convey all kinds of feelings—both positive and negative.

You'll probably need to give the students more than one session to finish their poems. Assign the final poems as homework and have the students turn them in on another day. They can turn them in for a grade, or they can read them out loud to the group, if they are comfortable with that.

WRAPPING IT UP

Assessment

Ask the students to select one of their own poems and to copy it onto paper without their name attached. Collect the poems and then randomly redistribute them (making sure, of course, that the students don't get their own poems). Have students become "peer coaches" and write a commentary on the poem they received. In their reviews they can describe the mood the poem creates, the way the words and phrases convey the mood and meaning, what the poem means to them, strengths of the poem, suggestions for improvement, and the connection between the poem and biodiversity. Assess the reviews on the students' ability to interpret and analyze the poem, to synthesize their critical review and personal reactions, and to apply critical thinking skills.

Unsatisfactory—The student is unable to analyze the poem and fails to synthesize his or her feelings or connect the poem to broader biodiversity issues.

Satisfactory—The student analyzes the poem and draws connections between the poem and broader biodiversity issues.

Excellent—The student provides a thoughtful analysis of the poem, including its mood and meaning. The student also makes connections between the poem and broader biodiversity issues.

Portfolio

The poems and critiques of the poems can be used as part of the student's portfolio.

Writing Ideas

- Before beginning to write poetry, invite your students to use one of the following leads to help them identify an idea or feeling they would like to capture in poetry:

1. I'll always remember . . . (Describe an experience with nature.)
2. I don't understand why some people feel _____ around animals.

Extensions

- Collect the students' poems and publish them. Distribute copies of the poetry book to the student-authors, as well as to others in your school and community.

- Have a Biodiversity Poetry Jam. Students can invite other classes to listen as they read their biodiversity poems out loud. They can also recite the biodiversity poems they read as part of this activity. You can even suggest that they bring refreshments to make the jam like a real coffeehouse event.

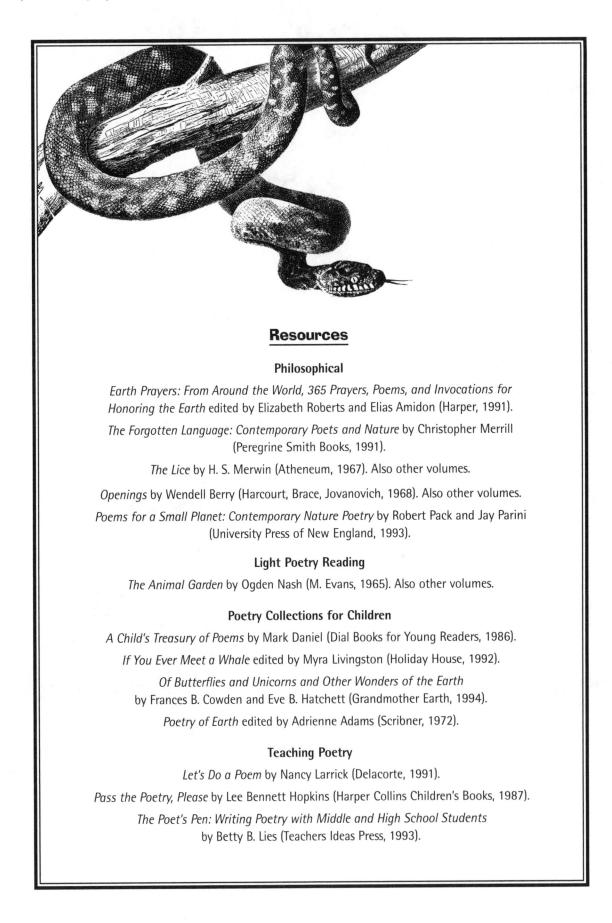

Resources

Philosophical

Earth Prayers: From Around the World, 365 Prayers, Poems, and Invocations for Honoring the Earth edited by Elizabeth Roberts and Elias Amidon (Harper, 1991).

The Forgotten Language: Contemporary Poets and Nature by Christopher Merrill (Peregrine Smith Books, 1991).

The Lice by H. S. Merwin (Atheneum, 1967). Also other volumes.

Openings by Wendell Berry (Harcourt, Brace, Jovanovich, 1968). Also other volumes.

Poems for a Small Planet: Contemporary Nature Poetry by Robert Pack and Jay Parini (University Press of New England, 1993).

Light Poetry Reading

The Animal Garden by Ogden Nash (M. Evans, 1965). Also other volumes.

Poetry Collections for Children

A Child's Treasury of Poems by Mark Daniel (Dial Books for Young Readers, 1986).

If You Ever Meet a Whale edited by Myra Livingston (Holiday House, 1992).

Of Butterflies and Unicorns and Other Wonders of the Earth by Frances B. Cowden and Eve B. Hatchett (Grandmother Earth, 1994).

Poetry of Earth edited by Adrienne Adams (Scribner, 1972).

Teaching Poetry

Let's Do a Poem by Nancy Larrick (Delacorte, 1991).

Pass the Poetry, Please by Lee Bennett Hopkins (Harper Collins Children's Books, 1987).

The Poet's Pen: Writing Poetry with Middle and High School Students by Betty B. Lies (Teachers Ideas Press, 1993).

*"In the end, the poem is not a
thing we see; it is, rather, a light by which we
may see—and what we see is life."*

–Robert Penn Warren, writer

17 The Spice of Life

World Wildlife Fund

SUBJECTS

language arts, science, social studies (ethics)

SKILLS

organizing (prioritizing), analyzing (discussing), presenting (articulating), citizenship (debating, evaluating a position, taking a position, defending a position)

FRAMEWORK LINKS

37, 40, 59, 60

VOCABULARY

ecological processes, insect, economics, **extinct, pollination**

TIME

one session

MATERIALS

chalkboard or pieces of flip chart paper, thick markers

CONNECTIONS

This activity works best once the students have become fairly familiar with biodiversity issues. It is a good culminating activity for a biodiversity unit.

AT A GLANCE

Explore beliefs and values about why biodiversity is important and why it should be protected.

OBJECTIVES

Explain personal beliefs and values about protecting biodiversity. List several reasons why people believe it is important to protect biodiversity.

People's feelings about biodiversity issues, including the importance people place on wild species and spaces and whether they think biodiversity should be protected, do not depend on just their knowledge of these issues and the sciences that relate to them (ecology, biology, sociology, political science, economics, and so on). People's feelings also depend on personal belief systems and values. This activity is designed to give your students a chance to examine their values and to sharpen their own thinking by sharing their opinions and feelings with their peers. The students first discuss their points of view in small groups and then talk about biodiversity conservation as a group. We've provided a series of numbered questions to get the students thinking about a range of biodiversity-related concerns, as well as additional guiding questions to help direct the discussions (see pages 209–210). Some groups may need a lot of prodding to carry on the group discussion; others may need only occasional questions or statements for direction. You may want to add your own questions to these lists and think about other ways to engage your group.

Before You Begin

Write each of the six "Why Care About Biodiversity?" statements (page 211) on separate pieces of flip chart paper. (You can adapt, shorten, add, or combine as needed.) Also write the word "other" on a seventh piece of flip chart paper. Use thick markers and write large enough so the students can read the statements from all areas of the room. Be sure to read through and familiarize yourself with the "Valuing Biodiversity" questions on pages 209–210. When you get to step 5, your knowledge and understanding of these questions will help you to guide the students' discussion.

What to Do

1. Ask your students whether protecting biodiversity is important and why they feel the way they do.

Explain that many people feel that it's important to protect biodiversity and that they have diverse reasons for thinking so. Ask your students how *they* feel. What reasons can they give to protect biodiversity? (These may be reasons they have read, reasons they have heard others express, or their own, personal views.) Write their ideas on a chalkboard or piece of flip chart paper. (It might also help to give them a few minutes to write their ideas before talking.)

2. Put up the statements and read each one out loud.

Using a different location for each one, tape the seven statements you copied earlier around the room. Place each one high enough for everyone to see. (Or you can put up the papers before class, folding each bottom half up and taping it in place so the students can't read the words until you uncover them.) Explain that the statements represent many of the key reasons people have given for why it is important to protect biodiversity. As a group, go over each of the statements. Compare the ideas represented in the statements with the lists that the students generated.

3. Students choose a statement to stand near.

Ask your students to carefully consider all of the statements. Have each of the students pick one of the statements and then go and stand near it. Explain that the statement the students choose should be one that they feel strongly about—either because they think it is an important reason to protect biodiversity or because they disagree with it. If they don't see a sign that reflects their viewpoint, they can stand at the sign marked "other." Explain that there is no correct answer and that it's OK to stand either alone or with a group.

Bruce Bunting/WWF

4. Discuss the choices that students made.

After everyone has made a selection, have the students at each statement discuss among themselves why they chose that particular statement. Remind them that each person will have personal reasons for making the choice he or she made and that they should explore some of those reasons.

Give the students about five minutes to discuss their thoughts before asking one person from each group to summarize the discussion. You might want to record each group's points on a chalkboard or piece of flip chart paper.

5. Open up the discussion to the entire class.

After all the groups have given their summaries, use the questions on pages 209–210 to spur a group discussion of some of the arguments that biologists, conservationists, ecologists, economists, and others have put forth for protecting biodiversity. Read one of the numbered questions and have the students react to it. You can use the guiding questions to challenge the students' thinking and to help direct their discussion as needed. Refer to the background information (pages 30–37) for more help in leading the discussion.

You do not need to ask the class all the guiding questions, and the students do not need to discuss each of the numbered questions in turn. The discussion may naturally flow from one topic to another. However, during the course of the discussion, make sure that the students confront the issues highlighted by each numbered question and that they explain why they feel the way they do. Have them give examples whenever they can, and be sure to challenge their ideas—especially when the students reach answers quickly or all of them seem to be agreeing with each other. Allow enough time for the students to fully discuss their points of view. Also give them an opportunity to research issues that come up.

Valuing Biodiversity

1. Is it important to conserve the diversity of life for medical and economic reasons?

Guiding Questions

- Do people actually need wild plants and animals for either medicinal or economic reasons?
- Can't people synthesize in a laboratory all the medicines they need?
- If genetic material is what's important, wouldn't it be sufficient if people froze wild plant and animal tissue samples, didn't worry about the actual organisms, and then used the samples when needed?
- If a plant or animal species is not known to have any medical or economic benefit to people, is it then OK to let the species die out?

2. Is the argument a good one that it's important to protect the diversity of life because biodiversity helps maintain important ecological processes that help support life on Earth?

Guiding Questions

- What sorts of ecological processes does biodiversity help maintain?
- People have developed an amazing array of technologies to deal with particular problems—everything from water treatment plants that purify sewage water to scrubbers that can take pollutants out of the air from factory smokestacks. Isn't it fair to assume that people will be able to develop technologies that can perform essential ecological processes in place of biodiversity?
- Are there any down sides to technological solutions?

3. Would your life be affected in any way if we lost species such as ladybugs, bears, tigers, and eagles?

Guiding Questions

- Is there anything about these species that makes them special?
- Would you feel the same way if the species we lost were venomous snakes, biting insects, and other species that are harmful to people?
- Are there species that you think are more important to protect than others? Which ones? Why?

4. Do all species have a right to exist?

Guiding Questions

- Do people have the right to use any of the world's resources as they see fit? Why or why not?
- Does the right to exist apply to ugly, obscure species that are of no use to people?
- Some species have been around for millions of years—and have survived incredible periods of destruction and change on the planet. Should that influence whether we decide to protect a species?
- Do people have any responsibilities to other living things?
- Do people have the right to drive species to extinction?

Valuing Biodiversity (Cont'd.)

5. Some people argue that no generation has the right to destroy the environment and resources that future generations will depend on. Do you agree or disagree with this idea?

Guiding Questions

- Why should people today do without things they want when we don't even know what future generations will need or want?

- How do you feel about the state of the world? Do you feel that past generations have left you with the environment and resources you need to live?

- There used to be millions of passenger pigeons in the United States. Today these birds are extinct. Has your life been affected in any way by the lack of passenger pigeons in the world? Will future generations really care about species that disappeared before they were born?

6. Is the diversity of life important for inspiring inventors and artists and for spurring curiosity and imagination?

Guiding Questions

- What human pursuits look to nature for inspiration?

- What inventions, stories, or works of art can you think of that were inspired by living things? Could these have been produced without the inspiration of nature?

- Isn't it reasonable to assume that all the photographs and films that have been made of wild plants and animals can provide inspiration to future writers and artists?

7. Is the diversity of life important for recreational activities?

Guiding Questions

- What kinds of recreational activities rely on wild spaces or species?

- Can well-tended golf courses and manicured parks provide the outdoor green space people need?

- Is it right to save an area so people can hike and fish if it means that other people lose their jobs?

- Does the fact that someone has done a particular job all his or her life—and perhaps his or her parent or grandparent also did the same job—give the person a right to keep doing that job even if it means wiping out a species or harming the environment?

- Should people be allowed to take part in any recreational activity (such as some off-road vehicle races) even if it harms the environment? How do we balance the rights of individuals and the rights of society as a whole?

WHY CARE ABOUT BIODIVERSITY?

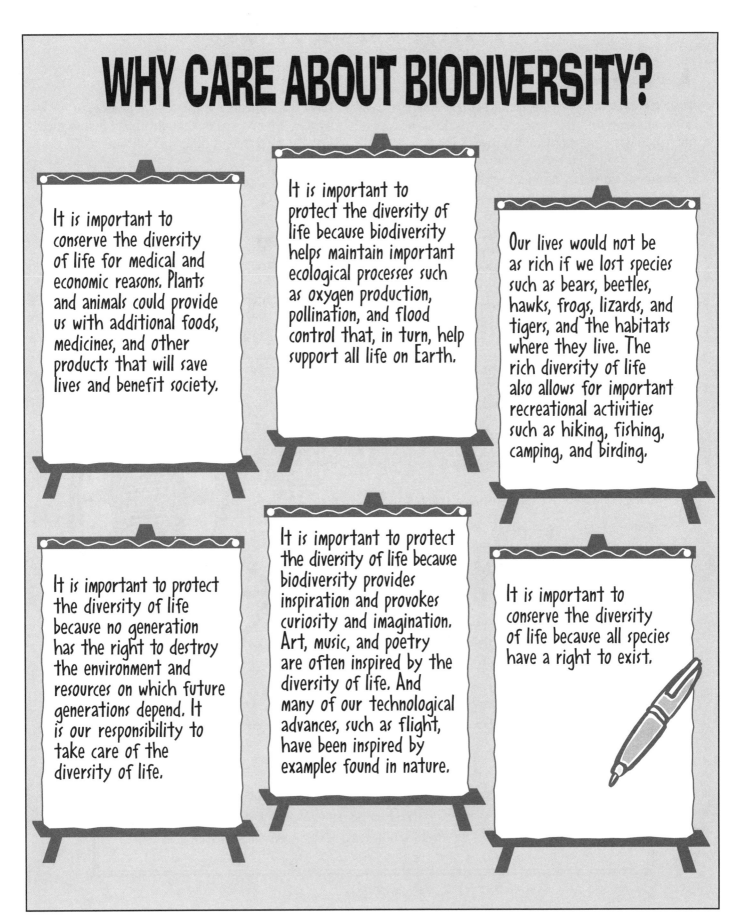

It is important to conserve the diversity of life for medical and economic reasons. Plants and animals could provide us with additional foods, medicines, and other products that will save lives and benefit society.

It is important to protect the diversity of life because biodiversity helps maintain important ecological processes such as oxygen production, pollination, and flood control that, in turn, help support all life on Earth.

Our lives would not be as rich if we lost species such as bears, beetles, hawks, frogs, lizards, and tigers, and the habitats where they live. The rich diversity of life also allows for important recreational activities such as hiking, fishing, camping, and birding.

It is important to protect the diversity of life because no generation has the right to destroy the environment and resources on which future generations depend. It is our responsibility to take care of the diversity of life.

It is important to protect the diversity of life because biodiversity provides inspiration and provokes curiosity and imagination. Art, music, and poetry are often inspired by the diversity of life. And many of our technological advances, such as flight, have been inspired by examples found in nature.

It is important to conserve the diversity of life because all species have a right to exist.

WRAPPING IT UP

Assessment

Have the students write a personal statement about the importance of protecting biodiversity. Explain that there is no right or wrong answer to this assignment—and that they don't even have to think protecting biodiversity is important at all. However, they should carefully consider everything they've learned about biodiversity as well as all of the points made during their discussion in order to make a well-reasoned and well-supported statement. Encourage the students to consider medical, economic, and ecological implications of biodiversity protection, as well as recreational activities, artistic inspiration, and any obligations of present generations to future ones. Tell them to use examples to illustrate their points.

Unsatisfactory—The student is unable to use examples to illustrate personal beliefs about protecting biodiversity. The student cannot make connections between the concepts discussed in the activity and personal beliefs in the statement.

Satisfactory—The student uses examples to support personal beliefs. The student makes connections between the concepts discussed in class and personal beliefs.

Excellent—The student uses examples to support personal beliefs. The student clarifies or challenges concepts from class using his or her individual belief system.

Portfolio

The student's biodiversity protection statement (created in the assessment) can be included in the portfolio.

Writing Ideas

- Have the students use the following as a journal starter: "Some ideas or thoughts I had before the activity are different now. They include . . . "

- Have the students write a dialogue between two people who have different viewpoints on protecting biodiversity.

Extension

Have each student or small group of students choose one of the "Why Care About Biodiversity?" statements on page 211 to use as a theme for a collage. Afterward have the students make a display of the collages under a title they've created.

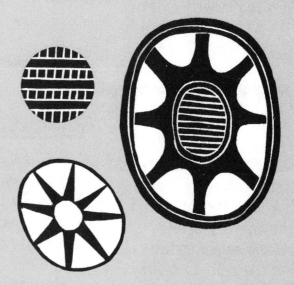

Resources

The Diversity of Life by Edward O. Wilson (Harvard University Press, 1992).

The Idea of Biodiversity: Philosophies of Paradise by David Takacs (Johns Hopkins University Press, 1996).

The Value of Life: Biological Diversity and Human Society by Stephen Kellert (Island Press, 1997).

WWF Atlas of the Environment by Geoffrey Lean and Don Hinrichsen (World Wildlife Fund, 1996).

*"The twentieth century has been extra-ordinarily successful for the
human species–perhaps too successful. As our population has grown from 1 billion to
6 billion and the economy has exploded to more than 20 times its size in 1900, we
have overwhelmed the natural systems from which we emerged and created the dangerous
illusion that we no longer depend on a healthy environment. As a result, humanity now
faces a challenge that rivals any in history; restoring balance with nature
while expanding economic opportunities for the billions of people whose basic
needs–for food and clean water, for example–are still not being met."*

–Lester Brown, President, World Watch Institute

SUBJECTS

language arts, social studies

SKILLS

gathering (researching), analyzing (comparing and contrasting, discussing), presenting (acting)

FRAMEWORK LINKS

5, 18, 40, 41, 42, 43, 58

VOCABULARY

cultural diversity, **introduced species, native species**

TIME

three or more sessions

MATERIALS

Part I—*copies of "The Nature of Culture Worksheet" (pages 85–90 in the Student Book)*
Part II—*maps, encyclopedias, and other reference materials on global cultures*
Part III—*copies of "The Cast of Characters" (page 91 in the Student Book); copies of "What Is Amaranth?" (page 92 in the Student Book)*

CONNECTIONS

For more on the ways humans depend on biodiversity, try "Something for Everyone" (pages 90–93), "Ten-Minute Mysteries" (pages 98–101), "Biodiversity Performs!" (pages 195–197), "The Nature of Poetry" (pages 198–205), or "Diversity on Your Table" (pages 230–237).

 **AT A GLANCE**

Through a role-play, explore some of the intriguing connections between culture and biodiversity.

 OBJECTIVES

Describe connections between nature and human culture. Explore personal heritage and traditions that are connected to the natural world. Examine culture/nature connections by taking part in a role-playing activity.

In the Middle Ages, European farmers believed that some animals, including bears, badgers, and hedgehogs, left their dens each year on the second of February to inspect the sky. Based on an old belief that the weather that day was the opposite of the weather to come, the farmers determined that if the animal could see its shadow, planting should be delayed by several weeks. If the animal could not see its shadow, then the planting could begin right away.

Sound familiar? Our Groundhog Day is a North American version of this European tradition. With no hedgehogs in sight, early European settlers in the United States chose groundhogs as the next best animal forecaster.

Around the world, every culture has traditions, foods, songs, art, games, and more that incorporate local plants and animals. Coyote stories are a popular part of many Native American traditions. Clambakes are a special feast in many New England communities. State birds, flowers, and other mascots contribute to people's regional identity. (See "Celebrating Nature" at the right for a list of some biodiversity-related festivals.)

CELEBRATING NATURE

Every year around the country there are festivals that celebrate locally important plants and animals. The festivals listed below will give you an idea of their diversity.

All About Alligators (Omaha, Georgia)

Arbor Day (nationwide celebration of trees)

Blueberry Festival (Montrose, Pennsylvania)

Buzzard Sunday (Hinkley, Ohio)

California Prune Festival (Yuba City, California)

Cape May Autumn Weekend (Cape May Point, New Jersey)

Catfish Days (East Grand Forks, Minnesota)

Cherry Blossom Festival (Washington, D.C.)

Festival of the Cranes (Socorro, New Mexico)

Florida Citrus Festival (Winter Haven, Florida)

Gilroy Garlic Festival (Gilroy, California)

Great Spangled Butterfly Days (Muskegon, Arkansas)

Lobsterfest (Mystic, Connecticut)

Maine Potato Blossom Festival (Fort Fairfield, Maine)

Marigold Festival (Pekin, Illinois)

Miss Crustacean USA Beauty Pageant and Ocean City Creep (Ocean City, New Jersey)

Moose Dropping Festival (Talkeetna, Alaska)

Morel Mushroom Festival (Muscoda, Wisconsin)

National Hard Crab Derby and Fair (Crisfield, Maryland)

Pumpkin Show (Circleville, Ohio)

Swan Days (Swan Quarter, North Carolina)

Washington State Apple Blossom Festival (Wenatchee, Washington)

Watermelon Thump (Luling, Texas)

Concern About Two Kinds of Diversity in Decline

A growing number of linguists, anthropologists, and sociologists have been exploring the connections between the loss of cultural diversity and biodiversity. Many of these scientists think language is the best indicator of cultural difference. And they point out that, like species, languages are experiencing a rapid decline worldwide. More than half of the world's languages have disappeared in the last 10,000 years.

These scientists also argue that similar forces are reducing both species diversity and language diversity. Many species cannot adapt quickly enough when their habitat undergoes rapid change. Similarly, many traditional languages disappear when communities change quickly and when residents no longer find the old language suitable for conveying modern needs and ideas. Also, just as introduced species may wipe out many native species, major languages from the outside may cause people to abandon their traditional languages.

Languages and cultures carry a legacy of knowledge, experience, and ideas—much of it rooted in specific places. For example, some indigenous names for plants and animals describe the uses, the ecological relationships, and other information about the species. This knowledge may be lost if the language dies. Or some groups may have culturally derived methods of farming that do more to accommodate specific local conditions than modern farming practices. For these and other reasons, many people believe we'd do well to avoid cultural homogeneity, and instead respect and support a diversity of knowledge, ideas, values, and ways of expressing our connections to planet Earth.

World Wildlife Fund

In the past, communities' cultural practices usually related to the specific plants and animals in their region. Differences in regional biodiversity therefore contributed to differences between cultures from one region to the next. But what happened as people started to move and cultures started to come into contact with one another?

In some cases, immigrants held on to their old traditions, even if the plants and animals their traditions were based on were no longer present. Others modified their cultural traditions to reflect the new species in their lives—as the European settlers did in the groundhog example previously mentioned. Other immigrants dropped their traditions in favor of those of their new neighbors. Today vast numbers of Americans subscribe to the same dominant culture—where mice are universally associated with Mickey Mouse and lush green grass is widely considered the proper ground cover for backyards.

Such changes have captured the attention of anthropologists, biologists, and sociologists alike, many of whom believe that there is a decline in both biodiversity *and* cultural diversity worldwide, and that the two may be linked. (See the box, "Concern About Two Kinds of Diversity in Decline.") But before we can begin to grasp these complex issues, it is important simply to appreciate some of the many ways that cultures and biodiversity are interconnected. This activity is designed to give your students that first step.

In Part I, students will read several examples of culture/nature connections from different places and compare these to things they have or do in their own lives. In Part II, they'll search their ancestral heritage to see when and how a culture/nature connection existed in another place and another time, and what has happened to it. And in Part III, they'll do a role-play to explore a set of characters' attitudes toward cultural differences. Each activity should open your students' eyes to the fascinating cultural dimensions of biodiversity.

Make copies of "The Nature of Culture Worksheet" (pages 85–90 in the Student Book) for each student.

1. Write the phrase "human culture" on the board.

Divide the students into groups of four or five. Ask the students to brainstorm about all the words that come to mind when they think of human cultures. Choose a student to record each group's responses on the board. If the students need any coaching, ask them to consider some of the elements of human cultures—rituals, language, and art, for example.

Ask the groups if they can think of any ways that human culture and biodiversity may be linked. Give them several minutes to discuss their ideas. Afterward tell them that they'll be doing an exercise to explore the wealth of connections between nature and culture.

2. Pass out copies of "The Nature of Culture Worksheet."

Give the students time to complete the worksheets, or assign them as homework. Afterward go over their responses as a group. We've included possible answers below to get you started.

Question 1

a. Only certain people could collect or wear resplendent quetzal feathers. What effect might that have had on quetzal populations? *(Restrictions on who could collect quetzal feathers probably helped protect these birds. Restrictions meant that fewer birds would be caught and injured. In addition, the practice of pulling out the feathers and letting the birds go didn't harm quetzal populations as much as killing the birds would have.)*

b. Are there animal products today that we have restricted use of to protect the animal populations? *(Bans on ivory, restrictions on harvesting certain species of fish and shellfish, and restrictions on using products from threatened and endangered species [such as feathers from eagles and fur from wild cats] are all examples of how we have restricted the use of animal products to protect animal populations. [For more information on laws protecting endangered species, see page 415.])*

Question 2

a. Why did the settlers choose buckskin and elkskin for their pants? *(Animal skins were probably one of the few readily available sources of clothing material that could protect settlers from cold temperatures and thorny shrubs. When the settlers killed deer and elk for meat, it was efficient to use the skins for clothing instead of letting them go to waste.)*

b. Where do your clothes come from? Do they have anything to do with nature? *(Today, our clothes come from natural and synthetic materials. Two of the most common natural materials are cotton and wool. The production of both natural and synthetic materials affects biodiversity. [See "The Many Sides of Cotton" on page 278 for more about the relationship between cotton and biodiversity.] Manufacturing processes for both natural and synthetic clothing can also produce pollution. People's clothes can also reflect their values. Even though leather can be made into clothes, shoes, and other items, some people won't wear it because they don't believe it is right to wear clothes made from animals.)*

Question 3

a. Why was purple reserved for high-ranking and wealthy people in ancient Rome? Why can all people wear purple today? *(In many societies, the use of rare products is restricted to those who can afford to buy them. Because the snails were the only source of purple dye, and because each snail produced only a little bit of dye, the dye was considered very valuable. Today purple dyes [natural and synthetic] are readily available and are no longer any more expensive than other dyes.)*

b. Are there things that only high-ranking and wealthy people wear today? What makes these things so valued? *(Valuable things include precious gems [diamonds, rubies, emeralds, and so on], precious metals [gold, silver], certain types of clothing [angora sweaters, suede leather, and so on], and other materials that are scarce or expensive to produce.)*

Question 4

a. Why do you think grenadiers first wore hats covered in bear fur? *(Wearing hats made of fur was considered to be a sign of strength and victory in battle. Killing wild bears was also a sign of power and virility.)*

b. Do people in your community use animals as symbols of strength? If so, in what ways? *(Answers will vary.)*

Question 5

a. What characteristics of lions do you think Mali mask-makers respected and wanted to imitate? *(The mask-makers probably wanted to imitate power, strength, dominance, and control.)*

b. How are animals or plants used as symbols in art or entertainment in your life? What characteristics of these animals do people prize? Give examples. *(Answers will vary. Examples could include cheetahs as symbols of speed; owls as symbols of wisdom; ants as symbols of hard work; and swans as symbols of beauty.)*

Question 6

a. If a truffle hunter did not have the help of a dog or pig, what skills would he or she need to find truffles? *(A truffle hunter would have to have knowledge of the type of habitat where truffles grow and skill at locating these mushrooms in the wild.)*

b. Have you ever collected food from the wild? If so, what kinds? Where do you get most of your food? If it comes from a store, where does the store get it? *(Answers will vary.)*

You may also want to use the following questions to broaden the discussion:

- What are some of the similarities and differences between your own cultural experiences and the practices described on the worksheets? *(Students may not be able to see how their own festivals, clothing, toys, and other cultural elements are tied to nature. Or if they do, they may find that they make use of global species and resources, whereas the examples on the worksheets describe the use of local resources. This "globalization" may make the students less directly familiar with the species they use as well as where each one comes from, how people obtain each one, and so on.)*

- Select one or more examples that show sharp differences between the people and situations described on the worksheets and the students' own lives. What do students attribute these differences to? *(Answers will vary. Possible conclusions include the following: modern or urban cultures do not choose to or do not need to be as tied to biodiversity as rural ones are or as traditional ones were; we don't have "cultural traditions" in the old sense; and wildlife is no longer a big enough part of people's lives to so directly influence cultures.)*

- What might happen to the cultures in these examples if the species they used disappeared? *(The people may have to, or choose to, replace that species with another one or with something manufactured. They may gradually lose the tradition altogether.)*

- Can you think of any ways that these cultural traditions might threaten biodiversity? *(Use could lead to overuse, especially if the human population grows or the species dwindles for other reasons.)* Can you think of any ways the traditions might help preserve or enhance biodiversity? *(Cultures may have the incentive to sustain the species, or they may be more attuned to the species status because they know it well. Some traditional management practices may increase the viability of species [through seed dispersal, game population control, and so on].)*

Before You Begin • Part II

Have maps, encyclopedias, and other reference materials on global cultures available. You may wish to set up a time for your students to use the library and Internet for researching their backgrounds.

What to Do • Part II

1. Discuss family heritage.

Ask the students if they know which countries their parents or ancestors came from. Or, if your students are Native American, ask them in which region of the country their ancestors lived. Have they ever been there? Do they still feel strong cultural ties to that region or to the people who live there? You may want to have the students put pushpins on a map to illustrate the geographic distribution of their ancestral homelands.

2. Research family heritage.

Tell the students they're going to find out more about the environment and cultural heritage of their ancestors. Tell them they can consult family members, encyclopedias, or other resources to find out (a) the characteristics of the natural environment which their ancestors originally came from (climate, topography, plants, animals, and so on), and (b) one tradition or aspect of their ancestors' culture that was tied to local biodiversity. The students should record the information they find and bring it back to class. (Note: It is not essential that students research their entire family heritage. Those students with complex ancestry should choose just one ancestor's heritage to explore in depth; those students who are adopted can choose the ancestry of their adoptive parents or any other culture that they feel an affinity for.)

3. Discuss the students' findings.

Ask the students to share their results. What kinds of environments did their ancestors come from? What traditions did the students find out about and how were the traditions tied to local biodiversity?

Then ask the students to reflect on the role those traditions now play in their families. Are they still practiced? If so, how have they been passed along? Are the same resources used? Do those resources occur naturally?

If the tradition has been lost, why might that have happened? (Was it lost when people moved? Was it gradually forgotten? Was it discouraged by mainstream culture?) Is the tradition still being practiced in the region where it originated?

4. Conclude the discussion.

Ask the students to describe how they felt about the culture/nature connections they discovered. Do they feel a greater sense of connection to their ancestors? To their ancestral homeland? Why or why not? If the traditions have been lost, how do they feel about that? Do any of them think it would be fun to try to revive them? Could they?

The Amaranth Tradition

Many years ago, the Tohono O'odham (TOE-ho-no AH-ah-tum) people of southern Arizona ate a wild plant called amaranth. The greens of this plant, which sprouted early in the spring before the people's own crops had grown, were tender and sweet for just a few weeks. The greens were also extremely high in vitamins, which gave a big boost to the diets of desert dwellers at that time of year.

When people from other cultures first came into contact with the Tohono O'odham, they didn't value the practice of eating amaranth greens. Little by little, the practice was lost. Nowadays people rarely admit to eating amaranth greens at all. But every once in a while, Tohono O'odham people living in Arizona, or even as far west as Los Angeles, still head out in the spring to gather amaranth greens and enjoy the sweet flavors of this plant.

amaranth

Before You Begin • Part III

Make copies of "The Cast of Characters" (page 91 in the Student Book) and copies of "What Is Amaranth?" (page 92 in the Student Book).

What to Do • Part III

1. Discuss cultural diversity in your area.

Ask the students if they think there is a lot of cultural diversity in the area where they now live. Why or why not? Do they think people in their community value diversity? Or do they discourage it? How?

2. Prepare for the role-play.

Explain to the students that they're going to do a role-play to see how different people in a family and community respond to one example of cultural diversity. Tell them you're going to read a brief general description of a biodiversity-related tradition (see the box on amaranth on the previous page), then assign them roles of people living in Los Angeles who are involved in this tradition today. Afterward, they'll act out a series of mini-dramas involving different combinations of these characters.

3. Pass out character descriptions, assign roles, and hand out "What Is Amaranth?"

There are six different role-play scenarios, and all the scenarios use a different set of the characters. Make sure each student has a part to play by passing out the "The Cast of Characters" slips. If you use the role play scenarios on the next page, you'll need four grandmothers, four mothers, six Ofelias, five Dannys, one father, one Mr. Williams (teacher), several friends of Ofelia and Danny, and a highway patrol officer. Feel free to make up new role-play scenarios to give every student a part. Also give students a copy of the "What Is Amaranth?" page in case they need more information about the uses of the plant.

Have the students read their character description. Explain that even though the character descriptions are written as quotes, they are not lines to be used in the role-play. They're character statements that describe the attitude of each character and will determine how the student will act in his or her particular scenario. To help the students understand their character descriptions, you can place all the same characters (mothers, fathers, Ofelias, Dannys, and so on) into "character groups" to discuss their roles.

Read aloud each of the "Role-Play Scenarios" on the right. Gather a set of actors as needed for each scenario, and give them a few minutes to discuss how they'll portray their scenario. After they've had time to prepare, have each group perform their role-play for the rest of the class.

4. Discuss the dramas.

After you have finished the mini-dramas, ask the students to reflect on what they saw. Could they identify with any of the characters? Which ones? Do they think the family would continue the tradition after the grandmother died? Why or why not? What would make it hard for Ofelia and Danny to keep gathering greens? What might make it easier? Name some of the benefits of keeping the tradition intact. Have the students had any experiences that are similar to the role-plays? Encourage them to think about how it might feel to be judged for the way they do things. Have they ever felt pressure to conform to what everyone else is doing? Have they ever made fun of someone because he or she dressed differently or had different traditions? Do they know anyone who has? Ask students to comment on what it means to them to value and respect cultural diversity.

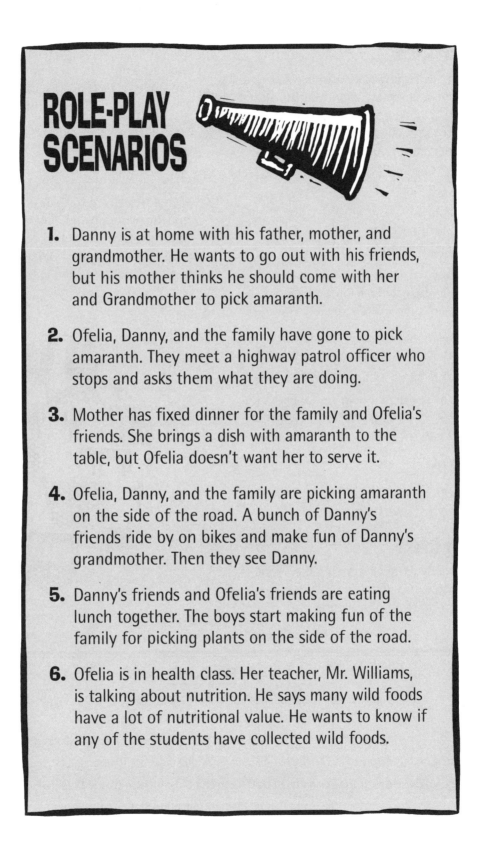

ROLE-PLAY SCENARIOS

1. Danny is at home with his father, mother, and grandmother. He wants to go out with his friends, but his mother thinks he should come with her and Grandmother to pick amaranth.

2. Ofelia, Danny, and the family have gone to pick amaranth. They meet a highway patrol officer who stops and asks them what they are doing.

3. Mother has fixed dinner for the family and Ofelia's friends. She brings a dish with amaranth to the table, but Ofelia doesn't want her to serve it.

4. Ofelia, Danny, and the family are picking amaranth on the side of the road. A bunch of Danny's friends ride by on bikes and make fun of Danny's grandmother. Then they see Danny.

5. Danny's friends and Ofelia's friends are eating lunch together. The boys start making fun of the family for picking plants on the side of the road.

6. Ofelia is in health class. Her teacher, Mr. Williams, is talking about nutrition. He says many wild foods have a lot of nutritional value. He wants to know if any of the students have collected wild foods.

WRAPPING IT UP

Assessment

Have students write down at least one example of how plants and animals are important to cultural identity—their own and other cultures—in each of the following categories. Also have them describe each culture/nature connection.

- food
- clothing
- entertainment
- status in the social structure
- values and religion

Unsatisfactory—The student is unable to give at least one example for each category and fails to describe each culture/nature connection.

Satisfactory—The student is able to give at least one example for each category and describes several of the culture/nature connections.

Excellent—The student is able to provide multiple examples for each category and describe different culture/nature connections.

Portfolio

"The Nature of Culture Worksheet" can become part of the student's portfolio.

Writing Idea

Have the students write an essay on their feelings about, and experience with, cultural diversity. Have they ever felt intolerant of someone else's way of doing things? Have the students ever felt that their way of doing things was not tolerated by others? Do the students have any friends who are not tolerated because they participate in traditional cultural activities?

Extension

Make a bulletin board of culture/nature ties. Pass out a large index card to each student. Then have all the students write up and illustrate the family culture/nature connection they investigated. Place the cards around a map of the world and link the tradition and the region with a piece of yarn or string.

+---+
| **Resources** |
| |
| *Biogeography: An Ecological and Evolutionary Approach* by |
| C. B. Cox and P. D. Moore |
| (Blackwell Scientific Publications, 1983). |
| |
| *Gathering the Desert* by Gary Paul Nabhan (University of |
| Arizona Press, 1985). |
| |
| *The Natural History of Domesticated Animals* by J. |
| Clutton-Brock (Cambridge University Press, 1987). |
| |
| *WWF Atlas of the Environment* by G. Lean and D. |
| Hinrichsen (World Wildlife Fund, 1996). |
| |
| Visit the North Central Plant Introduction Station's Web |
| site at |
| <www.ars-grin.gov/ars/MidWest/Ames/crops/amaranth.html>. |
+---+

*"We sang the songs that carried in their
melodies all the sounds of nature—the running waters,
the sighing of winds, and the calls of the animals.
Teach these to your children that they may
come to love nature as we love it."*

–Grand Council Fire of American Indians

SUBJECT

science

SKILLS

gathering (simulating), organizing (manipulating materials), interpreting (identifying cause and effect, inferring, making models), presenting (demonstrating, explaining), citizenship (working in a group)

FRAMEWORK LINKS

2, 12, 25, 28, 33

VOCABULARY

atmosphere, evaporation, heavy metals, impurity, mineral, **pesticides**, **photosynthesis**, sediment, toxic substance, transpiration, **wetland**

TIME

two sessions

MATERIALS

materials listed in each simulation

CONNECTIONS

"Biodiversity Performs!" (pages 194–197) works well before doing this activity. Try "Super Sleuths" (pages 182–187) to help your students explore some other important services—either before or after this activity. To learn about other ways that humans benefit from biodiversity, try "Something for Everyone" (pages 90–93), "Ten-Minute Mysteries" (pages 98–101), "The Culture/Nature Connection" (pages 214–225), or "Diversity on Your Table" (pages 230–237).

AT A GLANCE

Perform simulations that demonstrate some of the important ecosystem services that biodiversity provides.

OBJECTIVES

Perform a series of simulations that demonstrate ecosystem services. Identify and discuss the services illustrated in the simulations.

As the activity "Biodiversity Performs!" points out (see page 194), ecosystems and the variety of species within them provide many important services that help make life possible or at least more livable. These services are happening all the time—they are so common that we often don't notice them or think about how important they are. This activity is a series of five simulations that help illustrate a variety of these services. (More advanced students can try to develop their own simulations after learning more about ecosystem services.)

Before You Begin

There are a number of ways you can use this activity with your kids. We suggest that students be grouped into five secret service teams. Assign each team the task of setting up and testing one of the simulations on Day 1. On Day 2, have each team present its secret service simulation to the class. After watching each presentation, students will use the handout "The Secret's Out!" to identify the ecosystem service being demonstrated in the simulations.

You will need to arrange "stations" for each team's simulation. Put a copy of the directions and the necessary materials at each station (see pages 93–102 in the Student Book). Label each of the five stations. Also make one copy of "The Secret's Out" (page 103 in the Student Book) for each student. (Please note that Station #2 is shorter than the others, yet it still requires two days. It can be combined with Station #3. Stations #2 and #3 require some preparation ahead of time. Stations #1, #4, and #5 require activated charcoal, cobalt chloride paper, and elodea, respectively. Activated charcoal and elodea can be found in most pet stores that sell fish. Cobalt chloride paper can be ordered through science supply catalogues.)

What to Do

1. Day 1: Setting the stage.

Divide your students into five teams and assign one team to each station. Explain that the students will be working together to complete a simulation. Each team will be responsible for a different simulation. Students should not discuss their simulation with other members of the class. The simulations illustrate various ways that ecosystems provide important services for us and the environment. Identify the five stations around the room.

When they arrive at a station, all the members of the team should read the directions completely before setting up the simulation. Students should then set up and run their simulation. Tell them that on Day 2, each team will run its simulation for the class. Each team should discuss the expected outcome of the simulation. Each member of the team should also answer the questions listed under "Think about it."

Note: Remind the students that after they try their simulation out, they have to get it ready for the next day, so they might have to dry their equipment and/or supplies off or replace some of the parts. Stations 2, 4, and 5 require 24 hours to run. Let the students at these three stations know that they will not need to run the simulation again on Day 2, but that they'll have to explain what they did on Day 1.

2. Day 2: Presenting the simulations and matching the analogies.

Distribute the handout entitled "The Secret's Out!" to each student. Explain to the students that each team will have a few minutes to explain their simulation to the class. Ask each team to briefly review its procedures, perform the simulation (or explain the results of an overnight simulation) and discuss the results. Students should provide information to the class that answers the "What happened?" and "Think about it" sections on their handout.

After watching each presentation, have the students use the "The Secret's Out" handout to identify the ecosystem service being demonstrated in the simulation. Discuss student responses. (Answers are a–2, b–3, c–1, d–5, e–4.)

When all the teams have completed their presentations, review and summarize the different ways ecosystems provide important services to people and to the planet. The list should include flood control, water filtering and purification, erosion control, oxygen production, and climate control. (This final step can be used as the assessment—see page 228.)

WRAPPING IT UP

Assessment

Use the last step of the activity (page 227) as the assessment. Encourage the students to include on each list how the service is conducted in the "real" world and to give local, regional, or global examples of where the ecosystem services are taking place (e.g., local marsh, Everglades).

Unsatisfactory—The student's response is perfunctory reporting of one or more of the demonstrations.

Satisfactory—The student's list incorporates several of the demonstration stations and gives sound references to real applications.

Excellent—The student's list is a synthesis of the demonstrations and gives clear indication that the student is able to transfer the concepts of the services demonstrated to the real world.

Portfolio

Have the students keep lab reports for each station. The reports can include discussion notes as well as the list of ecosystem services (see Writing Idea). Place the lab reports in the portfolio.

Writing Idea

Students may keep a lab manual and make a report for each station. Lab reports should include an overview of the simulation, a description of what happened, and an analysis of the secret service that was simulated.

Extensions

- Identify places in your community where the ecosystem services that you simulated are occurring.

- Have students propose or create simulations that model other ecosystem services.

- Ask each student to illustrate one or two ecosystem services through sculpture, photography, or another art form. The student can use words to clarify points, but words should not be the focus of the illustration.

Resources

Discover Wetlands (Washington State Department of Publications Office, 1996).

Ecosystems (Science is Elementary—A Science Teaching Resource Publication) by Maureen Oates (Museum Institute for Teaching Science, 1995).

Environmental Science: Ecology and Human Impact by Leonard Bernstein, Alan Winkler, and Linda Zierdt-Warshaw (Addison-Wesley Publishing Company, Inc., 1995).

Project WET (The Watercourse and Western Regional Environmental Education Council, 1995).

The Wonders of Wetlands—An Educator's Guide by Alan Kesselheim (Environmental Concern, Inc., and The Watercourse, 1995).

Supply Catalogues

American Biological Supply, 288 E. Green St., Westminster, MD 21157, (410) 876-8599, Web site <www.qis.net/~ambi>.

Carolina Biological Supply, 2700 York Rd., Burlington, NC 27215, (800) 334-5551, Web site <www.carolina.com>.

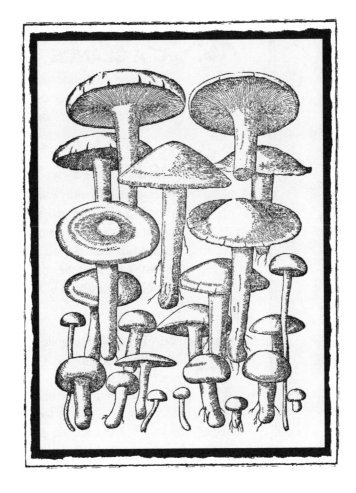

Nearly half of a tree's biomass is hidden in a vast
tangle of roots under the ground. And scientists have found
that these roots, in turn, are usually woven into an even bigger
web made of fungi. Fungi and trees have a symbiotic relationship that benefits
both. Fungi help trees absorb important nutrients like nitrogen and phosphorus
from the soil. And trees provide fungi with carbon, which the fungi absorb from the
trees' roots. Scientists are also finding that this tangled relationship is even more
complex than they realized and that the fungi may actually be helping to
"manage" the forest by giving some trees more nutrients than others.

–Adapted from "The Web Below,"
by Carl Zimmer
in *Discover*, November 1997.

20 | Diversity on Your Table

SUBJECTS

science, social studies (economics)

SKILLS

gathering (reading comprehension, simulating), analyzing (calculating, identifying patterns), interpreting (inferring), citizenship (working in a group, evaluating the need for citizen action, planning and taking action)

FRAMEWORK LINKS

4, 19, 20, 28, 36, 47

VOCABULARY

gene bank, **genetic diversity**, genetic engineering, **trait**, variety

TIME

two sessions

MATERIALS

Each group needs one copy of the "Planting Grid" (page 107 in the Student Book) and an empty bag labeled "Planting and Yield." The instructor needs examples of several varieties of potatoes and/or corn, one to two bags each of three different kinds of dry beans (e.g., kidney, black, and pinto), an empty bag labeled "Beans Sold or Not Salable" and a set of "Crop Crisis Cards" (page 108 in the Student Book). On Day 2, each student needs a copy of "Raiders of the Lost Potato" (pages 40–44 of WOW!–A Biodiversity Primer). OPTIONAL— copies of "Real Life Scenarios" (pages 104–106 in the Student Book) and copies of the "Food Journal" (page 109 in the Student Book)

CONNECTIONS

To extend learning about the importance of genetic diversity, use "The Gene Scene" (pages 158–167) or "Biodiversity—It's Evolving" (pages 168–179).

AT A GLANCE

Play a game to understand the importance of protecting the genetic diversity of food plants and learn about the ways scientists are working to protect plant diversity.

OBJECTIVES

Role-play a farmer and make decisions about what crops to plant. Identify several reasons that protecting genetic diversity in plants is important, including the consequences to food production when genetic diversity is lost. List examples of how people are protecting genetic diversity in plants today.

F ood, glorious food! While everyone eats it, most people don't know much about where it comes from or how it got to be what it is today. In this activity, your students will learn how preserving the genetic diversity of wild plants is essential to ensuring the long-term health of food crops. After you introduce the topic of crop varieties and genetic diversity, your students will play a game in which they act as farmers, making decisions about what and how many bean varieties to plant. Students will base their decisions on information about the plants' genetic traits, potential yield, and market value. In the process, they will learn why protecting genetic diversity in plants is so important.

For centuries, farmers have selected and bred plants that express certain desirable traits. Over time, this process has evolved into some of the modern farming techniques that allow us to produce vast quantities of tasty and nutritious crops. However, modern farming has also meant that crop plants have become more genetically uniform—and many people believe they lack enough genetic diversity to survive changing environmental conditions. For example, to be most profitable, all the tomatoes in a farmer's field have to mature at about the same time and be the same size and shape so the farmer can harvest them all at once using a mechanical harvester. There is no incentive for farmers to plant other varieties of tomatoes under these market conditions, so the less popular varieties die out, as do varieties of potatoes, corn, beans, and other crop plants for the same reason.

This activity illustrates the danger of losing genetic diversity and also demonstrates ways that farmers, plant breeders, and others are acting to preserve genetic diversity in plants. Your students will see why many people believe we need to protect genetic diversity while we still have the opportunity.

Before You Begin

Collect several varieties of potato and corn, as well as one to two bags each of three different kinds of dry beans (e.g., kidney, black, and pinto). You will also need to label some empty bags with "Planting and Yield" (one for each group) and "Beans Sold or Not Salable" (one for you). If you have purchased your bean varieties in separate packages, mix about 60 beans (about two handfuls) of each variety (about 180 beans total) into each group's Planting and Yield bag. Leave the remaining beans separated by variety. They can serve as the Gene Bank and can be used to replenish a team's bag as needed. As the game progresses, the groups will be able to purchase replacement beans from the Gene Bank if they need to. Placing purchased beans in the "Beans Sold" bag will remove them from genetic circulation. Copy one "Planting Grid" (page 107 in the Student Book) for each group, "Raiders of the Lost Potato" (pages 40–44 in *WOW!—A Biodiversity Primer*) for each student, and one set of "Crop Crisis Cards" (page 108 in the Student Book) for you. On a chalkboard or flip chart, draw the two tables on page 233. You may need to adapt the "Traits, Yield, and Value" table depending on the varieties of food you purchased. You can also make copies of the "Real Life Scenarios" on pages 104–106 in the Student Book.

What to Do

1. Briefly discuss the concept of genetic diversity in food plants.

Display different varieties of corn, potatoes, and/or beans that you have purchased from the grocery store. Explain to the students that thousands of potato and corn varieties have been grown around the world, but today most of these varieties are no longer being grown and, in many cases, have been lost. Also explain that each variety has different traits that make it especially good for different uses. For example, russet potatoes (with a dark brown skin) are good for baking while Atlantic potatoes were developed primarily for making potato chips. Sweet corn contains lots of sugar and is a favorite for summer eating. Flour corn gives us—you guessed it—flour. And dent corn, which contains a lot of starch, makes the best animal feed. Also mention that certain varieties of food may have cultural significance to certain groups of people. For example, blue corn (a type of flour corn) is important to the Hopi Indians who use it extensively in traditional foods and drinks.

Do your students know what makes the varieties of each food unique? Explain that while these varieties are closely related, each contains different genes that produce different shapes, colors, flavors, and other characteristics. In explaining how different members of the same species can have very different traits, compare the varieties of plants to breeds of dogs (which are all members of the same species). For example, the bean species called common bean *(Phaseolus vulgaris)* has hundreds of varieties currently used in farming. They are all members of the same species, yet the individual varieties can be as different as a Chihuahua and a St. Bernard. Explain to your students that they will now play a game to help them understand why genetic diversity in plants is important to us all.

Foods from Around the World

Agriculture began 10,000 years ago when people first realized that the seeds of wild plants could be collected, planted, and cultivated to produce a reliable source of food. All around the world, wild species were domesticated and then selectively bred to produce improved varieties.

The foods we eat today can be traced back to those original wild species. For example, barley and wheat originally came from wild grasses of the Near East; corn, squash, peppers, and beans first grew wild in Mexico; and today's potatoes descended from wild Peruvian species.

Even what we think of as a traditional "American" breakfast is really an international affair: Orange juice is squeezed from a fruit that originated in Southeast Asia, toasted wheat bread comes from those wild Near Eastern grasses, our hash browns can be traced to those Peruvian tubers, and our coffee has descended from a wild Ethiopian bush. And, if you like peach jam on your toast, you have Chinese farmers to thank!

But despite such a wide variety of food origins, only about 12 basic crops, including wheat, rice, and corn, feed most of the world today.

2. Divide the class into groups of four or five students and hand out materials.

Each group needs a 20-square "Planting Grid" along with its own Planting and Yield bag. Explain that each group of students represents a family of farmers and the grid represents the family's 20-acre plots of land. Each family has a variety of beans that may be planted (one bean per square on the grid) and each family will be given information about the beans to help decide which varieties and how many of each to plant. No group is obligated to plant all three varieties of bean. The object is to bring in the most money after several years of harvesting.

3. Discuss the rules.

Draw the two tables (at the right) where everyone can see them. Then explain the rules as you show the tables to the students. (You can also copy the page and pass it out to the students.)

©Keren Su/Tony Stone Images

World Wildlife Fund

VARIETY*	TRAITS	YIELD	VALUE
Kidney Beans	High yielding, resistant to spot fungus.	**TWO** beans for every planted bean	**$10.00** per bean
Black Beans	Low yielding, drought tolerant. Leaves poisonous to rabbits.	**ONE** bean for every planted bean	**$5.00** per bean
Pinto Beans	Low yielding, resistant to bean mold and bean-eating slugs.	**ONE** bean for every planted bean	**$1.00** per bean

*The names in the "Variety" column will depend on the varieties purchased.

PROFITS

	Round 1	Round 2	Round 3	Round 4	Round 5	Total
Group 1						
Group 2						
Group 3						
Group 4						

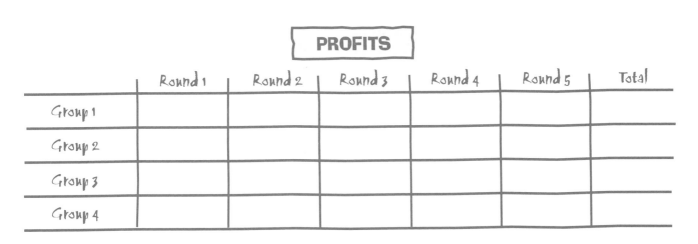

Rules

- Each round of the game represents a different growing season. During each round, through reading the "Crop Crisis Cards," you will learn about the environmental conditions for that particular season.

- Every black bean you plant will yield one additional new black bean. For example, if you plant five black beans, you will gain five new black beans, giving you 10 beans at the end of the round. The same is true for pinto beans.

- Since kidney beans are the higher-yield variety, every kidney bean you plant will yield two new kidney beans at the end of the round. For example, if you plant five kidney beans, you will gain 10 new kidney beans, giving you 15 beans at the end of the round.

- After each round, you must decide what to plant for the next season and what to sell (at the prices listed in the table).

- You can buy bean varieties from the Gene Bank (a place where important seeds and plants can be preserved in an artificial setting) for $20 per bean at any time during the game.

4. Play Round 1 of the game.

Have each group plant one bean from its Planting and Yield bag in each square of its planting grid. Each square represents one acre of land. Students should consider the information in the "Traits, Yield, and Value" table and the rules you have gone over as they make their planting decisions. What seems to be the most economically sound plan? Tell them also to think carefully about the different characteristics of each type of bean and about what they will gain from each one at harvest time.

When the groups have finished planting, tell the class that it has been a good year—not too hot or too cold, not too dry or too wet, no insect or slug infestations—and that all the farmers will get a full harvest.

Have each group count its yield according to the "Traits, Yield, and Value" table by removing beans from its Planting and Yield bag. Advise the group to be very careful to make an accurate count, because each bean represents the group's profits. And remember that for every high-yield kidney bean, the students should get two extra beans at harvest, while the lower-yield varieties bring in only one extra bean per bean planted.

Students should then "plow their fields" by removing all the beans from the grid. Now they must decide which of those beans to replant. The remaining beans will be sold. Keep track on the board of how much money each farm makes on the beans they decide to sell (based on the beans' values described in the table). Have each group place the beans it has marketed in the Beans Sold or Not Salable bag. (The beans in this bag either represent beans that were sold or that were diseased, damaged, or not of high enough quality to sell.) Explain to the class that these beans have been removed from circulation.

5. Play Round 2 of the game.

In this round the students will begin to see the benefits of a diverse planting scheme. They should prepare for the second season by replanting with their group's remaining seeds. (They can no longer take beans from their Planting and Yield bags.) After

the fields are planted, read one of the Crop Crisis Cards (page 108 in the Student Book) to the class. Is it a blight of bean-eating slugs? An unprecedented drought? A destructive mold? Whatever the crisis, it will affect different groups in different ways. Only those beans resistant to the crop crisis survive. All the other beans should be cleared from the field and placed in the Beans Sold or Not Salable bag. Place the Crop Crisis Card back in the deck and shuffle.

Students who have fared poorly do have an option. Direct their attention to the Gene Bank and explain that a gene bank is an important conservation tool used to preserve plant parts—including seeds—and, therefore, genes. (Students may already be familiar with how gene banks are used in research and breeding.) Tell the groups that they may purchase their choices of replenishment beans in any combination of varieties from the Gene Bank, but that each bean costs $20. A group may not be able to afford to replant its entire field for the next round, but explain that this is part of the recovery process from the Crop Crisis. Remind the groups to think carefully about the varieties of beans they choose from the Gene Bank. Distribute beans as necessary, and subtract the purchase price from the profits of those groups buying replenishment beans.

6. Play three more rounds of the game.

Continue to plant seeds and draw crisis cards for as many rounds as time will allow. After drawing, remember to place the card back in the deck and reshuffle. After each round, help students determine how they fared based on varieties preserved, and record their profits or losses on the Profits table. After the last round no purchases of seeds can be made, and whatever beans the students have left should be sold. Add up the winnings on the board to determine which farm has been the most successful. Discuss the events that led to either a profit or loss by each group. If the groups have planned independently of one another and have made different decisions, those who planted diverse crops from the beginning should come out ahead (although results depends on the specific conditions encountered).

7. Wrap up with a discussion.

At this point your students should be able to discuss some of the pros and cons of the farming decisions they made. The activity demonstrates how farmers are often better off economically if they plant the highest yielding, most popular crops instead of less common varieties. But short-term profit may mean sacrificing the genetic diversity needed to withstand an environmental crisis. Your students can see the long-term value of protecting genetic diversity, even when it may not always seem economically beneficial in the short term.

The real-life scenarios (pages 104–106 in the Student Book) will help you guide a discussion as well. (You can hand these out, read them to the group, or use them as research starters.) However, it's also important to emphasize that the game was a very simple simulation. And there are many other factors (politics, social equity, and so on) that influence what farmers plant and why. Encourage your students to conduct more research to find information about this. You might also want to stage a debate with local experts to highlight the complexities of the issue.

8. What can we do?

Ask your students to think of ways that people can help preserve plant diversity. For example, they can purchase (or encourage their families to purchase) nontypical varieties at their grocery stores. Consumer demand can encourage grocery stores to carry more varieties and, in turn, farmers to supply those varieties. By using their power as consumers, people everywhere can help others realize the value of food diversity. Also, they can shop at local farmers' markets, co-ops, and whole-food stores for unusual varieties of foods. Students can save and plant the seeds of unusual varieties from the produce they consume, or they can buy seeds and plant them. In this way, like the seed savers (on page 106 in the Student Book), they'll follow the footsteps of farming families who have passed down seeds from one generation to the next. (See page 237 for seed companies that specialize in unusual varieties.)

9. Hand out a copy of "Raiders of the Lost Potato" to each student.

Have your students read "Raiders of the Lost Potato" to summarize and expand on the first part of this activity. After everyone has read the article, divide the class into small groups to discuss the following questions:

- Why are scientists collecting plants from around the world? *(They are trying to save some of the diversity of crop species.)*

- What are scientists doing with the material? *(They are storing it in gene banks.)*

- In what sort of place might scientists look for important plant material? *(They might look in areas where genetic diversity is high, such as natural habitats and areas where traditional agricultural methods have been used for generations.)*

- Are there ways that scientists can preserve genetic diversity other than in gene banks? *(Although gene banks are important, it is even more important to protect natural habitats as a means to conserve crop species. By setting aside and protecting the natural habitats of plants, we allow the natural process of evolution to continue to produce new combinations of genes that may someday prove valuable.)*

©D. Cavagnaro/DRK Photo

WRAPPING IT UP

Assessment

Have each student write down the pros and cons of the farming decisions that his or her group made. (This can be done either before or after the wrap-up discussion on page 235.)

Unsatisfactory—The student cannot explain both the benefits and weaknesses in increasing specific traits of their crops. The student is unable to show how increasing desired traits can minimize diversity.

Satisfactory—The student is able to explain both the benefits and weaknesses in increasing specific traits. The summary statement provides a connection to biodiversity.

Excellent—The student clearly explains how increasing traits are concurrently beneficial and problematic. The summary statement reveals individual thinking and reflection.

Writing Ideas

- Ask students to keep records in their journals of the plant-derived foods they eat, noting which ones they depend on the most. (See "Food Journal" on page 109 in the Student Book.)
- Have students write about how their lives would be different if tomato (or corn or potato) crops were suddenly decimated by disease or environmental crisis.
- Have students create their own recipe that includes at least five ingredients. Have them find out from which species each ingredient originates. Also have them research the origins of three or four key ingredients. For a list of foods and their origins, see *Global Diversity— Status of the Earth's Living Resources* (Chapman and Hall, 1992), pages 332–38.

Extensions

- Obtain some drought-resistant and non-drought-resistant beans. (See Resources at the right for sources of unusual varieties of seeds.) Plant the seeds in sand mixed with a little soil.

Keep the seeds watered. When the plants sprout, water them only once a week. Have students monitor the progress of their beans. Do the drought-resistant beans grow better? Why or why not? *(The drought-resistant beans should grow faster because they can withstand the dry conditions.)* What role might the sand play? *(Water drains more quickly through sand, creating drought-like conditions.)* Explain that drought resistance is a trait that people have found advantageous and therefore have favored. Which bean would you prefer if you lived in the southwestern United States? If the other bean grows faster and produces more in non-sandy soil, which bean would you prefer if you lived in the midwestern United States?

- Have students visit farmers' markets, health food stores, and small urban markets to look at crop varieties that are sold. Have students compare the varieties at those locations to the ones sold at supermarkets. Did they find differences in the varieties offered? Why or why not? What different groups of people do the markets appeal to? Ask them at which market they would prefer to shop. (Often you will find different crop varieties being sold at different stores. Many times, the smaller markets cater to different groups of people than do the large supermarkets, demonstrating that consumers from different cultures prefer different crop varieties. Cost is also a factor and smaller markets sometimes charge more for variety.)

- Ask students to choose an unusual food item from the store, prepare it at home, and bring it in to class (if that is permitted). Then have your own feeding frenzy. How are the foods unusual? Where did the students purchase the food? How many parts of the world are represented in the class meal? Are there different varieties of the same species of plant represented? What have the students learned about food origins and diversity from this meal?

Resources

Conserving the Wild Relatives of Crops by Erich Hoyt (IUCN/WWF, 1992).

Enduring Seeds: Native American Agriculture and Wild Plant Conservation
by Gary Paul Nabhan (North Point Press, 1989).

Global Biodiversity: Status of the Earth's Living Resources
by Brian Groombridge (Chapman and Hall, 1992).

Green Inheritance by Anthony Huxley (Anchor Press/Doubleday, 1985).

The Last Harvest by Paul Raeburn (Simon and Schuster, 1995).

Visit the Ohio Corn Growers Association *Corn World* Web site at <www.ohiocorn.org>.

Rain Forest in Your Kitchen by Martin Teitel (Island Press, 1992).

Seeds of Change: The Living Treasure by Kenny Ausubel (Harper, 1994).

"The World's Food Supply at Risk," *National Geographic* 179, no. 4 (April 1991).

WOW!—A Biodiversity Primer (World Wildlife Fund, 1994).

Conservation Groups

The Center for Plant Conservation, P.O. Box 299, St. Louis, MO 63166. (314) 577-9450.
This group is a national network of botanical gardens and arboretums dedicated to saving
native plants of the United States. <www.mobot.org/CPC>.

Native Seeds/SEARCH, 2509 North Campbell Avenue #325, Tucson, AZ 85719.
(602) 327-9193. <www.azstarnet.com/~nss/>.
This organization promotes the use of seeds to preserve the traditional crops and their wild relatives
of the southwestern United States. A seed catalogue that lists over 200 varieties is available.

Seed Savers Exchange, Rural Route 3, Box 239, Decorah, IA 52101. (319) 382-5990.
This network is dedicated to cataloguing and preserving the diversity of nonhybrid vegetable seeds.

Seeds of Change, P.O. Box 15700, Santa Fe, NM 87506. (888) 762-7333. <www.seedsofchange.com>.
This group was the first major national grower and supplier of organically grown seeds.
A seed catalogue is available.

Education Materials

Rural Advancement Foundation International-USA (RAFI-USA), P.O. Box 640, Pittsboro, NC 27312.
(919) 542-1396. <www.rafiusa.org>. Available materials include a wall map, "The Seeds Map:
Dinner on the Third World," which illustrates the centers of genetic diversity and plant genetic resources;
and a "Community Seed Bank Kit," which describes the importance of crop diversity as well
as explaining how to establish a community-based program to preserve crops.

What's the Status

The activities in this section explore biodiversity in today's world—what's happening to it, why it's threatened, where it's thriving, and how it's being studied. For background information, see pages 38-49.

World Wildlife Fund

# of Biodiversity?

"We consider species to be like a brick in the foundation of a building. You can probably lose one or two or a dozen bricks and still have a standing house. But by the time you've lost 20 percent of species, you're going to destabilize the entire structure. That's the way ecosystems work."

**–Donald Falk,
restoration ecologist**

SUBJECTS

science, language arts, social studies

SKILLS

gathering (collecting, researching), analyzing (comparing and contrasting), interpreting (defining problems, generalizing), applying (creating, designing), presenting (illustrating, writing)

FRAMEWORK LINKS

3, 12, 21, 23, 26, 35, 45

VOCABULARY

endangered species, *Endangered Species Act*, **habitat, introduced species,** *range map*

TIME

two to three sessions

MATERIALS

list of endangered species (see pages 113–114 in the Student Book), poster board, magazines with pictures of endangered animals, reference materials, Internet access (optional)

CONNECTIONS

For more on endangered species, try "The Case of the Florida Panther" (pages 246–251). To look more closely at the link between species and their habitats, try "Space for Species" (pages 286–301) or "Mapping Biodiversity" (pages 252–265). Use "The Spice of Life" (pages 206–213) to help your students examine their feelings about endangered species protection.

AT A GLANCE

Conduct research to create a poster about an endangered species, then take a walk through a poster "gallery" to find out more about endangered species around the world.

OBJECTIVES

Identify groups of animals that are endangered and name some species within each group. Research one species, describe why it's endangered, and compare its problems to those of other endangered species.

Have your students ever heard of an iriomote cat? Do they know what a kakapo looks like? How big is the Galápagos marine iguana? By researching one endangered species, your students will put a face on species biodiversity and gain insight into some of the problems that threaten all living things. As they share their research, they'll begin to understand broader issues of biodiversity loss—such as the HIPPO dilemma.

HIPPO is an acronym that represents the five major threats to biodiversity, which are caused by human activity: **H**abitat loss, **I**ntroduced species, **P**ollution, **P**opulation growth, and **O**verconsumption. See "What's the Status of Biodiversity?" (pages 38–49) for more about each threat.

This activity provides a way to help students understand the threats to biodiversity and what the word "endangered" means. It also gives your students a chance to use their creativity to design a poster and share information with their peers.

Before You Begin

Make photocopies of the list of endangered species (pages 113–114 in the Student Book). Copy the Poster Parameters below onto a chalkboard or a flip chart. Have a piece of poster board available for each student.

In addition, gather magazines that the students can use to cut out pictures of their endangered-species. Past issues of *National Wildlife, International Wildlife, Natural History, Audubon, Ranger Rick,* and *National Geographic* are all good sources. CD-ROMs and the Internet may also have information and pictures students can download. If your school has a computer lab with these resources, you may want to set up a time for your group to use the lab for their research.

What to Do

1. Discuss the term "endangered."

Show students the list of endangered species. Explain that every animal on this list is endangered. Discuss what "endangered" means (see box on page 242).

2. Explain the task and have students choose an animal to research.

Hand out the list of endangered species and tell students that each of them should pick an animal from the list that they would like to research. There are more than 100 different endangered animals from all over the world on the list. Encourage the class collectively to choose as wide a variety of species as possible in order to have a broad range of posters at the end of the activity. Encourage students to pick an animal they are unfamiliar with. Each student will need to research his or her animal and then create a poster designed to teach others about it. Point out the Poster Parameters you copied earlier, and review what the students should be trying to find out about their species—this is the kind of information they should include on their poster. Hand out poster board and give students time to conduct research. (You can also add plants to the list and adapt the poster parameters accordingly. For a complete list of endangered animal and plant species, see the U.S. Fish and Wildlife Service Web site listed on the next page.)

POSTER PARAMETERS

Your poster should include the following:

- Natural history information about your endangered species (what type of habitat it lives in, what it eats, and other natural history information such as how it gets its food, how long it lives, where it breeds, and when it is active [night? day? year round?])

- Why your species is in trouble

- What people are doing to help your species

- What your species looks like (pictures or drawings)

- Where your species lives (a range map)

The poster should not be a report. It should present the required information in easy-to-read chunks arranged in an informative and creative way that will capture people's attention. Pictures should have captions that explain what is shown.

Legal Lingo

*In the United States, many species are protected by the Endangered Species Act. Under this act, species may be listed as **threatened** or **endangered**. Populations of threatened species are generally low or declining but not in immediate danger of extinction. Endangered species are in immediate danger of becoming extinct. Typically, their populations are critically low and require high levels of protection.*

The U.S. Fish and Wildlife Service (USFWS) maintains a list of all the animals and plants in the United States that are threatened or endangered. The list can be found by visiting the USFWS Web site at <www.fws.gov/r9endspp/endspp.html>.

> *"Endangered species are sensitive indicators of how we are treating the planet, and we should be listening carefully to their message."*
>
> —Donald Falk, restoration ecologist

3. Take an endangered species gallery walk.

When the students have finished their posters, use them to create a scavenger hunt. Your scavenger hunt might include statements like the following:

- Name three endangered marine animals and state why they're endangered.
- Find two cats that are endangered. Record their names and why they're endangered.
- List three different kinds of animals that are threatened by loss of habitat.
- Name an endangered animal that lives in Asian tropical rain forests.
- Name two animals that winter in one part of the world and breed during spring or summer in another part.
- Name three animals that live on islands and state why they're endangered.

The questions themselves will need to come from the students' posters and should be designed to get students to read each other's posters and to draw conclusions about why different kinds of animals are in trouble. You should also be sure to include questions that cover the full range of animals your students researched. Next hang the posters where everyone can see them and hand out copies of your scavenger hunt clues to the group. Explain that the students should "tour the gallery," reading each other's posters to find the answers to the scavenger hunt sheets.

4. Share scavenger hunt results.

After students have finished their scavenger hunt, review their answers as a group. What seemed to be the biggest problem(s) facing the species in the gallery? What other generalizations can the students make about endangered species? For what reasons are island animals in trouble? Rain forest animals? Birds of prey? Students should notice, for example, that pollution has been a big problem for many birds of prey, that many island species are in trouble because of introduced species, that many big cats are in trouble because they are hunted for their furs, and that animals that live in tropical rain forests are threatened by habitat loss. What are people doing to help endangered animals? Is there anything that students can do? Why do people care about losing species? (See "The Spice of Life" on page 206 for more on examining personal feelings and beliefs about endangered species and biodiversity.)

WRAPPING IT UP

Assessment

Posters should be assessed separately. To determine students' understanding of why different species are endangered, have them give examples of species that are endangered and describe why. Have each student choose a group of species from the endangered species gallery (e.g., a group of marine animals) and list major reasons the animals are endangered. Identify and match endangered animals to each reason. Can the students recognize a common reason or reasons for endangerment of different groups of animals?

> **Unsatisfactory**—The student is unable to recognize why groups of animals are endangered and is unable to give examples of species that are endangered for different reasons.

> **Satisfactory**—The student is able to recognize why groups of animals are endangered and is able to give two or three examples of species that are endangered for different reasons.

> **Excellent**—The student is able to recognize animal groups that are endangered and is able to give many examples of species that are endangered for different reasons.

Portfolio

The student's endangered species poster can be added to the portfolio.

Writing Idea

Have students write a report that describes the main threats facing a group of animals or a habitat type.

Extension

Have students create a group database using the information they found about their animals. (They can also do a Web search to add to their information—see page 244). The students can then use the database as a long-lasting reference for information about endangered species. It could include the following entry fields: name of animal, type of animal (mammal, bird, fish, reptile, amphibian), type of feeder (herbivore, carnivore, omnivore), size (less than 5 pounds, 6–25 pounds, 26–100 pounds, 101–500 pounds, more than 500 pounds), habitat type (tropical rain forest, desert, grassland, ocean, wetland, and so on), main reason it's endangered (habitat destruction, introduced species, pollution, overhunting, overcollecting), and the second most important reason it's endangered.

After students have entered information on their animals, they can use the database to answer questions about the species they researched. The database can be updated and expanded by other classes during the year and in future years. It can be used as an up-to-date resource on endangered animals of interest to the school.

jaguar

Magazines

*Audubon, International Wildlife, National Wildlife,
National Geographic, Natural History, Ranger Rick.*

Books (General)

The Atlas of Endangered Species by John A. Burton (Macmillan, 1991).

The Collins Guide to Rare Mammals of the World by John A. Burton
(Greene Press, distributed by Viking Penguin, 1988).

The Doomsday Book of Animals by David Day (Viking Press, 1981).

The Encyclopedia of Mammals edited by David Macdonald (Facts on File, 1995).

The Encyclopedia of Reptiles and Amphibians edited by Tim Halliday
and Kraig Adler (Facts on File, 1986).

IUCN Red Data Lists of Threatened and Endangered Species edited by Jonathan Baillie and
Brian Groombridge (World Conservation Union, 1996).

The Official World Wildlife Fund Guide to Endangered Species of North America, Vol. I–II
edited by Charles Moseley (Beacham Publishing, 1990).

Vanishing Animals by Kurt Benirschke (Springer-Verlag, 1986).

Books (Middle School)

Endangered Mammals of North America by Victoria Sherrow (Twenty-First Century Books, 1995).

Endangered Species of the World by Laura O'Biso Socha (Mallard Press, 1991).

Saving Endangered Mammals: A Field Guide to Some of the Earth's Rarest Animals by
Thane Maynard (The Zoological Society of Cincinnati, Inc., 1992).

Vanishing Wildlife of Latin America by Robert M. McClung (William Morrow and Co., 1981).

Web Sites

There are many Web sites with great information and pictures of endangered species—and there
are sites that have information and pictures for each of the animals on the list we provided.
Have your students learn how to search for and find these sites. To find information on a
particular species, students should type in the full name of the animal and have the search
engine look for any relevant Web sites. For example, a student researching the
northern hairy-nosed wombat would type in the full name, not just "wombat."

black-footed ferret

*"One of the greatest challenges
of conservation is preserving the incredible
diversity of life while ensuring that people's
needs are respected and protected."*

**–Henri Nsanjama, Vice President,
WWF's Africa and Madagascar Program**

SUBJECTS

science, social studies, language arts

SKILLS

gathering (reading comprehension), analyzing (comparing and contrasting, discussing), applying (proposing solutions)

FRAMEWORK LINKS

1, 21, 22, 23, 29, 35, 38, 44, 46, 47, 48, 52, 55

VOCABULARY

endangered species, habitat loss, **introduced species, over-consumption,** panther

TIME

two sessions

MATERIALS

one copy of the "Panther Problem" summary (page 115 in the Student Book) and one set of "Panther Cards" (pages 116–118 in the Student Book) for each group; one copy of "The HIPPO Dilemma" in WOW!–A Biodiversity Primer (pages 23–29) and one copy of "Panther Solutions" (page 119 in the Student Book) for each student

CONNECTIONS

Use "Endangered Species Gallery Walk" (pages 240–245) to learn about other endangered species around the world. To investigate the importance of habitat to species, try "Space for Species" (pages 286–301). "The Gene Scene" (pages 158–167) can help your students understand the importance of genetic diversity in small populations. And "The Spice of Life" (pages 206–213) can help your students think about why it is important to protect species like the Florida panther.

AT A GLANCE

Work in small groups to discover how the Florida panther's decline is tied to the major causes of biodiversity loss around the world, and discuss what people are doing to help protect the panther.

OBJECTIVES

Describe how habitat loss, introduced species, pollution, population growth, and over-consumption are threatening Florida panthers and biodiversity in general. Discuss ways people are trying to protect the Florida panther.

There are only 30 to 50 Florida panthers left in the wild, and that makes these cats one of the most endangered animals in the world. Habitat loss has pushed the small remaining population into the southern tip of Florida. People's increasing demand for water and space, as well as the effects of non-native species introduced by people, continue to threaten panthers even in this last refuge. Pollution, in the form of mercury, has also been found in the tissues of many panthers, and mercury poisoning has been the cause of at least one panther's death.

In this activity, your students will learn about the threats that the Florida panther faces. Along the way, they will discover that Florida panthers are beset by the same problems that threaten biodiversity around the world—something we call the HIPPO dilemma. "HIPPO" is an acronym for the five major problems threatening the Earth's biodiversity: **H**abitat loss, **I**ntroduced species, **P**ollution, **P**opulation growth, and **O**ver-consumption. Students will also learn some of the ways people are trying to protect the Florida panther and to slow other kinds of biodiversity loss in Florida and around the world.

Before You Begin

For each group of four to five students, copy one "Panther Problem" summary (page 115 in the Student Book) and make one set of "Panther Cards" (pages 116–118 in the Student Book). For each student, make one copy of "Panther Solutions" (page 119 in the Student Book). If you don't have a copy of *WOW!—A Biodiversity Primer* for every student, make copies of "The HIPPO Dilemma" article on pages 23–29.

What to Do

1. Divide the class into groups and describe the assignment.

Divide the class into groups of four or five students and explain that they're going to be learning about one of the most endangered animals in the world—the Florida panther. Give each group a copy of the "Panther Problem" summary and have one student in each group read the summary to the rest of the group.

When the students have finished, give each group a set of "Panther Cards." Tell the students to read each card out loud in their group. Next have them try to organize the cards into four or five major categories of threats to the Florida panther. Tell the students that it's OK if each group organizes the cards differently and if some categories have only one or two cards. Explain that some cards may seem to fit into more than one category. In that case, students should pick the category that seems most appropriate to them. Then have each group make a list of the categories of threats that it developed.

2. Discuss as a class the threats to the Florida panther.

Have each of the groups name the threats it came up with and the problems that fit into those threats. Record the ideas on the board. After all the groups have participated, have the students compare the categories the groups came up with. Are there categories that can be lumped together? (You may want to draw lines to connect similar categories.)

3. Explain and discuss the HIPPO dilemma.

Explain that one way to think about the major threats to biodiversity worldwide is by creating broad categories that characterize the threats. Have the students compare their categories of threats to the Florida panther with those threats we've included, which are based on the thinking of many conservationists around the world. The categories are easy to remember by the acronym HIPPO.

Florida Panther Habitat

A Note to Educators

This activity focuses on efforts to protect the Florida panther–a subspecies of the cougar. As your students will discover, there are many different efforts taking place that are designed to protect these wild cats. But it's also important that your students understand that protecting endangered species and subspecies can be controversial and that not everyone agrees on how best to conserve those plants and animals that are in the most serious trouble–or even if we should protect them. Although the issues are complex, you might want to have students explore some of the more controversial ones–including the Endangered Species Act–in more detail. See the extension on page 249 for ideas about getting your group to examine some of the controversial issues surrounding endangered species.

dodo

Review each of the categories with the students: H–habitat loss, I–introduced species, P–pollution, P–population growth, O–over-consumption. Can the students assign each of the panther problems they read about to one of the HIPPO categories? (See answers below.) Ask your students to compare the HIPPO categories to the categories they came up with. How are they similar? Different? Can they think of any other endangered animals and plants that are affected by one or several of the HIPPO problems? Ask students to describe ways that one type of threat can be related to another. For example, introduced species can cause habitat loss and human population growth can increase pollution. Next have students read "The HIPPO Dilemma" article from *WOW!–A Biodiversity Primer*. As a class, talk about any new insights and information.

Answers:

Habitat: A, C, G, H, J, K
Introduced species: F
Pollution: E
Population: D, G
Over-consumption: B, I, K

4. Discuss possible solutions and distribute copies of "Panther Solutions."

Once your students have a better understanding of the HIPPO dilemma and the threats to the Florida panther, have them brainstorm ways to protect the panther. (Coach them a bit to draw out more concrete ideas than "increase habitat" or "stop pollution.") Write their thoughts on the board until they run out of ideas, then have them read the "Panther Solutions" handout to discover some innovative ways people are trying to protect the panther. When they have finished, ask them if they learned any new alternatives. Did the students find out any more about the problems faced by the Florida panthers by reading about the ways people are trying to protect them? *(Students should point out that Florida panthers are also threatened by having such a small population size and, therefore, limited genetic diversity. Review this problem with the students, making sure they understand that low numbers may mean a disease or a natural disaster could easily wipe out the population and that low numbers can lead to problems of inbreeding.)*

WRAPPING IT UP

Assessment

Pass out five index cards or pieces of paper to each student. On each card, have the students write the first letter of one of the following: **H**abitat loss, **I**ntroduced species, **P**ollution, **P**opulation growth, and **O**ver-consumption. Ask them to bold the first letter on each card. (You might want them to write "Pop" on one card to separate population from pollution.)

Then, one at a time, read each of the statements in the list below. Ask students to hold up the card that they think portrays the threat to the species or natural area described in the statement. (More than one answer may be correct!) You can do a quick visual assessment of the class or have your students keep track of their individual scores.

- Leopard skin coats were popular in the 1960s. **(O)**
- The demand for beachfront property near large cities is reducing wetlands and mangrove forests. **(Pop, O, and H)**
- When industries use water from rivers and bays, the water may go back into waterways at a higher temperature, killing organisms that can't tolerate warmer water. **(P)**
- The multiflora rose, an introduced species in the United States, destroyed many native plants in Midwestern fence rows. **(I)**
- More people are moving from urban areas to the "country" where there is more "natural scenery." **(Pop, O, and H)**
- Demand for mahogany and teak wood destroys acres of rain forests each year. **(O)**
- More than 60 percent of the animal life on the Hawaiian Islands is not native. **(I)**
- In the United States, the deer population in many states is larger now than in any other period of recorded history because natural predators have been destroyed by people and because people have created more deer habitats. (Deer like disturbed forests.) **(Pop and O)**

- The numbers of eagles, peregrine falcons, and ospreys dropped steeply in the 1960s because pesticides that farmers sprayed on crops moved through the food chain and caused the birds to lay eggs with soft shells. These soft-shelled eggs broke before the young could develop and hatch. **(P)**

Portfolio

Have the students compare and match HIPPO to the list of threats their small group identified. Have them write their comparison on a piece of paper and include it in their portfolio.

Writing Idea

Have the students write the text of a brochure for the Florida Department of Natural Resources that explains how Florida's increasing human population is affecting the state's natural resoures. (Allow time for research.)

Extensions

Species in Peril: Telling Their Stories

Have each student research and write a short report or prepare a poster on a species that has gone extinct or is currently endangered because of people's activities. Students can use the following examples, or they can look for their own examples of over-exploited species:

a. extinction of the passenger pigeon and heath hen in the United States because of overhunting

b. reduction of elephant populations in Africa for ivory

c. reduction of egret and tern populations for plumes at the turn of the 20th century

d. reduction or near extinction of whale species because of overhunting for meat, oil, and baleen

e. reduction of populations of wild American ginseng for herbal supplements and medicines

f. reduction of snow leopard populations because of commercial hunting for pelts

WRAPPING IT UP (Cont'd.)

Protecting Endangered Species: Debating the Issues

For older students, you may want to explore through a debate the controversy surrounding the Florida panther and other endangered species that live in your area. Although most people recognize the value of biodiversity, much controversy centers on the following issues:

Saving Species Versus Saving Habitat

In the past, many conservation programs focused on saving individual species. The current Endangered Species Act is an example of the species approach to protecting biodiversity. Under the act, species and subspecies are listed as threatened or endangered. The U.S. Fish and Wildlife Service, which oversees the act, is required to develop a recovery plan for each endangered species or subspecies. Although protecting habitat is a key component of most species recovery plans, the emphasis is on individual species and subspecies.

Today, many people would like to see the Endangered Species Act include more of a focus on protecting habitat than individual species. By protecting habitat, many argue that more species will be protected in the long run. They also feel that too much time and money are being spent on individual species and subspecies—and that we need to protect larger tracts of habitat if we want to protect biodiversity. Others argue that we need to do both, and that there are some keystone species (see page 21 for definition and examples) that need special protection if they are to survive.

Saving Species Versus Subspecies

Another debate centers on the difference between species and subspecies and the importance of each. Although the Endangered Species Act currently protects species and subspecies, some people feel that subspecies are so genetically similar to their relatives (which are often not endangered) that little genetic information will be lost if a subspecies becomes extinct. This is the case with the Florida panther. Although there are only 30 to 50 Florida panthers left in the wild, its relatives throughout much of the West are doing fine.

Other conservationists argue that it's just as important to save a subspecies as it is to save a species. Subspecies develop when a small population is isolated from the main population and, over time, develop distinct characteristics that help them adapt. For example, many subspecies form in isolated valleys and islands where breeding with the main population can't occur. Because the Florida panther is genetically distinct from its relatives, many conservationists think the panther is worth protecting. They feel that it's important to protect as much of the world's genetic diversity as possible. They also point out that, in many cases, a subspecies like the Florida panther is very important to an area's ecological health. Florida has a higher number of subspecies than most other states because of its isolation, climate, and diverse habitats. And many feel that the Florida panther is part of what makes Florida unique.

Losing Genetic Diversity

Another issue focuses on efforts to protect genetic diversity within populations. Again, using the Florida panther as an example, many scientists fear that because the population of Florida panthers is so low, continued inbreeding will weaken the small number that remains. To improve the genetic diversity of the panther population, Texas cougars have been released in Florida with the hope that they will mate with the Florida panthers and improve the overall genetic variability of the Florida panther population. But other scientists have raised questions. Is the new hybrid panther still a Florida panther that needs to be protected? And where does this new hybrid fit on the list of priority species and subspecies?

Setting Priorities

Another controversial issue centers on how to set priorities for protecting species. Which species and subspecies are most important and why? In many cases, politicians, not scientists, decide how to spend national or state dollars. And that often means those species that have public appeal take precedence over those species that are less cute and cuddly but just as important ecologically. In the case of the Florida panther, some scientists argue that other species are more important to protect and that all the effort expended protecting the Florida panther is not based on good science.

Saving Endangered Species and Ensuring Economic Development

Your students will find a variety of articles that look at the economic issues of species and subspecies protection. There are many differing views regarding how much money we should spend on protecting endangered species, how to resolve differences between economic growth and species protection, and other issues related to the implementation of the Endangered Species Act.

To help your students explore these issues, you can have them research different aspects of the Endangered Species Act or some of the issues outlined here and make presentations to the rest of the group. Or you can stage mini-debates and have several students take different sides of an issue. If your students decide to investigate the economic or policy issues surrounding the Endangered Species Act, we suggest they write to a number of organizations so that they will get a more balanced view of how scientists, economists, and others feel about the importance of the act and the value of protecting biodiversity.

The Dusky Seaside Sparrow: A Subspecies Lost

Older students may also research the case of the dusky seaside sparrow, a species that became extinct in 1987. The story of this endangered species is very similar to the panther's because the dusky seaside sparrow was also a subspecies and its protection was hotly debated. Ultimately, it became extinct. How is the case of the dusky seaside sparrow the same as that of the Florida panthers? How is it different? How do your students feel about the loss of this subspecies?

Resources

"The Everglades: Dying for Help" by A. Mairson. *National Geographic* 185, no. 4 (April 1994).

The Florida Panther (brochure). Florida Game and Freshwater Fish Commission, Bureau of Wildlife Research, 620 South Meridian Street, Tallahassee, FL 32399-1600.

Florida Panther Habitat Preservation Plan: South Florida Population, November 1993. Contact Florida Panther Recovery Coordinator at U.S. Fish and Wildlife Service, 117 Newins-Zeigler Hall, University of Florida, Gainesville, FL 32611-0307.

The Florida Panther Society. Route 1, Box 1895, White Springs, FL 32096. Telephone: (904) 397-2945. Visit their Web site at <www.atlantic.net/~oldfla/panther/panther.html>.

Our Endangered Species: A Citizen's Guide to Protecting Parks, Wildlife, and Plants. (National Parks and Conservation Association, 1995).

WOW!–A Biodiversity Primer (World Wildlife Fund, 1994).

23 Mapping Biodiversity

AT A GLANCE

Take part in mapping activities to explore the world's ecoregions and learn how experts make decisions about which natural areas to protect.

OBJECTIVES

Give several examples of how maps help scientists protect biodiversity. Compare the following terms: biome, ecoregion, and habitat. Give several examples of how plants and animals are adapted to specific ecoregions.

Conservation biologists know that if we want to protect species over the long term, we need to protect the places where species live. Sound simple? Maybe. But species protection is complicated work for biologists, social scientists, and planners. It's not enough to know the life cycle of one species. Experts need to understand the relationships between a species and its ecosystem, among different species in the same ecosystem, and among adjacent ecosystems. They also need to know how people affect and are affected by species and ecosystems. And they need to decide which areas are most important to protect.

None of this is easy. And beyond that lies the challenge of putting all available information together in a way that helps create effective conservation strategies in the short term and long term. The good news is that two of the most effective planning tools— maps and computers—have come a long way in the past decade.

In Part I of this activity, your students will learn more about how biologists survey and map critical ecoregions by taking a close look at World Wildlife Fund's Global 200 map of the world, which highlights more than 200 ecoregions that WWF believes represent the most important conservation priorities for the twenty-first century. By playing a mapping game with the Global 200 map, the students will learn more about the species, ecological processes, and landscapes that make these ecoregions so important. In Part II, they'll conduct their own gap analysis in an area they create. Depending on the resources you have in your school or community, you can introduce them to Geographic Information Systems (GIS) technology and give them a chance to see how it can be used locally to study and plan for the protection of biodiversity.

Make one one-sided copy of each of the "Ecoregion Species Cards." (pages 123–137 in the Student Book). Make 20 copies of "Ecoregion at-a-Glance" (page 121 in the Student Book), which is one for each ecoregion covered in this activity. Make seven copies of the "Secret Message Cards" (page 138 in the Student Book). Gather scissors, clear tape or glue, string, and world maps or atlases (optional).

THE GLOBAL 200

WWF's Global 200 map highlights 233 of the richest, most diverse, and most threatened terrestrial, freshwater, and marine ecoregions. Ecoregions are broad habitat types that are tied to specific geographic regions. For example, the *Namib and Karoo Deserts and Shrublands* is an ecoregion in western Africa. And the *Everglades Flooded Grasslands* is an ecoregion in the southeast corner of the United States. The ecoregions highlighted on the map include a broad representation of the world's biodiversity and provide a blueprint for WWF's conservation priorities in the next century. (See the box on page 256 for more about ecoregions, ecosystems, habitats, and major habitat types.)

In creating this map WWF and its partners first looked at which regions support the greatest total species diversity. Areas such as tropical rain forests in South America and the barrier reefs of Costa Rica, Belize, and Australia support an enormous number of species. In fact, tropical rain forests as a whole occupy less than seven percent of the Earth's total land mass but may contain more than half the species on the planet. (See page 264 for more about how biodiversity is distributed on the planet.)

But numbers of species are just one measure of biodiversity. Scientists also gave a high priority to areas that house *endemic* species—those found nowhere else on Earth. Examples of endemic species include the lemurs of Madagascar, the snail kites of the Florida Everglades, and the Cape sunbirds of South Africa.

The scientists working on WWF's map also assessed several other factors, including areas that are essential to migrations, breeding, and other natural phenomena, those areas that are facing severe threats, and those areas that are most likely to be protected.

Finally, using GIS layering, the scientists created a finished map that conservationists can use to make decisions about how to protect biodiversity worldwide. In this part of the activity, students play a game and use the map to learn more about some of the amazing plants and animals that live in different ecoregions around the world.

1. Discuss ways to protect species.

Ask your students if they have any ideas about what scientists do to protect threatened plants and animals. After students give a few suggestions (e.g., enforcing the Endangered Species Act, running captive breeding programs, protecting wildlife habitats as parks and wildlife refuges), explain that most scientists today believe that the most effective way to protect species is to protect the places where they live. That means protecting forests, lakes, deserts, coasts, and other types of habitat so that the greatest number of species and ecological processes are protected.

Show your students the Global 200 map you've posted (or the one on page 122 in the Student Book) and explain that it shows some of the most important and threatened ecoregions around the world. (See introduction at the left for more about the process.) Describe ecoregions according to the information in the box "Talk the Talk" (page 256). You might want to give examples to the students to explain the difference between major habitat types, or biomes, and ecoregions. Explain that in this activity they'll find out more about these amazing—and in many cases, threatened—places.

A Powerful Conservation Tool

Maps are a great way to make complex data easier to understand and analyze. Conservation biologists use maps to show land forms, fluctuations in human populations, patterns of ecological change, species' ranges, and many other ecological phenomena. And in the last two decades maps have become even more valuable thanks to a computer technology called Geographic Information Systems, or GIS.

With GIS, scientists make one map on the computer for each batch of data they collect. For example, they can create a map showing all of the bodies of water in an area, the location of every nesting site of a rare bird species, or the location of every surrounding village. Then GIS allows the scientists to combine the different map layers into a single image so they can analyze how the different pieces of data relate to one another.

One of the most important applications of GIS for conservation biologists is something called gap analysis. After scientists have identified areas especially high in biodiversity, or areas where species or ecological processes are highly threatened, they have to decide where to focus their protection efforts. Using GIS, they can determine which areas are already protected and where there are gaps in protection.

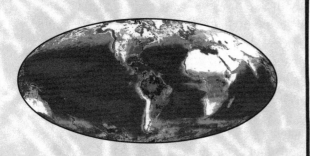

World Wildlife Fund

2. Pass out "Ecoregion Species Cards."

In this activity, there are 60 "Ecoregion Species Cards," which represent 20 ecoregions. This demonstration will be done in rounds to cover all 60 species. To determine how many ecoregions will be explored in each round, count the number of students in your group and divide by three. The number you come up with is the number of ecoregions your group will cover in each round of this activity. If you have 30 students, for example, you will select the ecoregion species cards for 10 ecoregions. Then scramble and pass out the 30 cards (there are 3 species for each ecoregion)—1 to each student. (To learn which species live in which ecoregions, see pages 260–261.)

Explain to the class that each of the cards depicts one species of three that live together in a specific ecoregion. The students' goal is to find the other two members of their ecoregion using the information on their card. Clues as to the identity of one or both of the ecoregion partners are given within each species' description. Once the students have located their two teammates they should identify the ecoregion that all three organisms belong to.

Each species card includes information on the natural history of the animal or plant along with its common name and scientific name. Explain to the students that scientific names are used to distinguish one species from another, and that no two species have the same scientific name. The first word of the scientific name tells which genus the animal or plant belongs to. The second word is the species name. In the "Ecoregion Species Cards," some of the plants and animals represented include more than one species. In these cases, instead of stating a species name, we've listed the genus followed by "spp.," indicating that the animal or plant represents numerous species. (For more information on scientific names, see page 110 in "Sizing Up Species.")

3. Match organisms with ecoregions.

After the students have located their partners, ask them to tape their three species cards to a piece of paper and write the name of their ecoregion on the paper. Ask them to post their paper next to the map and link the ecoregion to its location on the map with a length of string. Explain that the names of the ecoregions are listed by habitat type at the bottom of the map. Each habitat type has a different color that corresponds to shaded areas on the map.

As each group finishes (or if others are waiting for their turn), hand the students an "Ecoregion at-a-Glance" worksheet (on page 121 in the Student Book) to fill out on their own. When all the groups have finished posting their species cards, pass out the next round of cards and repeat the process. If necessary, run a third round to cover all 20 ecoregions.

4. Give presentations.

When all of the ecoregion cards have been posted, have the students do a brief presentation on their ecoregion(s) to the rest of the class. To expedite this process, you might have a spokesperson from each group come to the front of the room, point out and name the ecoregion, and give a few interesting bits of information about it. The "Ecoregion at-a-Glance" worksheet should help the students pick out and organize the information for their presentation.

5. Decipher secret messages.

Have the students form seven groups. Distribute copies of the secret message and bonus question cards to each group. Explain that, for some ecoregions, letters in the text of the species cards are boldfaced and underlined. To find the answers to the secret message questions, the students must first locate these letters in each ecoregion mentioned in their secret message cards. Then by unscrambling the letters they'll come up with the answer. The bonus question uses all the boldfaced and underlined letters on the species cards.

For example, the first secret message card poses the question: "The island of Kauai (part of the Pacific Ocean's Hawaiian Island chain), has more of this than any other place on earth." To find the answer, students will need to search the "Ecoregion Species Cards" from the Namib Desert, the Scandanavian Alpine Tundra, the Klamath-Siskiyou Coniferous Forests, and the Southwest Australian Shrublands and Woodlands for boldfaced, underlined letters. The letters found on these cards are *a, i, n,* and *r* and these letters can be rearranged to find the answer to the question: "rain." (For answers see pages 260–261.)

6. Discuss the importance of ecoregional mapping.

After students have completed the mapping activity, review the purpose of the map and why it's a useful conservation tool. Ask the group to think about how this type of map is different from other maps they've seen. (The concept of using ecoregions as the basis for conservation planning is fairly new.) Today many scientists believe that it's more effective to conserve ecoregions than to focus on protecting individual species or small patches of habitat, and that by conserving the diversity and richness of ecoregions we can protect more species and important ecological processes.

Explain that WWF and many other conservation organizations are working together to protect biodiversity by using an ecoregional approach. They're looking at protecting marine and freshwater ecoregions as well as terrestrial ecoregions. You can ask the students to think about the types of threats that these areas are facing and how maps like this can help the public better understand conservation issues. (*Many ecoregions are threatened by pollution, global climate change, development from housing, road building, farming, and other human activities, as well as an overall lack of habitat protection. Maps like the Global 200 show that there are important ecoregions all around the world and that many are threatened and vulnerable. They also show that marine, freshwater, and terrestrial ecoregions are all important and that many areas are in trouble.*)

sample ecoregion card

Windows on the Wild: Biodiversity Basics

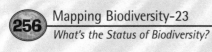
TALK THE TALK

Definitions of Ecoregions, Biomes, and Habitats

Before starting the game in Part I, you might want to help your students learn the "lingo" that conservation biologists use to describe natural areas. The two easiest-to-confuse terms are **ecoregion** and **biome** (or major habitat type). But many people also get confused when they try to explain the difference between habitats and ecoregions. Here are definitions and examples of all three words.

Biome (Major Habitat Type)

Biomes, which are also called major habitat types, are large areas characterized by the types of plants that dominate the area. For example, temperate coniferous forests, and savannas and grasslands are biomes. Coniferous forests are dominated by conifers (trees with needles), and savannas and grasslands are dominated by grasses and small shrubs.

Ecoregion

Ecoregions are geographically distinct areas that are characterized by the types of communities that live there. Areas that make up ecoregions have similar plant and animal species and similar environmental conditions such as climate, terrain, altitude, and soil type. In the example provided below, the Klamath-Siskiyou Coniferous Forest in the Pacific Northwest is one ecoregion in the temperate coniferous forest biome. But it has a certain set of plants and animals and certain environmental conditions that separate it from other temperate coniferous forest ecoregions in other parts of the world. The tallgrass prairie of the Midwest is an example of an ecoregion in the temperate grasslands and savannas. It has a certain set of plants, animals, and environmental conditions that make it unique within the grasslands biome.

Habitat

Habitat refers to a place that provides whatever a species needs to survive, such as food, water, and living space. "Black bear habitat" and "sagebrush habitat" mean places that support those specific organisms. There are many different types of habitats within ecoregions and biomes.

Note: See the background information on page 28 for more on these and other ecological terms.

Biome (Major Habitat Type)
Temperate Coniferous Forest

Ecoregion
Klamath-Siskiyou Coniferous Forest

World Wildlife Fund

Before You Begin • Part II

Make copies of the "Key to Map Symbols" (pages 139–140 in the Student Book) and the "Wild Hills County Gap Analysis" worksheet (page 141 in the Student Book) for each team of three students. Gather plain white paper for students to use for base maps, three transparencies (the same size as the plain white paper) for each group of three students, and colored transparency markers (four colors per group).

What to Do • Part II

GAP ANALYSIS

This part of the activity introduces your students to the work of conservation biologists and how they use a process called *gap analysis* to figure out if we're protecting the areas that support the most important, diverse, and unique examples of biodiversity. By making their own maps of a fictional county called Wild Hills, the students will be able to see how gap analysis can be a useful tool in making local and national conservation decisions. You might also want to check with a local conservation organization, state department of natural resources, or university to see if your students can get an introduction to GIS (page 254) and observe firsthand how conservation biologists, geographers, demographers, and other scientists use GIS to organize data, make conservation decisions, and overlay science and social science information to plan for the future.

1. Introduce gap analysis.

Hang the Global 200 map where everyone can see it. Then ask for volunteers to explain how the priority ecoregions were selected *(number of different kinds of species, the number of endemic species [species that live in only a small geographic area], the level of threat facing each ecoregion, and the number and types of outstanding ecological phenomena and processes that occur in each area)*. Explain that many of the areas shown on the map are protected as parks, wildlife refuges, and reserves and through a

variety of other local, state, national, and international efforts. But others are not protected. To find out which areas still need to be protected, scientists conduct an evaluation called a gap analysis. Explain that in this activity the students will conduct their own gap analysis to determine how much land they need to protect in an area called Wild Hills County.

2. Read the scenario.

Tell your students that the following scenario sets up the work they'll do in this activity.

You live in a place called Wild Hills County, known for having some of the last intact wetland and forest ecosystems in the state. Some extremely rare plants and animals also live in your county. But lately, a boom in housing development has made some people concerned that rare species and ecosystems in Wild Hills County may be threatened.

WWF scientists working in the GIS lab

Matt Finarelli/WWF

Protecting Biodiversity Worldwide and in the United States

Around the world, one of the ways that countries protect biodiversity is by setting aside land as parks, wilderness areas, national forests, wildlife refuges, and other specially designated areas. Some countries, like Bhutan, Ecuador, Venezuela, Austria, and Denmark, have set aside more than 20 percent of their land for protection. Others, like Cameroon, El Salvador, Jamaica, Uruguay, and Lebanon, have set aside less than 1 percent. All together, less than 5 percent of the world's total land mass is protected.

How does the United States stack up? About 13 percent of the land in the United States is under some form of federal, state, or local protection. These protected areas are managed by a variety of government agencies, each of which has different restrictions on the kinds of activities that can take place on the land. Most of our protected lands fall within national parks, national forests, national wildlife refuges, national seashores, national recreation areas, military bases, and other designations. Some land is also protected by state and local governments. However, a big part of our country's biodiversity is in the hands of private landowners. For example, private citizens own and manage about 59 percent of the nation's forests. Many conservation biologists believe that, to protect more of our terrestrial, freshwater, and marine ecosystems, we need to better educate private landowners and private companies so that they can do a better job of protecting the country's biological wealth.

World Wildlife Fund

A local conservation organization called The Diversity Trust has just received a large donation from a member interested in helping to protect the county's natural biodiversity. The Diversity Trust plans to use this money to purchase land, protect it from development, and manage it carefully to sustain the diversity of species and ecosystems. But first the trust has to find out where the rare species and ecosystems are located and which species and ecosystems are already protected by existing parks and reserves.

3. Organize the class to conduct a gap analysis.

Divide the class into teams of three. Explain that each team has been hired by The Diversity Trust to conduct a gap analysis of Wild Hills County. Distribute to each team a plain sheet of white paper, three transparency sheets, four colors of transparency markers, and the "Key to Map Symbols—Step 1" (page 139 in the Student Book).

4. Create a base map.

Ask each team to work together to create its own map of Wild Hills County on the plain white sheet of paper. Have the team use the whole sheet of paper and assume the county is rectangular. The teams' maps should contain one to three items from each of the following categories: (1) mountains or hills, (2) lakes, rivers, or streams, and (3) areas developed by people (cities, towns, farms, roads, and so on). It works best if your students use blue ink or another dark color for their base maps. (See page 259 for a sample base map.)

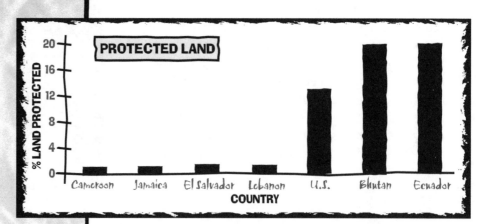

5. Draw map layers.

Now explain that each member of the team will draw his or her own transparency—one member should show where ecosystems of concern are located, another where rare species live, and another where protected areas exist.

Hand out a copy of "Key to Map Symbols—Step 2" and a different colored marker to each team member. Have the students take turns using the base map. One by one, they should place the transparency over the base map and draw in the symbols they want. When one student finishes, he or she should pass the base map to the next student without sharing what he or she drew. (See page 261 for sample overlays.)

Here's a summary of what each person should draw (see symbols on pages 139–140 in the Student Book)

#1 Ecosystems of Special Concern (The student can choose from ancient forests or wetlands but should limit the number of symbols from three to five of each.)

#2 Rare Species (The student can pick from plants, land mammals, aquatic animals, or birds but should limit the number of individual species to 10.)

#3 Protected Areas (The student can choose from parks, reserves, or recreational areas but should limit the number of areas to five or fewer, and each should be about the size of quarter.)

6. Combine the layers and fill out the worksheet.

After all the students have finished, give each team a copy of the "Wild Hills County Gap Analysis" sheet (page 141 in the Student Book). Explain that the members of each team should place all three transparencies on top of their base map and discuss the patterns they see. Then have each team work together to fill out the worksheet.

7. Discuss the results.

Discuss the questions on the worksheet as a group. (See possible answers at the right.) You might also want to invite a speaker who uses GIS or has worked on a local gap analysis project to come in to talk to

your students. He or she can explain how this process is being used locally and around the world to help identify areas that need more protection.

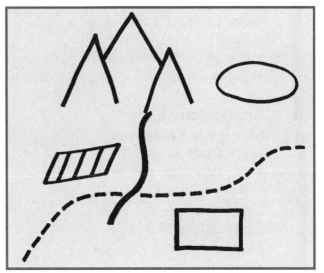

base map (see page 261 for examples of overlays)

Possible Worksheet Answers:
(for questions see page 141 in the Student Book)

1. *Answers will vary.*

2. *Possible threats include development (homes, resorts, shopping centers, office parks, roads), logging activities, expansion of agriculture, collection or harvest of rare species, pollution, and traffic.*

3. *Answers could include (a) enlarging some of the parks and protected areas; (b) maintaining corridors between protected areas—especially for carnivores and other large mammals; (c) setting aside buffers, or border areas, around lakes, rivers, and streams to prevent erosion and reduce the amount of pesticides and other types of pollution that might run into them; and (d) setting aside new parks to protect areas with large numbers of species.*

4. *Other ways to protect species and ecosystems include pollution control, public education, restricting hunting and mining, and restricting certain types of development and road building.*

5. *Answers will vary.*

ANSWERS—ECOREGION SPECIES CARDS

Ecoregion #25
Southern Congo Basin Forests
- Okapi
- Bonobo
- Salongo Monkey

Ecoregion #57
Madagascar Dry Forests
- Giant Baobab
- Sicklebilled Vanga
- Panther Chameleon

Ecoregion #64
Mexican Pine-Oak Forests
- Monarch Butterfly
- Imperial Woodpecker
- Wild Maize

Ecoregion #66
Klamath-Siskiyou Coniferous Forests
- Cobra Lily
- Western Azalea
- Siskiyou Mountains Salamander

Ecoregion #78
Central/Southwest China Temperate Forests
- Fountain Bamboo
- Giant Panda
- Golden Pheasant

Ecoregion #85
Central and Eastern Siberian Boreal Forests and Taiga
- Wolverine
- Siberian Spruce
- Long-Eared Owl

Ecoregion #92
Scandinavian Alpine Tundra and Taiga
- Norway Lemming
- Reindeer
- Arctic Fox

Ecoregion #94
Tallgrass Prairies
- Prairie Dog-Tooth Violet
- Greater Prairie Chicken
- American Bison

Ecoregion #105
Terai-Duar Savannas and Grasslands
- One-Horned Rhinoceros
- Chital
- Tiger

Ecoregion #110
Everglades Flooded Grasslands
- Snail Kite
- Florida Tree Snail
- Red Bay

Ecoregion #114
North Andean Paramo
- Frailejón
- Andean Condor
- Vicuña

Ecoregion #125
Namib and Karoo Deserts and Shrublands
- Welwitschia Plant
- Darkling Beetle
- Sand-Diving Lizard

Ecoregion #136
Southwest Australian Shrublands and Woodlands
- Drosera Sundew
- Numbat
- Brown Goshawk

Ecoregion #154
Eastern Australian Rivers and Streams
- Platypus
- Freckled Duck
- Murray Cod

ANSWERS (Cont'd.)

Ecoregion #156
Varzea and Igapó Freshwater Ecosystems
- Amazon Dolphin
- Arawana Fish
- Black Caiman

Ecoregion #166
Lake Baikal
- Nerpa
- Golomyanka
- Gammarid Shrimp

Ecoregion #183
Danube River Delta
- White Stork
- Black Stork
- Corncrake

Ecoregion #195
Red Sea Marine Ecosystem
- Whitetip Reef Shark
- Giant Clam
- Green Turtle

Ecoregion #213
Icelandic and Celtic Marine Ecosystems
- Puffin
- Narwhal
- Gray Seal

Ecoregion #231
Bering and Beaufort Seas
- Walrus
- Polar Bear
- Bowhead Whale

Sample Overlays for Base Map

overlay #1—ecosystems of special concern

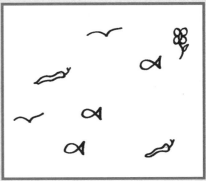

overlay #2— rare species

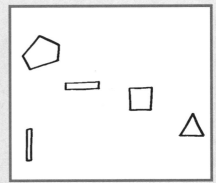

overlay #3—protected areas

base map with three overlays

ANSWERS–SECRET MESSAGE CARDS

The island of Kauai (part of the Pacific Ocean's Hawaiian Island chain)
has more of this than any other place on Earth.

 R   A   I   N

To find the answer, look in the Namib Desert, the Scandinavian Alpine Tundra, the
Klamath-Siskiyou Coniferous Forests, and the Southwest Australian Shrublands and Woodlands.

In the deciduous forests of the eastern United States, the elimination of large predators and
the increase of edge habitat has led to a dramatic increase in the population of these animals.

 D   E   E   R

To find the answer, look in the Madagascar Dry Forests, the Tallgrass Prairies of the United States,
the Mexican Pine-Oak Forests, and the Central and Eastern Siberian Boreal Forests.

In many areas of the world, the establishment of one of these can lead to vast amounts of
pollution as chemicals and sediments wash into nearby watersheds.

 M   I   N   E

To find the answer, look in the North Andean Paramo, the Southern Congo Basin Forests,
the Klamath-Siskiyou Coniferous Forests, and the Everglades Flooded Grasslands.

A portion of soil the same size as this object might contain millions of
microorganisms including fungi, bacteria, and tiny animals.

 D   I   M   E

To find the answer, look in the Klamath Siskiyou-Coniferous Forests, the Madagascar
Dry Forests, the Southern Congo Basin Forests, and the Mexican Pine-Oak Forests.

This phenomenon is created by the gravitational pull of the sun and moon on planet
Earth and is an important force that constantly changes the shape of the land's surface.

 T   I   D   E

To find the answer, go to the Mexican Pine-Oak Forests, the Central/Southwest China Temperate
Forests, the Klamath-Siskiyou Coniferous Forests, and the Central and Eastern Siberian Boreal Forests.

In the summer, you may find one of these normally seafaring birds hundreds of miles from the
ocean, breeding in vast colonies in places such as the high plateaus of Tibet in central Asia.

 T   E   R   N

To find the answer, head to the Scandinavian Alpine Tundra, the North Andean Paramo, the Central/
Southwest China Temperate Forests, and the Southwest Australian Shrublands and Woodlands.

When people do this to wetlands to build houses or create agricultural land, many valuable services the wetlands provide (such as flood-control and serving as a nursery for sealife) are lost.

<u>D</u> <u>R</u> <u>A</u> <u>I</u> <u>N</u>

To find the answer, head to the Klamath-Siskiyou Forests, the Tallgrass Prairies of the United States, the Terai-Duar Savannas and Grasslands, the Mexican Pine-Oak Forests, and the Everglades Flooded Grasslands.

Bonus Question

This habitat type holds 20 percent of the Earth's plant species, and every ecoregion within it is in critical danger of becoming extinct. What is it?
(The answer uses all the letters highlighted in the Ecoregion Species Cards.)

<u>M</u> <u>E</u> <u>D</u> <u>I</u> <u>T</u> <u>E</u> <u>R</u> <u>R</u> <u>A</u> <u>N</u> <u>E</u> <u>A</u> <u>N</u>

How many ecoregions are there within this habitat type?
Five—and all of them are critical priorities for conservation.
Does this habitat type exist in the United States? If so, where?
Yes—the California Chaparral and Woodlands in Southern California.

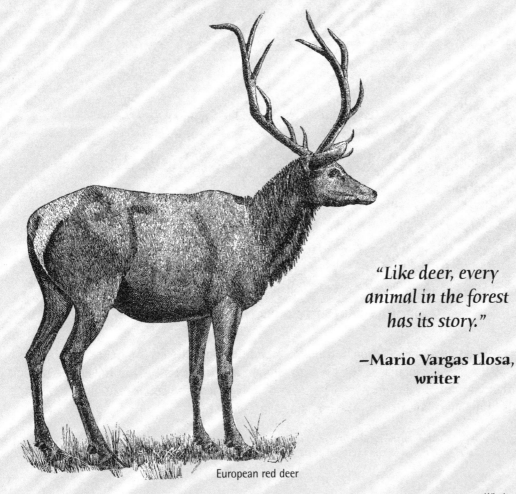

"Like deer, every animal in the forest has its story."

–Mario Vargas Llosa, writer

European red deer

WHERE ON EARTH IS BIODIVERSITY FOUND?

While almost every environment on Earth supports life, certain areas have more biodiversity than others. In general, the diversity of species tends to increase from the poles to the equator. The tropical regions, with their rain forests, coral reefs, and large tropical lakes, have an average of twice as many mammal species per square mile as temperate areas. And flowering plants are even more abundant in the tropics: A tropical forest in the Amazon basin, for example, might have over 80 species per acre, while the same amount of temperate forest in the United States would typically have less than 12. Amazingly, more than half the world's species live in tropical forests, which make up less than 7 percent of the planet's land area.

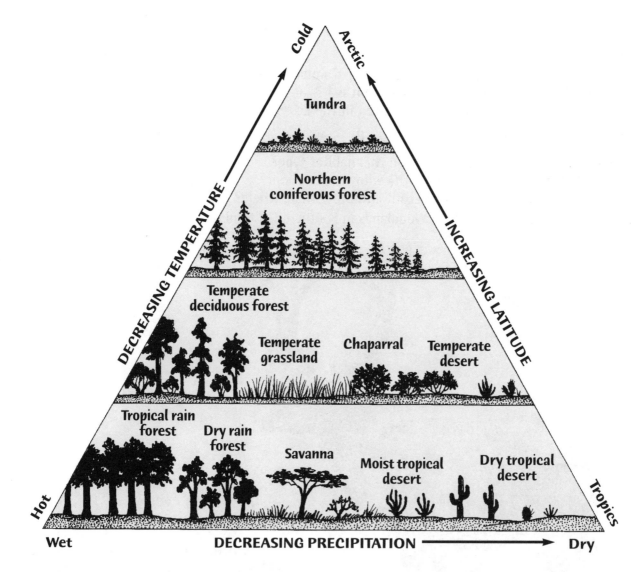

This simplified diagram shows how precipitation, latitude, temperature, and the Earth's major land biomes are related. For example, as the latitude increases, the average temperature decreases. If you follow the triangle, you'll see biomes shift from the hot rain forests, savannas, and deserts near the Equator to more temperate forests and grasslands in the mid-latitudes to cold, icy tundra in the Arctic.

Diagram adapted from *Environment (Second Edition)* by Peter H. Raven, Linda R. Berg, and George B. Johnson (Saunders College Publishing, 1998); and *Environmental Science: Earth as a Living Planet (Second Edition)* by Daniel B. Botkin and Edward A. Keller (John Wiley and Sons, Inc., 1998).

WRAPPING IT UP

Assessment

This activity can be assessed in two steps. Step One has two parts: (1) using the Global 200 map, have students identify three ecoregions and the major habitat type in which the ecoregions are found; and (2) using examples from their maps and the habitat types they've identified, have students individually or in small teams work to develop a definition for ecoregion.

Step Two also has two parts: (1) have students select one ecoregion and identify three different types of information they could use that would help them prioritize conservation efforts for that region; and (2) have students list three strategies they would pursue to protect their selected ecoregion.

Unsatisfactory—The student is unable to complete at least one component of each step of the assessment.

Satisfactory—The student completes at least one component of each step of the assessment.

Excellent—The student completes both components of each step of the assessment and is able to list more than three strategies for protecting his or her selected ecoregion.

Portfolio

The map and individual definitions of *ecoregion* can serve as portfolio evidence.

Writing Idea

Have students write species cards for three wild animal or plant species in your area. They should include a drawing or a photograph.

Extensions

- Conduct a gap analysis for an ecosystem of concern in your area, and propose a conservation plan for its protection. Obtain maps for your county (try county government offices or planning commissions). Contact the U.S. Fish and Wildlife Service for maps of wetlands, forests, or other ecosystems of concern, as well as for the locations of threatened or endangered species. Using tracing paper or overhead transparencies, trace the locations of relevant items or areas such as protected areas, housing developments, specific animal habitats, and so on. (Different scales may add to the complexity of this project.)
- Contact a local office of the Natural Heritage Network or the U.S. Fish and Wildlife Service to see if someone would be willing to speak to your students about local gap analyses (if they have been performed) or strategies for protection.

Resources

Biodiversity in the Balance: Approaches to Setting Geographic Conservation Priorities
by Nels Johnson (World Wildlife Fund, Biodiversity Support Program, 1995).

A Conservation Assessment of the Terrestrial Ecoregions of Latin America and the Caribbean by
Eric Dinerstein, David M. Olson, Douglas J. Graham, Avis J. Webster, Steven A. Primm,
Marnie P. Bookbinder, and George Ledec (The World Bank, 1995).

Conservation Design for Subdivisions: A Practical Guide to Creating Open Space Networks
by Randall Arendt and Holly Harper (Island Press, 1996).

Endangered Ecosystems: A Status Report on America's Vanishing Habitat and Wildlife
by Reed F. Noss and Robert L. Peters (Defenders of Wildlife, 1995).

Global 200: A Blueprint for Saving Life on Earth (poster) by World Wildlife Fund, Marketing
Membership, and Communications Department, 1250 24th St., NW, Washington, DC 20037.

AT A GLANCE

Read an article about re-introducing wolves to Yellowstone National Park as you learn about new thinking in natural resources economics.

OBJECTIVES

Discuss economic and non-economic values of natural resources and how personal economic values about natural resources influence our decisions.

How much do you value safe drinking water? How much would you be willing to pay to see a clear view of the Grand Canyon? What is it worth to you to know that there are tigers alive in the wild?

These are the kinds of difficult questions being asked today by a growing number of economists in the field of natural resource economics. As you can imagine, it's tricky and quite controversial to try to find a specific dollar amount for each of the above circumstances. But many people think we need to try. Why? Because many environmental decisions today—about how to grow food, how to use a forest, how to protect the lives of animals—are made on the basis of economics. And many people believe that plants, animals, and resources typically haven't had an assigned value that represents their true worth. So a group of progressive economists has taken a unique approach. They're trying to determine dollar values for such hard-to-quantify things as scenic views and the existence of endangered species.

While you'll scare away most kids (and most adults) if you start discussing economics in all its complicated lingo, giving them a jargon-free introduction to some basic economic concepts can grab their attention. Most kids understand that we use price to denote the value of something. So they should be ready to learn why we might need to give natural resources their own price tags and why doing that can be difficult. This activity provides a user-friendly introduction to valuing natural resources.

Before You Begin

Make copies of "How Much Is a Gray Wolf Worth?" and "Computing the Costs of Wolf Reintroduction" (pages 142–144 in the Student Book). You'll need one copy for each student.

What to Do

1. Discuss how things are priced.

Have the students imagine they've gone to a store that sells two kinds of basketball shoes. One pair costs $40, and the other costs $135. What reasons can they come up with that might explain the difference in the two prices? *(The first pair may be made from cheaper materials or its craftsmanship may be poor; the second pair may have higher labor, advertising, and transportation costs, it may be popular but hard to find, or it may have symbolic status—such as being the same kind of shoe a famous athlete wears, and so on.)* Fill in any gaps in the students' ideas. Point out how the prices for things represent many values from quality to trendiness.

2. Begin the tree exercise.

Tell the students that they're going to determine the value of something very different from a pair of sneakers: a half acre of 60-year-old oak trees. Read them the following scenario:

> *You live in a house bordered on one side by half an acre of oak trees. You understand that the wood in the trees is worth about $8,000 at the local mill. One day a woman approaches you and offers to pay you $12,000 if you'll let her harvest the trees. She'll also give you enough tree seedlings to replant the forest. Would you accept her offer?*

Ask for a show of hands: Who would take the money? Who wouldn't? Ask for some explanations from the group for their answers.

3. Read additional pieces of information.

Now tell the students you're going to read some additional pieces of information about the trees, and ask them again if they would accept the offer. After reading each of the questions below, ask for a show of hands and record the number of "yeas" and "nays" on the board.

- What if the oak trees had been planted by your great-grandfather?

- What if you and other neighborhood kids love to play in the woods?

- What if you were told that, left standing, the oak trees would provide a hundred or more years of ecological benefits? For example, they would provide shelter and food for birds and other small animals, shade, oxygen, soil stabilization, and so on. And simply planting new seedlings would not replace these benefits for many years.

4. Review the idea of "valuation."

Explain to the students that in all of the above examples, some value has been added to the stand of trees that would, for many people, increase its price. By how much? That would vary from person to person. But it's clear that a person's decision to sell off the trees would be significantly affected by the type and number of values that are attached to them as well as the person's financial situation. Can the students think of any cases when it might be useful to find the value of a natural resource—not just a stand of tress, but, say, a river? A mountain view? A species of wild animal? After soliciting some ideas, hand out "How Much Is a Gray Wolf Worth?" and "Computing the Costs of Wolf Reintroduction" (pages 142–144 in the Student Book) and have the students read it for homework.

©Erwin and Peggy Bauer/Bruce Coleman Inc.

5. Review and discuss the article.

Use the following questions to review and discuss the article with your students.

- What are some of the positive and negative effects of reintroducing wolves?

  (Positive effects could include increased income from tourists for the park and surrounding businesses; better balance between predators, prey, and vegetation within the park ecosystem with the return of a major native predator; and the value of just knowing the wolves are there. Negative effects could include loss of livestock, increased cost of providing services such as safety, and maintaining public areas for a greater number of people.)

- If you chose to reintroduce wolves, from which groups would you expect resistance or opposition and how would you deal with it?

  (Potential opposition could come from ranchers, animal rights advocates, or local communities that might not want increased numbers of people passing through. Possible ways of dealing with the opposition could include educating people about the ecological importance of reintroducing wolves and/or providing reimbursement to the groups that suffer economic losses as a result. Make sure your students understand that any successful reintroduction program should include a way for citizens to be involved in making decisions.)

- If you were the head park ranger, would you have introduced wolves? Why or why not? What additional information might you need to make the decision?

  (Answers will vary.)

WRAPPING IT UP

Assessment

Have students identify a natural resource they value. Then have them develop a chart that identifies considerations they could use to place a dollar value on the resource. Have students also compare how much they value that particular resource in comparison with other resources.

Unsatisfactory—The student does not complete the exercise or does not present logical and sound arguments for his or her position.

Satisfactory—The student is able to present sound arguments and rational reasons for his or her valuations.

Excellent—The student is able to compare how he or she valued the selected resource with other resources.

Portfolio

Have the students write down their answers to the discussion questions for the story. The answers can be placed in their portfolio.

Writing Idea

Have students write a persuasive argument for or against the idea of attaching a dollar amount to natural resources such as rivers, wild animals, and rain forests.

Extensions

- Simulate a public meeting with your students. Suppose that the park ranger is trying to make a decision about whether to request funding from the National Park Service for a wolf reintroduction program and wants to hear different arguments for and against the proposal. The following interest groups have assembled to participate in the meeting: ranchers, animal rights advocates, environmental groups, the local chamber of commerce, and conservation biologists. Divide your students into teams to represent each interest group. Have teams research their positions and make short presentations at the public meeting. Then conduct a debate. Emphasize that not all groups think alike—for example some ranchers might favor the reintroduction and some environmental groups might take different positions on the reintroduction.

- Have the students put a "price tag" on a species in your state or community. Ask them to think about all the ways it might benefit or not benefit people economically and then to come up with a dollar value.

SUBJECTS

social studies (economics, ethics, geography), science, mathematics

SKILLS

gathering (simulating), analyzing (comparing and contrasting, identifying patterns), interpreting (relating, identifying cause and effect), applying (decision making, restructuring, proposing solutions), citizenship (working in a group)

FRAMEWORK LINKS

39, 45, 51, 59, 65

VOCABULARY

ambassador, foreign aid, gross national product (GNP), immigration, per capita, resource distribution

TIME

one to two sessions

MATERIALS

188 ft. of yarn or string, copy of "Ambassador's Cards" (pages 145–146 in the Student Book), five large placards for signs, approximately 140 peanuts in shells, 160 individually wrapped candies, 100 toothpicks, 15 self-sealing clear plastic sandwich bags, labels for bags

CONNECTIONS

Before doing this activity, you might want to try "The Spice of Life" (pages 206–213) to get your students thinking about why biodiversity is important. After exploring some of the social and economic issues affecting biodiversity, use "Future Worlds" (pages 324–329) to help your students think about their future and what's important to them.

AT A GLANCE

Play an interactive game to compare and contrast how population density, distribution, and resource use affect biodiversity.

OBJECTIVES

Explain how the distribution of population and resources varies among different regions of the world, and how population size and growth combine with resource distribution and use to affect biodiversity.

e know that an increasing human population affects the environment. Having more people means more demand for resources, especially food, water, and energy. The number of people already in the world (5.92 billion as of 1998) and the population growth predicted over the next 25 years can be mind boggling. (The world population is projected to be 7.7 billion people by 2020.) But calculating the effects of human population on the environment isn't as simple as looking only at numbers. How the world's population is distributed geographically and how people among different geographic regions use resources also have a big effect on the environment. This activity will give your students a global perspective on population distribution, resource use, and the effects of both on the environment. (Refer to "Population Pointers" on page 275 for definitions and information about how population growth affects the quality of life.)

This activity is adapted from *People and the Planet: Lessons for a Sustainable Future* (Zero Population Growth, 1996). Visit ZPG's Web site at <www.zpg.org>.

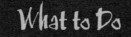
Before You Begin

You'll need to gather and organize materials and locate a large space (such as a gym or other open area) for this activity. Although a large, open space works best, if you don't have access to one, simply move the desks to create one in your classroom.

Measure the yarn or the string in the lengths indicated on the "Distribution Chart" (page 276) for each region. Tie the two ends of each length of yarn together to form a circle, and mark each circle in some way so you'll know which region it represents. Wind each circle into a bundle to keep it from getting tangled. When you are ready to do the activity, unroll the yarn and arrange the circles on the floor to represent each world region. (The regions in this simulation are those defined by the United Nations, so Mexico is included in Latin America rather than North America. The sixth world region, Oceania, is not included because its population is so small relative to the others that it cannot be accurately represented in this simulation.)

Make a large placard for each of the world regions (North America, Latin America, Europe, Asia, and Africa) and place the placard inside the boundary of its region when you are ready to start. You'll also need to make one copy of the "Ambassador's Cards" (pages 145–146 in the Student Book) and cut them into sections.

Count out the toothpicks that will represent energy consumption for each region. Put them in individual bags or bundles using the information from the chart (page 276). Label each bag with the region, number of toothpicks, and energy consumption (for example, North America—Energy Consumption: 58 barrels/toothpicks). Count out and bag the peanuts, labeling the bags as above. The peanuts represent protein intake for each region. Do the same for the candy, which represents the wealth for each region.

Note: This activity works best with about 58 students, but it can be done with a smaller group. The "Distribution Chart" (page 276) contains information for a group of 58 and a group of 29.

What to Do

1. Appoint ambassadors.

Explain that this activity illustrates how differences in population distribution and resource use among five world regions combine to affect the quality of life for the people who live in each region. Ask for five volunteers to be ambassadors—one from each region. Give each ambassador the appropriate "Ambassador's Card," and ask the ambassadors to stand inside their region's boundary with the placard prominently placed.

2. Divide the students into each of the regions.

Explain that your students need to populate each region with the appropriate number of people. If you are playing with 58 participants, each player represents about 100 million people. (With a group of 29, each player represents 200 million people.)

Using the "Distribution Chart" (page 276), appoint each student to a region and ask the student to stand inside the region's boundaries. (You could also pass out slips of paper with a region's name and have students go to the region named on their paper.)

If you don't have enough people, you can use chairs as "citizens." If you have a few too many people, distribute them to Asia and, if necessary, one each to Europe and Africa.

3. Look at the population and land area around the world.

Once your students populate the regions, it will be apparent which regions have the largest populations and which have the highest population densities. Ask your students to look around and name the regions with the highest population densities.

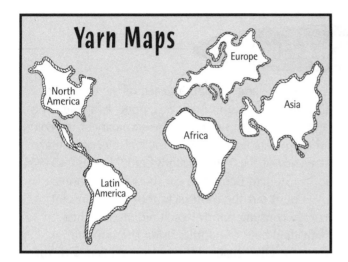

Yarn Maps

Now ask the ambassadors to look on the "Ambassador's Cards" to find out how many years it will take for their region's population to double. Have each ambassador announce their respective doubling time to the class. Then ask students to describe what it might be like to have a doubled population within the same size border of yarn. Where would everyone live? Would there be enough food to go around? Would more or less energy be needed? Where would it come from? How might nonhuman species and the environmental systems that support them be affected? (Of course, your students won't know exactly but the questions will help them think about issues.)

4. Look at energy consumption around the world.

Hold up the bags of toothpicks and explain that they represent energy use, such as electricity and fuel used for heating homes, running factories, and operating vehicles. While energy is provided in many ways, including wood, coal, natural gas, and nuclear power, in this activity all these sources have been combined and are expressed in terms of barrels of oil (that is, each toothpick represents one barrel of oil). The bags of toothpicks represent the average amount of energy used per person per year in each region. For example, on average each European uses 26 barrels of oil in a year, so Europe gets a bag with 26 toothpicks in it.

Pass out the appropriate bag to each ambassador, and ask the ambassadors to hold up their bag and announce the number of toothpicks

they have. Then ask students: "In which region does the average person use the most energy? How does this compare to other regions?" Discuss with your students how, even though North America's population is smaller than that of other regions (and it is growing more slowly than most), the average North American uses far more energy than does the average person in most other regions of the world. Tell the ambassadors that they can distribute the energy (toothpicks) however they see fit.

5. Look at protein consumption around the world.

Explain that the basic protein requirement of each person in this simulation is about 100 grams of protein a day and that three peanuts represent 100 grams of protein. If you obtain fewer than three peanuts, you are not getting sufficient protein in your daily diet. Then pass out the bags of peanuts to the appropriate regions, and ask the ambassadors to distribute the bags as they see fit.

At this stage of the activity, interesting things may occur. Some students, realizing they are being "shortchanged," might demand a fairer distribution system. Others might, without warning, simply jump into a circle containing more resources. Others might offer to share what they have. Still others might threaten to forcibly remove resources either within their own or from adjacent circles.

Be prepared for such student reactions and, as they occur, call them to the group's attention. You might even shout "Freeze!" and have students observe and describe what they see happening. Ask them if they can describe what they see their peers doing in "real world" terms. Are they emigrating? Are they being charitable (offering foreign aid)? Are they starting a revolution? A war?

6. Look at the distribution of wealth around the world.

Now hold up the bags of candy and explain that they represent each region's gross national product (GNP), which is the value of all goods and services produced by a country's citizens in one year. The bags represent the amount of money each citizen would get if that region's GNP were distributed equally among all people.

Each candy represents $300. So Africa, with its per capita GNP of $660, gets two candies. Distribute the plastic bags of candy to the ambassadors, asking them not to open the bags yet. Then have the ambassadors hold up their bags so everyone can see the relative wealth per person in the different regions. Remind the students that the bag represents one person's share of the wealth in each region.

Have the ambassadors distribute the candy as they'd like. As in step 5, a variety of responses might occur.

7. Discuss the activity.

Use student responses to questions such as those in step 5 to find out how students felt about being citizens of a particular region. Other questions might include the following:

- How do the people in Asia, Africa, and Latin America feel about getting only a few candies, while the North Americans get 90 per person? How did you respond? What would you like to do about the situation? *(Students might suggest moving to North America or getting foreign aid from countries in the northern hemisphere.)*

- In what ways do those things happen in the real world? Have students recall and discuss some of their reactions. *(Many people immigrate or want to immigrate to North America in search of opportunity. Sometimes they find it, but many times people also experience poverty here. Some countries receive foreign aid. Providing food, medicine, and money during famines is one obvious example. But governments also provide regular foreign aid to some countries. Although this aid is helpful, it is not enough to meet all people's needs, and it can also be challenging to find ways to make sure the aid is getting to the people who really need it. In addition, aid can disempower people who want to make it on their own. In some countries, governments work very hard to improve the domestic situation by supporting new industries, restructuring the government, or providing more support for social programs.)*

- How did those students who had a lot feel when others had only a little? How did they respond? *(Students might have found it uncomfortable to have a lot of resources when others had little, and they might have shared their "wealth." Others might have felt bad for those with less but still have wanted to protect the resources they had.)*

HANEL
FRANKFURTER ALLGEMEINE
Frankfurt
GERMANY

Hanel. ©Cartoonists & Writers Syndicate

- What would happen if the regions possessing the smallest amounts of the resources distributed in this activity experienced some environmental catastrophe, such as increased pollution or a large-scale natural disaster? *(Many people would die. As above, some people might receive foreign aid.)*

8. Link the activity to biodiversity.

In small groups or as a class discuss the following questions:

- Based on this activity, what are some ways that the human population can affect biodiversity? *(As the human population expands, there is more pressure to develop land and to extract resources, which destroy natural habitat. Having more*

SORTING OUT TOUGH ISSUES

This activity can raise some very complex and controversial issues regarding family size, birth control, religion, culture, poverty, and resource use. We encourage you to work with other educators in the community to discuss the best ways to address the links between population, resource use, and biodiversity, and to have local experts come in to talk about specific issues that you believe are most relevant for your students. For example, it's important that students explore some of the reasons why population growth is so much higher in some areas than others. (It's linked to health care, education, poverty, religion, and many other factors.) Your students could investigate the growing research on the importance of education in reducing birth rates throughout the world and on the links between resource use and population growth. We also encourage you to use the resources developed by a variety of groups, including the National Audubon Society, the Population Reference Bureau, Zero Population Growth, World Resources Institute, the Centre for Development and Population Activities, the United Nations, and other conservation and development organizations to get more ideas about how to integrate population and environment into your teaching (see page 277 for additional resources).

people also means more demand for food, energy, and other products. Increased use of energy and fossil fuels means more pollution and possible changes in climate because of temperature increases. Increased demand for food will require the protection of arable lands for agriculture. You might want to compare your students' ideas to the HIPPO Dilemma [page 48].)

- What are some of the social issues that can result from an unequal distribution of resources? *(Oppressive living conditions can occur when resources aren't available or don't reach certain groups within a country. Social unrest, conflict, and war can occur as the people with resources try to protect them and those people without the resources try to get access to them. Immigration pressures, stealing, and demand for foreign aid can also arise.)*

- In what ways can the unequal distribution of resources affect biodiversity? *(People need to have their basic needs met—food, water, and energy to cook and heat with. When many people don't have access to plentiful food, water, and energy, they often have no alternative but to cut down local forests for firewood or for land to grow crops. They also might have little choice but to overharvest certain plants or animals for food because they don't have other resources to draw upon. These situations may lead to loss of habitat and population declines of certain species— even outright extinction. In addition, people living in poverty often don't have the resources needed to catalyze action on their behalf. Environmental experts recognize that because poverty and uneven resource distribution contribute to many environmental problems, both locally and globally, protecting biodiversity means working to meet people's basic needs.)*

- What knowledge and skills would you like to learn to help address these issues in the future? *(Answers might include conflict resolution, economics, geography, and so on.)*

- What is the relationship between environmental justice, health, and poverty? *(In many parts of the world, economically deprived and under–represented communities suffer from the worst pollution and lack of green space, clean water, health care, and educational opportunities.)*

POPULATION POINTERS

Population Definitions:

Population: the number of people living in a region

Birth Rate: the number of births per 1,000 people per year

Death Rate: the number of deaths per 1,000 people per year

Rate of Natural Increase: growth caused by having more births than deaths in a year (does not include immigration or emigration)

Per Capita Energy Consumption: the total amount of energy used by each region per year divided by the population of that region (includes industrial use)

Population Growth Rates and Quality of Life

- A population grows whenever its birth rate is higher than its death rate.

- The growth rate is determined by the size of the difference between the birth and death rates. The closer these rates are, the lower the growth rate.

- Where birth and death rates are equal, the population's growth rate is zero.

- The world's current birth rate is over 2.5 times its death rate.[1]

- The world's women bear an average of 3.0 children.[2]

- The average human life expectancy at birth is 66 years[1] and is expected to be 73 years by 2025.[2]

- About half of the world's population now lives in urban areas.[1]

- By 2025, the number of the world's people living in urban areas is expected to be almost double the current number and to reach 5 billion.[3] (About 75% of the population of the developed world already lives in urban areas.[1])

- The 20% of people living in the most developed countries account for 86% of the world's consumption.[4]

- About 97% of the world's annual population increase is in developing countries.[5]

[1] Population Reference Bureau. "1998 World Population Data Sheet." <www.prb.org/pubs/wpds98/wpds98.htm>.

[2] World Health Organizaiton. "Fifty Facts from the World Health Report 1998."<www.who.org/whr/1998/factse.htm>.

[3] Roberts, Leslie, ed. 1998. *World Resources 1998-99.* New York: Oxford University Press, p. 39.

[4] United Nations Development Programme. "Overview of the Human Development Report 1998." <www.undp.org/undp/hdro/e98over.htm>.

[5] United Nations Population Fund. "United Nations Population Fund Moves Day of Six Billion Based on New Population Estimates" press release. <www.unfpa.org/NEWS/RELEASES/D6bnew.htm>.

DISTRIBUTION CHART

	North America	Latin America	Europe	Asia	Africa
Land Area (sq. miles)	7,099,448	7,922,197	8,782,320	11,927,208	11,443,826
Length of Yarn	28 feet	32 feet	35 feet	48 feet	45 feet
1998 Population (with 58 players) (with 29 players)	301 million 3 students 1 student	500 5 students 2 students	728 7 students 4 students	3,604 36 students 18 students	763 7 students 4 students
Per Capita Energy Consumption (1 toothpick = 1 barrel of oil)	58 toothpicks/ person	7 toothpicks/person	23 toothpicks/person	11 toothpicks/person	2 toothpicks/person
Per Capita Protein Consumption (3 peanuts=100 grams of protein)	100 grams protein 3 peanuts/ person	67 grams protein 2 peanuts/person	100 grams protein 3 peanuts/person	67 grams protein 2 peanuts/person	67 grams protein 2 peanuts/person
Per Capita GNP Number of Candies	$27,100 90 candies	$3,710 12 candies	$13,710 46 candies	$2,490 8 candies	$650 2 candies

Note: The regions in this simulation are those defined by the United Nations so Mexico is included in Latin America. Oceania is not included because of its relatively small population. The energy consumption, protein consumption, and GNP figures are from 1996.

WRAPPING IT UP

Assessment

Have each student draw a diagram that illustrates how population distribution, energy, food, and wealth relate around the world. Then show how differences in any one of the variables affect the relationship among all the variables. Label each part of the diagram.

Unsatisfactory—The student is unable to complete the activity or does not include all components in the diagram.

Satisfactory—The student completes the diagram with all components and is able to demonstrate how the relationships between the variables change when one of the variables is altered.

Excellent—The student completes the diagram and demonstrates how the relationships between the variables change when one of the variables is altered. The work is complete, thoughtful, and reflects the discussions held in class.

Portfolio

The diagram completed in the assessment portion can be used in the portfolio.

Writing Ideas

- Have students describe how certain situations in the game can be compared to actual occurrences in the world.

- Offer the following journal starter to the students: "I found taking part in this simulation to be _____ because . . . "

Extensions

- Have the students work in groups to identify the countries that the United Nations groups together for the regions in this activity. Using maps or globes, students can locate and then make a list of these countries. Have them choose a country or region and conduct research to find out more about the culture, biodiversity, population, and social issues in that part of the world.

- Using their knowledge of current events or history, have the students describe how certain situations that occurred in the game can be compared to "real world" occurrences. Can they think of any historical events, either in the recent or distant past, that could have led to the unequal distribution of global resources we see in the world today?

Resources

Conserving Land: Population and Sustainable Food Production by Robert Engelman and Pamela LeRoy (Population Action International, 1995).

Global Environmental Change Series: Carrying Capacity by Margaret Edwards, Irwin Slesnick, Susan Plati, Linda Wygoda, and Brad Williamson (National Science Teachers Association, 1997).

People and the Planet: Lessons for a Sustainable Future edited by Pamela Wasserman (Zero Population Growth, 1996).

The Population Explosion by P. R. Ehrlich and A. H. Ehrlich (Simon and Schuster, 1990).

The Third Revolution: Population, Environment, and a Sustainable World by Paul Harrison (Penguin Books, 1993).

World Development Report, 1995 by The World Bank (Oxford University Press, 1995).

"1998 World Population Data Sheet" (Population Reference Bureau, 1998). For a free copy, call (800) 877-9881. Visit the Population Reference Bureau's Web site at <www.prb.org>.

26 The Many Sides of Cotton

AT A GLANCE

Investigate the pros and cons of conventional and organic cotton farming, analyze conflicting opinions, and develop a personal opinion.

OBJECTIVES

Compare methods for growing organic and conventional cotton. Analyze various essays on the pros and cons of organic cotton. Identify the problem, issues, parties involved, and interests, beliefs, and values of each party. Defend a personal opinion in support of either traditional or organic cotton or describe what additional information you would need to make a decision.

All of us have the ability to make changes that can help protect biodiversity. But how do we decide what's best to do? After all, environmental issues are complicated. Parents struggle to decide if it's better to use cloth or disposable diapers. Shoppers can't keep up with the latest information on dolphin-safe tuna, pesticides on produce, or paper versus plastic grocery bags.

In these and other issues, we hear conflicting arguments about what the problems are and the best ways to resolve them. Experts on opposing sides can provide reports, statistics, and plenty of commentary justifying their positions. So how do we know who is right? How do we decide what to believe?

When we learn more about an issue, we usually find that deciding on a solution is never simple. The "sound bites" offered by opposing sides often don't clarify the murky complexities of the issue. Because the issues are so complex, no one can *give* us the solution to any problem; we must become as informed as possible, then use our own best judgment. And there are often more than two sides to every issue, and a host of individual values and beliefs that influence our decisions.

It is essential that kids learn to think critically about environmental controversies because they will face such issues throughout their lives. One of the most important steps for making sound decisions and resolving conflicts is learning to analyze different perspectives. Students should be able to identify the different parties involved in an issue, their positions, and the beliefs and values that shape those positions. They should also be able to challenge the validity of each party's arguments. Where does each party get its information? Where might it be biased? And where might it be exaggerating or distorting the truth?

The process of analyzing different perspectives requires taking two different approaches. The first is one of empathy: You realize that all parties have concerns that should be heard and considered when you step into their shoes. But the second approach is one of skepticism: You can avoid being persuaded by inaccurate or false information by carefully assessing each party's arguments.

In this activity, your students will learn to apply both of these critical skills to the issue of cotton farming. In the process, they'll increase their ability to understand and make sound decisions about environmental controversies—something they can apply to other issues throughout their lifetimes. Because this activity involves quite a lot of reading, it can be used to develop different reading skills (see skills, pages 446-450). For students with lower reading levels or for students for whom English is a second language, you may want to summarize the reading to give them easier access to the information needed for the discussions.

Before You Begin

For each student make copies of "A Tale of Two T-Shirts" (pages 147–148 in the Student Book), "Cotton Concerns" (pages 149–159 in the Student Book), and "Sorting Out the Issues" worksheet and chart (pages 160–161 in the Student Book).

What to Do

1. Create a controversy "map" on the board.

To begin the activity, come up with a hypothetical or real controversy that you think will elicit a variety of responses from your students. For example, suppose your school has to cut something from the budget. Should it cut the music program or after-school sports?

Write the question on the board and ask for opinions from the group. Write the responses on the board and ask each student to explain his or her position. You may want to group similar responses.

2. Introduce students to issue analysis.

Explain that conflicts like the one you described usually originate with a *problem*. For example, a school has financial problems and therefore has to cut something from the budget. So it decides it must cut either music or sports or reduce funding for both.

Is it easy to make that kind of decision? Not at all. And when people disagree on the best way to solve a problem, some people say the problem changes into an *issue*. For example, in this case, it

became an issue when there were disagreements about whether or not to cut music or athletics. (Some people don't differentiate between problems and issues, and use the words interchangeably.)

Now tell the students that any time there is a conflict, they should be able to identify the different *parties* involved. Parties can be individuals or groups. For example, a parent may be a party, so may the PTA. Teachers, student musicians, athletes, coaches, and others may also be parties. Each of the parties may have a different view, or *position*. Ask the students to identify the different positions taken by members of their class. (This may simply amount to "music should be cut," "athletics should be cut," or "both should be cut some.") It's also important to remind the students not to stereotype groups, and that individuals within a party may have different views. For example, it's possible that not all members of the PTA will agree on an issue, not all athletes will agree, and not all students will agree. However, it sometimes helps to solve an issue by looking at what the interests are so that different groups and individuals can try to reach consensus.

The Components of an Environmental Issue

Understanding environmental issues can be complex. One way to investigate and analyze an issue is to look at the following components:

PROBLEM

What's at stake? Most environmental problems involve a risk or a threat created by some interaction between humans and their environment. For example, an environmental problem might be that migratory songbird populations are dwindling because people are destroying much of the birds' native habitat.

ISSUE

What's the conflict? More often than not, different parties disagree about how to address or resolve an environmental problem. This conflict turns an environmental problem (e.g., migratory songbird populations are dwindling) into an environmental issue (e.g., should more money be spent to save birds' habitats?).

PARTIES

Who are the individuals and groups involved in an issue? The parties in the songbird example might be environmentalists, wildlife biologists, local farmers, government officials, parents, and housing developers.

INTERESTS

What interests do each party have? (Some people may support habitat protection and restoration. Others may prefer to develop the areas to provide homes for people who need and want them.)

BELIEFS

What ideas about the issue does each party have? For any issue, you should be able to make a list of the beliefs of each party. These may or may not be based on facts, but they are usually closely tied to people's values. (Some people and environmental groups may believe that the world should have songbirds, that birds have as much of a right to exist as do humans, that humans should cut down on their consumption and reduce urban sprawl to protect habitats, and so on.)

VALUES

What are the ideals that motivate the different parties? The way people look at issues is deeply influenced by the values they hold. Less concrete than beliefs, values are a person's overarching priorities and principles. For example, there are ecological values, ethical values, economic values, and so forth. Other common environmental values are recreational, aesthetic, social, and utilitarian. (See "Bear in Mind" from WOW!–A Biodiversity Primer for more on environmental values.)

SOLUTIONS

What can be done to address complex issues? When it comes to personal decision making, think about creative ways to address your own and others' interests. A respectful search for common ground is more likely to lead to a lasting solution.

OTHER IDEAS

It is important to realize that there are many ways to analyze environmental issues. Some people, for example, do not distinguish between issues and problems. It's also important to find ways to size up social, economic, cultural, and political contexts when analyzing environmental issues. We encourage you to come up with your own way to help students explore complex issues.

Adapted from *Investigating Environmental Issues and Actions* by Hungerford, Litherland, Peyton, Ramsey, and Volk (Stipes Publishing Company, 1992).

World Wildlife Fund

Now review the reasons given by students for their different interests. These reasons, also called *beliefs,* shape parties' positions. (Beliefs are shaped by facts, information, and experiences.) Examples of beliefs may include the following: music is culturally more important than athletics, athletics can serve more students than music, high school athletics can lead to high-paying professional sports jobs, music develops artistic creativity and allows for personal expression, and community sports leagues can take the place of school sports.

Ask the students why, if they were in a disagreement with someone, it might be useful to understand that person's underlying beliefs. (*They may learn something they didn't know. The beliefs may be inaccurate. It may be easier to discuss beliefs than positions.*)

Explain to the students that by looking closely at a person's interests, they can start to identify the *values* that others bring to an issue. For example, the comment about sports leading to good jobs reflects economic values. Other comments may show how people may value equity, security, or fairness.

Finally, explain that the analysis can be applied to any issue. This kind of analysis helps clarify what's going on in a conflict and makes it easier to see how a solution can be reached. Tell the students that they're going to practice looking at another issue in this way so they'll become better at making their own decisions and working through group conflict.

3. Introduce the cotton issue.

Ask the students if they wear cotton clothing. Then ask them if they've ever thought about how cotton is grown. Explain that people disagree about the best way to grow cotton. Some people think that too many chemicals are used on cotton plants and that those chemicals cause problems for people, wildlife, and the global environment. Other people think that we can't afford to grow cotton without pesticides and that there aren't enough good reasons to stop using them. And there are many opinions in the middle.

4. Pass out copies of "A Tale of Two T-Shirts."

Tell the students that the pictures they're looking at show both organic (pesticide-free) cotton production and conventional cotton production. Give them a chance to read through the captions, then ask if they have any responses to the diagrams. Can they predict why some people might oppose conventional practices? Can they anticipate why others might be OK with conventional growing? (*Some people might have health or environmental concerns about the pesticides and other chemicals used in growing conventional cotton. Others may feel that conventional methods are more cost effective, faster, and easier, and that they produce more cotton than organic methods do.*)

5. Pass out copies of "Cotton Concerns" and "Sorting Out the Issues" worksheet and chart.

Now tell the students that they're going to read what certain individuals and organizations have to say about organic and conventional cotton growing. You'll probably want to assign this reading as homework. (You may want to assign the readings over two days or divide the class into groups—see variations below to lighten the reading load of individual students.) After they complete the readings, ask the students to fill out the "Sorting Out the Issues" worksheet and chart.

Note: The views expressed in the articles are those of the authors of the articles. World Wildlife Fund is not endorsing any of the companies or the views expressed. The articles simply represent different viewpoints on a controversial issue. Point out to your students that the readings represent seven different perspectives on organic cotton, but they do not represent every viewpoint on the issue or every party involved. Also point out that the e-mails were compiled from interviews with students to provide two diverse viewpoints. Carlos and Rachel are fictional.

Innovative Cotton!

by Sally Fox, founder and president of
Natural Cotton Colors, a producer of
colored cotton without chemical dyes

*In 1989 I started a company called
Natural Cotton Colors, Inc., to produce
and market FoxFibre™—naturally colored
cotton. Most cotton plants produce white
lint. This lint can be made into natural or
bleached white fabric for T-shirts, socks,
and other types of clothing or cotton
products. It can also be dyed to produce
every color imaginable—from bright
yellows to neon pinks to heather grays.
But making those colors usually requires
chemical dyes. And both the bleaches for
bright whites and the dyes for most colors
pollute the environment. This is one
reason I began experimenting with cotton
plants that produce colored cotton
naturally—sort of the way different kinds
of apple trees produce yellow or red or
green apples. My company succeeded in
growing cotton with lint that is green,
brown, khaki, auburn, and a variety of
other colors—using no dyes at all. In
addition, these naturally colored cottons
are generally more pest resistant than
white cotton. In introducing these hardier
plants to farmers, I hoped to help reduce
pesticide use, too. (To find out more about
Sally Fox and organic cotton see the
resources on page 285.)*

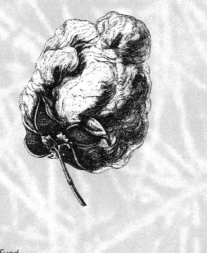

VARIATIONS:

Team Up—To decrease the amount of reading
for each student, divide the class into groups
of six and have each student read one article.
Then the students can share the parties, positions,
beliefs, values, and solutions presented in their
articles with the rest of the group. While filling
out the worksheet and chart, the groups should
compare and contrast the different points of
view presented in the essays. You could also have
the students stage a debate about organic versus
conventional cotton farming.

6. Review handouts.

Ask the students if they had any general reactions to
the essays they read. Then review their responses to
the worksheet and chart. Use the following questions
to help guide your discussion:

1. List several things that the different writers
   agree on regarding conventional and organic
   cotton growing. *(Cotton is an important
   industry in the United States; the health and
   safety of people are important; and so on.)*

2. List several things that the writers disagree on
   regarding conventional and organic cotton
   farming. *(Some authors think that when used
   properly, regulated agricultural chemicals are safe
   for people and the environment, but others think
   they aren't safe; some regard farms as living
   ecosystems while others think of farms as
   human-made, highly controlled environments;
   some authors think the higher prices of organic
   cotton are justified if it means the environment is
   protected, but others think the price for organic
   cotton is too high; and so on.)*

3. Which writers seem to have similar points of
   view? *(Daniel Imhoff, Carlos Vielma, Yvon
   Chouinard, and Patrick Leung are all strongly in
   favor of organic cotton. The National Cotton
   Council is in favor of conventional cotton, and
   Rachel Shutkin believes there are more important
   things to worry about than how cotton is
   produced. The Gap comes from a combined point
   of view, saying that organic cotton is a good idea
   but that it is more expensive and harder to
   obtain than conventional cotton.)*

4. Which writers have the most different points of
   view? *(Answers will vary, but the National Cotton
   Council's point of view is quite different from the
   views of Daniel Imhoff, Carlos Vielma, Yvon
   Chouinard, and Patrick Leung.)*

5. Fill out the "Sorting Out the Issues" chart, being sure to include every writer from the readings. *(Answers will vary, but you may want to discuss the following questions for each point.)*

a. Parties

Were you surprised by how many different parties are involved in an issue like this one? What other parties might be involved? *(Other parties might include chemical manufacturers, field workers, clothmakers, parents, the U.S. Environmental Protection Agency, and so on. It's important to emphasize that not all individuals in a party have similar views. For example, not all parents or farmers agree.)*

b. Interests

How easy was it to state the parties' priorities? Were the interests as simple as being for or against conventional cotton? Why do you think some of them were more complicated? *(Answers will vary.)*

c. Beliefs

Why do you think some of the beliefs contradicted each other? Where do you think the parties get their information? *(You may need to explain to the students that people often choose to interpret scientific data differently, depending on their values. That's why it's often useful to go to the source of the information to see what it says and how reliable it is.)*

d. Values

How did the parties' different values compare? In particular, you may want to talk about the different interpretations of the "bottom line."

6. What are some of your thoughts about conventional and organic cotton growing now that you've read these essays? *(Answers will vary with individual students.)*

7. What other types of information would you need to form your own opinion about the conventional versus organic cotton farming dilemma? Do you think you could find that information? What facts would you like to verify? *(Answers will vary.)*

7. For a discussion on bias, return to "A Tale of Two T-Shirts" diagram.

Ask the students to look at the diagram again. Tell them the diagram was supplied by Patagonia, Inc. Based on what they now know about Patagonia's position on organic cotton, do they look at the diagram differently? Ask them whether the information in the diagram was slanted in any way. Have them circle anything in the text or pictures that could be seen as biased.

Discuss with the students the concept of "bias" when evaluating environmental issues. If an organization or an individual holds a strong belief (based on values), it might be difficult for the party to present balanced information on a topic. It's important to point out that everyone is biased to some extent.

Ask the students what values were held by the authors of essays that were in favor of organic cotton. *(Supporters of organic growers place a high value on environmental and human health. They recognize that organic farming practices may be more expensive at this point, but hope that if consumers understand the benefits of the organic method they will be willing to pay more for the organic product.)*

How could the views of the organic-cotton supporters affect their opinions? *(They might place a lot of emphasis on the effects of pesticides on human and environmental health when describing the organic/conventional cotton debate. They might downplay the economic issues such as the fact that organic cotton is much more expensive than conventional cotton. They might also downplay problems such as the small supply of organic cotton and the use of non-organic dyes on organic cotton.)*

Cotton Clips

In 1993, 16.2 million bales of conventional cotton and 50,000 bales of organic cotton were produced in the United States. This is enough cotton to make 19,776,250,000 T-shirts. Now that's a lot of cotton!

The United States is the largest cotton exporter in the world. On the average, more than one quarter of the world's cotton originates in the United States.

Cotton is the single best-selling fiber in America, outselling all synthetic fibers combined.

How did the T-shirt get its name? Starting back in the 1880s, sailors in the U.S. Navy were issued an elbow- and hip-length undershirt that, when laid out on a flat surface, was made in the shape of a perfect "T."

Cotton has been grown and used to make fabrics for at least 7,000 years! Egyptians may have been using it all the way back in 12,000 B.C. Other cultures also have a long history with cotton. Cotton fabric scraps have been found by archaeologists in Mexico (from 3500 B.C.), India (3000 B.C.), and the southwestern United States (500 B.C.).

In the United States, cotton is grown in 17 states along the "Cotton Belt," which stretches from Virginia to California.

What values were evident in the essays in favor of conventional cotton? *(The conventional growers place a strong emphasis on economic values. With regard to their environmental values, they also point out that using pesticides and intensive growing practices may make better use of a smaller amount of land.)*

How could this thinking bias the conventional-cotton supporters' view of the subject? *(Conventional growers might emphasize the economic and environmental benefits of using pesticides to grow more cotton on a smaller amount of land. They might focus on the fact that the conventional method can currently produce a greater amount of cotton to meet the world's high demand for the product. They might also point out that, by offering cheaper cotton, they are serving underprivileged buyers who can't afford to buy organic cotton. They might not mention the health and environmental concerns associated with the use of pesticides.)*

How can you find out if an organization or an individual is presenting balanced information? What are some clues you can look for? *(Keep your eyes open for superlatives such as "always" and "never." These words can signal that the author is making a generalization or an assumption. If the statement is coming from an association or a company, think about where the organization gets its funding. You can also contact the organization, look for other publications it has produced, or visit its Web site to get a better idea as to what its views are on the topic.)*

How can biased information be useful? *(Some would say that all information is biased in some ways and that it's important to get information from a variety of sources. It's also important to realize that with most controversial issues there's never enough information to always know what to do and that people have to use their best judgment.)*

8. Assign a personal-opinion piece.

As a follow-up to the activity, have the students write what they now think about conventional versus organic cotton farming. Tell them to be sure to back up their statements with specific reasons. And tell them to include at least one sentence describing whether and how their values will affect their specific actions as a consumer and a citizen.

WRAPPING IT UP

Assessment

Have the students select (or supply them with) an article from the newspaper on an environmental issue. Then have them "map" the issue, delineating the problem, issue, parties, interests, beliefs, and values. Have them include their personal opinion on the issue.

Unsatisfactory—The student fails to adequately address any of the required elements of the project.

Satisfactory—The student presents a document that contrasts different perspectives, reveals a basic knowledge of issue analysis tools and use, and presents his or her personal opinion.

Excellent—The student presents a document that clearly delineates differing views on the issue, reveals skill in use of issue analysis tools, and has a clearly defined personal opinion that logically emerges from the analysis.

Portfolio

The issues worksheet and assessment can be used in the portfolio.

Writing Idea

Have the students interview consumers, retailers, manufacturers, and/or growers in their region about organic cotton and other organic products. Then have them write a report on the issues, parties, interests, beliefs, and values they encountered.

Extension

This activity focuses on just one aspect of cotton farming: conventional versus organic farming methods. There are other issues associated with cotton farming, including where the cotton is grown, how much water the plants need, and how the increased price of organic cotton affects who can afford to buy it. Have your students investigate these other issues. Or have your students investigate integrated pest management (IPM), a farming system that combines elements of both conventional and organic systems and results in a much lower dependency on pesticides and other chemicals for production.

Resources

Basic Guide to Pesticides: Their Characteristics and Hazards by Shirley Briggs (Taylor and Francis, 1992).

Cotton, Health, and the Environment by the National Cotton Council and Cotton Incorporated. For a copy, contact Cotton Inc. at (919) 782-6330.

"King of Fibers" by Jon Thomson. *National Geographic* 185, no. 6 (June 1994).

Organic Cotton Story (Patagonia, Inc., 1995). For a copy, contact Patagonia's public affairs department at P.O. Box 150, Ventura, CA 93002.

Visit FoxFibre's Web site at <www.primenet.com/cotton>.

---

27 Space for Species

SUBJECTS

science, mathematics

SKILLS

gathering (simulating, collecting), organizing (graphing, charting), analyzing (calculating, identifying patterns), interpreting (inferring), applying (proposing solutions), citizenship (evaluating the need for citizen action, planning and taking action)

FRAMEWORK LINKS

3, 12, 21, 23, 29, 35, 38, 44, 46, 47, 54

VOCABULARY

biogeography, *edge effect*, **fragmentation**, **habitat**, *immigration*

TIME

Part I—*one to two sessions*
Part II—*two sessions*

MATERIALS

See "Before You Begin" Part I and Part II.

CONNECTIONS

For an introductory outdoor activity, try "Backyard BioBlitz" (pages 134–143). Use "Insect Madness" (pages 144–157) for more on measuring biodiversity. Follow up with "The Case of the Florida Panther" (pages 246–251) to investigate the effects of habitat loss and other factors that can cause species to become endangered. For more on designing reserves, try the gap analysis activity in "Mapping Biodiversity" (pages 252–265).

AT A GLANCE

Play an outdoor game, conduct a survey of plant diversity, and analyze current research to explore the relationship between habitat size and biodiversity.

OBJECTIVES

Describe factors that affect the relationship between habitat fragmentation and biodiversity loss. Create a graph that demonstrates the relationship between biodiversity and the size of a habitat. Describe different strategies for designing reserves that could help lessen the effects of fragmentation.

Habitat loss is one of the biggest threats to biodiversity. Roads, shopping centers, housing developments, agricultural fields, and other types of development are breaking up our large forests and other natural areas into smaller and smaller chunks—a problem conservationists call *fragmentation*. Many scientists compare the remaining habitat fragments to islands because they are so isolated. And like islands, habitat fragments are often too small and isolated to support a large number or a wide variety of species. Conservationists have the tough job of trying to figure out how fragmentation is affecting biodiversity. They're asking questions like "How small is too small?," "Which species are we losing?," and "How can we balance our need for development with other species' need for space?"

Development and fragmentation can be difficult concepts for students to understand. Seeing the relationship between the two will help students realize there are certain tradeoffs that result from our decisions to develop natural areas. Most development occurs to fill people's needs for schools, homes, roads, food, and income. While most people recognize that developers are not trying to destroy biodiversity when they build roads or homes, many people also feel that the value of biodiversity is not factored into our decisions to develop. Many conservation biologists would like to see communities consider how development impacts biodiversity and work to accommodate natural systems as much as possible.

In Part I of this activity, your students will play a game that will allow them to explore some of the actions we can take to try to balance human need for development and species' need for space.

The students will become species trying to move between habitat fragments, and they'll begin to understand why animals have such a tough time living in fragmented landscapes. Then your group will try to come up with some ways they can help species move between habitat fragments more easily.

In Part II, the students will take a closer look at the relationship between the size of a habitat and the biodiversity it supports. By going outside and measuring plant diversity in habitats of different sizes, and then graphing their results, your group will see that, in general, smaller areas have fewer species.

The "Java Raptors" option in this activity will give your students the opportunity to look at some real research about fragmentation and its effect on birds of prey on the island of Java. They'll read about the fragmentation problem on Java, then use a graph to analyze research done by scientists on the island.

You can use the different parts of this activity to fit your needs. Each part can stand alone, or they can be used together to help build a unit on the effects of development and fragmentation on biodiversity.

Before You Begin • Part I

You'll need an open playing area about 60 feet by 40 feet, with plenty of extra room for students to work in groups outside the playing area. Use four traffic cones or other visible markers to mark the boundaries. Use two 25-ft. ropes to make two small islands with diameters of about 8 feet. Use two 40-ft. ropes to make two large islands with diameters of about 13 feet. Arrange the islands in the playing area as they are arranged in the diagram at the right. If possible, tape the ropes to the playing surface so that students can't kick them out of place during the game. We recommend that you do the activity outside, if possible. You'll also need two colors of tokens or poker chips (about 20 of each color), flip chart paper, markers, and a stop watch.

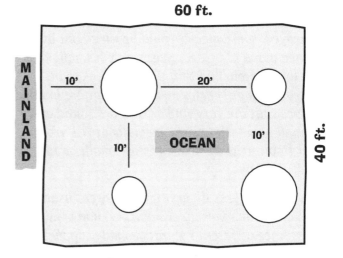

What to Do • Part I

Island Hopping

In the 1960s, two scientists, Robert MacArthur and Edward O. Wilson, studied how many species lived on islands of different sizes at different distances from the mainland. Their theory, called the theory of island biogeography, has helped ecologists think about the effects of habitat fragmentation. Although the theory's name sounds complicated, its point is very simple: More species can live on islands that are large and close to the mainland than on islands that are small and far from the mainland.

Why is this theory still important to scientists today? Because it relates directly to the study of the numbers of species in habitat fragments on the

mainland. In many areas, fragments of forest and other habitats are all that remain where the landscape used to be covered with vast areas of continuous natural vegetation. Scientists compare these fragments to islands because they are so isolated.

In this part of the activity, your students will learn the basics of island biogeography by imagining they are species trying to get to different-sized "islands" at different distances from the "mainland." Then they'll apply the concepts of island biogeography to habitat islands. They'll explore some of the threats facing species in habitat islands and think about ways we can reduce those threats by planning development with biodiversity in mind.

What's the problem with patches?

Habitat fragmentation is one of the most serious threats to biodiversity. A researcher studying birds in one part of Australia, for example, found that the numbers and ranges of almost half the birds native to the region have decreased since the early 1900s. He thinks that almost all the decline is a result of habitat fragmentation. Small, fragmented habitats, called habitat islands, usually can't hold as many species as large, more continuous ones. Here are some of the reasons we lose species, and biodiversity, in small patches of habitat:

Luck of the draw: When a piece of habitat is destroyed, some species could be wiped out by chance alone. If a species uses only a small part of a larger area, and that part happens to be destroyed, that species and its habitat are lost. Species that are very rare or that are found only in small populations are especially at risk when their habitats are broken up into smaller and smaller chunks.

Less habitat, less diversity: Large areas usually contain a wider variety of habitats than smaller ones. Since different habitats usually support different species, a fragmented area will often contain fewer habitats and fewer species than a larger area. Many scientists think this is the main reason diversity is lower in habitat patches.

Road blocks: Some species can live in habitat fragments if they can move from one area to another to get everything they need, such as food, shelter, and mates. Unfortunately, many fragments are surrounded by barriers that prevent species from moving between different areas. Roads are a common barrier that many species can't cross, but buildings, parking lots, and fences can also keep species from getting where they need to go. When a species is isolated from others of its kind, it can become subject to inbreeding and lose some of its genetic diversity. Species that need a lot of space or that spend a lot of time on the move can be very sensitive to these "road blocks."

On the edge: When we build developments and break a habitat into small chunks, we create more boundaries between the habitat and the outside world. Conditions at these boundaries, called edges, are very different than the conditions in the habitat's interior. There may be more sunlight and wind at the edge, and because there's no canopy overhead to keep the moisture in, the edge is often much brighter and drier than the interior. These different conditions can change the plant and animal species living in the area. There can be different predators and prey, making it harder for animals to find food and to avoid being food themselves. In small fragments, edge conditions can take up most of the habitat. Scientists call this problem the edge effect, and species that can't adapt to the edge often become threatened.

Fragmentation doesn't affect all species in the same way. Some are more sensitive to habitat loss than others. And some species can even benefit from fragmentation and the edge effect. All of the factors listed above affect different kinds of species in different ways, and that's what makes the problem of fragmentation so difficult for conservationists trying to protect a wide variety of species.

1. Introduce the activity.

Explain to the students that they'll be investigating a well-known ecological theory called the theory of island biogeography. Briefly explain that scientists Robert MacArthur and Edward O. Wilson wanted to study species that traveled from the mainland to nearby islands in the ocean. (You might want to introduce the scientific term "immigration" here, explaining that MacArthur and Wilson were studying species that immigrated to islands from the mainland.) The scientists wanted to know how many species from the mainland lived on islands of different sizes at different distances from the mainland. They were also interested in those species that became "locally" extinct, which means they were no longer living on the islands, but could still be found living on the mainland.

Tell students that they'll be doing a similar investigation outside. Some students will be animals immigrating to "islands" you've laid out in the yard. Other students will be playing predators, diseases, and different forces out in the "open ocean," that can cause animals to become extinct.

2. Explain the rules.

Familiarize the students with the playing area. Show them the islands and their sizes and distances from the mainland.

Select about 22 students (or three-quarters of the group) to be species immigrating to the islands and about 8 students (or one quarter of the group) to be taggers that represent threats that can cause immigrating species to become extinct. Explain that immigrating species will have one minute to run from the mainland to an island, but they'll have to avoid being tagged by the students in the playing area because being tagged will make them extinct on the islands. As you select students to be the extinction taggers, you can have them think of some of the causes of extinction (predators, diseases, pollution, severe weather, and so on) they might represent to species immigrating across the ocean.

Explain that once you give the signal, species on the mainland should begin immigrating to the islands by making a run for them. Species can be tagged out of the game only when they are out in the open ocean. If they are on an island or the mainland, they can't be tagged. Although they're safe on the mainland, tell students that at the end of the game you'll only count the species that successfully have made it to an island.

3. Play Round 1: Immigrate!

Tell the taggers to spread out in the playing field, and make sure they keep moving all the time that students are immigrating. Explain that, as in nature, threats to species are spread all around the landscape, so the taggers should also be spread out. Keep taggers from crowding around islands and not allowing any students to pass. Try to make the game fair for everyone.

Yell "Immigrate!" to let students know when to begin. Keep time and tell the students to stop after one minute. Ask any students who become extinct to help you monitor the game.

4. Evaluate the results.

Have the students count the number of animal species on each island. Keep track of the results on a piece of easel paper. You can make a chart or a graph, or you can write the number of species on each island in a diagram of the playing area.

Have the students gather around to go over the results of Round 1 and to talk about what they'll do in Round 2. Having them nearby will help them focus on you and not on the many distractions there can be outside. Figure out the percent of students who survived (divide the number of students who made it to an island by the total number of students who started on the mainland, then multiply by 100) and record the percentage on the easel paper.

Tell the students that, according to MacArthur and Wilson, the large island close to the mainland should have the most species. Is that what your group found? Why are there more species close to the mainland? Ask students to think about their experiences while immigrating. *(Those who tried to run to the farthest islands faced many more threats on their journey than those who traveled only to a nearby island.)* If your students found different results than MacArthur and Wilson found, talk about some reasons they may have had a different outcome. *(Your students may have been better at getting to islands than most species are. Or the extinction taggers could have made more species extinct than happens in nature.)*

Regardless of how many species made it to islands at different distances, more students should be on islands that are large than on islands that are small. Ask students why this is true. *(Small islands don't have the space or variety of different habitat types to support many different species, just as the small islands in the game were not big enough to hold many students. If a small island was overcrowded, a student could have been pushed out and, while moving to another island, would have been open to an extinction tagger.)*

5. Discuss habitat islands.

Ask the students to think about what's happening to many of our natural areas and what that may have to do with ocean islands. Why might conservationists use the MacArthur and Wilson model when they think about designing reserves in natural areas? *(Explain that many of our forests and other natural areas have been separated from each other. Only small patches of the continuous vegetation that once covered much larger areas still remain. And the things that separate these habitat islands, such as roads, buildings, and agricultural fields, are often even more difficult for species to cross than the ocean.)*

Ask the students why animals need to move between habitat islands. *(Many islands are too small for all the species living in them, and they can become crowded. Competition for resources may force animals to move to find more food or shelter. Some animals need to migrate. Others may be looking for mates.)* Then ask the students what kinds of barriers the animals might face. *(Animals are often killed trying to cross roads. Many animals also become easy targets for predators to spot when they leave their habitat. Animals traveling a long distance through developed areas may not be able to find enough food and could become pests to humans by rummaging through garbage cans or waiting for people to provide food.)* Record the students' ideas on easel paper if you can.

6. Play Round 2: Habitat island hopping.

Round 2 will demonstrate what it's like for species trying to move between habitat islands. Tell students that the playing area now represents habitat islands in a sea of development rather than an ocean.

For this round, you'll need two-thirds of the class as species in habitat islands and one-third as extinction taggers. (Have the students think about how the extinction factors might be different in habitat islands as opposed to oceanic islands.) You might make the species that were tagged out in the last round become taggers, and you might move some taggers into habitat islands so that the students get a chance to experience both roles. Have the students think of what the taggers might represent in the sea of development by going over the threats they came up with in their discussion in step 5.

Tell students that the two different colors of tokens you have represent some of the things that species need. The tokens may be food and water, shelter and mates, or any of the other needs you discussed in step 5. Tell the students that they'll be competing for these resources in the habitat islands. Count out enough tokens so that there is one of each color for every student in the habitat islands. Scatter the tokens throughout the four islands so that larger islands, which can hold more resources, have more tokens. State the following rules:

- Students must collect at least one token of each color to survive, but they can collect more if they like.

- Students can pick up only one token at a time from any island. So if a student picks up a token on the island he or she starts from, the student must run to at least one other island for another token. Students can return to their first island for tokens if they need to.

Shout "Immigrate!" to start the game again. This time give students as much time as they need to move between the islands. Stop the game when every student either has been tagged or has collected at least two tokens. Tell taggers they should spread out in the landscape, just as threats to species are spread out. They shouldn't stand in front of moving species and keep them from passing. Use your judgment about how to keep the game fair. After they finish the round, count up the number of students who survived and record it on easel paper.

7. Go over the results of Round 2.

Have the students gather together to figure out the percent of species that survived. Most likely, a large percent became extinct. Ask students why they think so many species didn't make it. *(There were more extinction taggers out to get them, and they were forced to leave their habitat to get all the resources they needed. Species that had to travel to several islands to get their tokens faced many threats, and most probably didn't survive.)*

Ask the students who didn't survive why they think they were tagged. Were the extinction taggers faster than they were? Did they have to go to several islands, leaving themselves open to taggers each time? Were they forced out of a small island that was too crowded? Ask them if real species are also affected by the same things. *(Species that are slow moving are often hit by fast-moving cars; some species are forced to travel between many habitat islands to find all the resources they need and are thereby open to threats when they leave their habitat; and many species can be pushed out of overcrowded habitat islands and forced to move to other habitats.)*

Ask the students who survived why they think they were never tagged. Were they faster than the taggers? Did they have to go to only one or two islands to get their tokens? Did the taggers see them? Tell them that in real habitats, just like in the game, not every species becomes extinct. Ask the students if real species can survive in the same way that they did. *(Some species can avoid many threats that other species face. For example, birds can fly over cars that might hit other species. Some species*

need only a very small amount of space to get what they need, so they might not ever need to leave their habitat island, or they might not have to look far outside their habitat. Still others might not be affected by the main threats in an area. If, for example, there was a predator killing many small animals in an area, a larger species might be too big to be eaten by that predator.)

8. Introduce edge effect.

Now tell the students that there is an additional threat that species face when living in habitat islands. Species are at risk not only when they travel between islands, but also while they're inside their habitat island. Many species can be lost to something called the edge effect.

The conditions at the edge of a patch of forest (as well as some other ecosystems) are quite different from those in the interior of the patch. Ask the students if they can think of some of the ways the edge of a habitat might be different from the middle. *(Often, more wind and sunlight make the edge much hotter and drier than the interior. The difference in conditions can change the plant species living in the area, and plants that are better suited to the edge might out-compete plants that would normally grow in the forest. Some of the plants that are lost could have been a source of food or shelter for animals in the forest, so these animals could be lost with the plants. Also, it is often easier for predators to find prey on the edge, so some prey species can have a hard time living on an edge.)*

The figure below illustrates the impact edge effects can have on habitats of varying sizes. Smaller habitats will have a larger proportion affected.

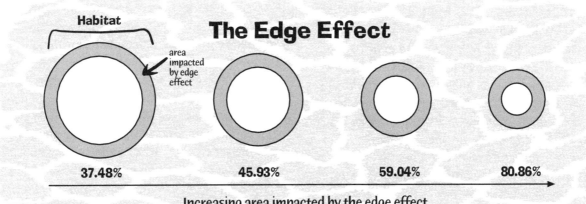

The Edge Effect

Habitat

area impacted by edge effect

37.48%   45.93%   59.04%   80.86%

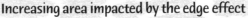

Increasing area impacted by the edge effect

9. Play Round 3: Life on the edge.

Tell students that they'll play another round to find out how the edge effect can affect species in habitat islands. Start the round with the same number of species and tokens that you started with in Round 2. Let some of the students who have been taggers since the first round become moving species. Collect the tokens and scatter them throughout the habitat islands as you did for Round 2. Tell the students that the rules for this round are the same as they were in the last round, except for one major change. Now the taggers will simulate the edge effect by reaching an arm's length into the islands to tag species. Their feet can't cross the rope that marks the island—they can only reach in.

Shout "Immigrate!" to start the round, then give the students as much time as they need to collect their two tokens. Once the round is over, count the students that survived and record the number on your easel paper.

10. Go over the results of Round 3.

Have the students gather together to calculate the percentage of surviving species. Did more or fewer species survive this round? What did the students expect would happen? Explain that in most cases some species can be lost to the edge effect, so they would expect that the percentage of surviving species is lower than in Round 2 when they didn't consider the edge effect. Ask students who were tagged if they were tagged on the edge or outside an island. Students tagged on the edge are much like species that can't adapt to the new conditions or like species that are easily spotted by predators, and so can't survive as well on the edge. Had any students who were tagged outside an island been pushed out of an overcrowded island? Did they notice that their amount of "safe," or healthy, habitat had been decreased by the edge effect?

Tell the students that although many of the original species didn't survive in the game, the total number of species in real habitat islands with lots of edges can actually go up. Ask if anyone knows how this could happen. *(Many species that are well adapted to the conditions on the edge can move in and take the place of a smaller number of species that were adapted to the interior and not to the edge.)* In the game, when a species was tagged out, it wasn't replaced with another species, but in nature

this can happen. Some disturbed habitats actually have more species than healthy ones, but they don't necessarily play the same ecological roles as these species do in a healthy habitat. And the disturbed habitats often contain widespread, non-native species rather than the more rare local species that are important in the area.

11. Discuss reserve design: making a good plan.

Tell the students that the challenge of understanding species' need for space is in building developments like roads, homes, and schools so that both people and wildlife can get the things they need. Because it's not always easy to do this, many species are in trouble. But it can be done. Ask the students to think about how we can help wildlife in fragmented areas, whether in reserves or in developments.

For example, one way to make sure that there's enough habitat for species is to set aside land in reserves. Ask the students if they can think of any potential problems with reserves. *(If they have trouble thinking about problems, lead them back to the concepts they learned in the game. Many reserves are like habitat islands—they are surrounded by developments and can become isolated. Small reserves might be too small to support many species. And wildlife moving between reserves can face many threats. Your students may come up with other problems not related to the concepts of the game, but be sure that the ones listed here are covered.)*

Draw the diagrams at the right on easel paper or on the chalkboard. Each diagram represents a possible reserve design, but in each set one is a better choice than the other. Ask the students which design in each pair they think is best, based on what they learned in the game. Ask them to explain their choices, and then give them the answer and explanation provided. Remember that there are many different ways of thinking about reserves, and your students may have many different issues in mind. The answers provided are the best choices according to what we know today about island biogeography. As long as the students can justify their answer with an explanation that demonstrates that they understand the material, they're right. But let them know that there are many ways of looking at the problem.

Reserve Design Choices

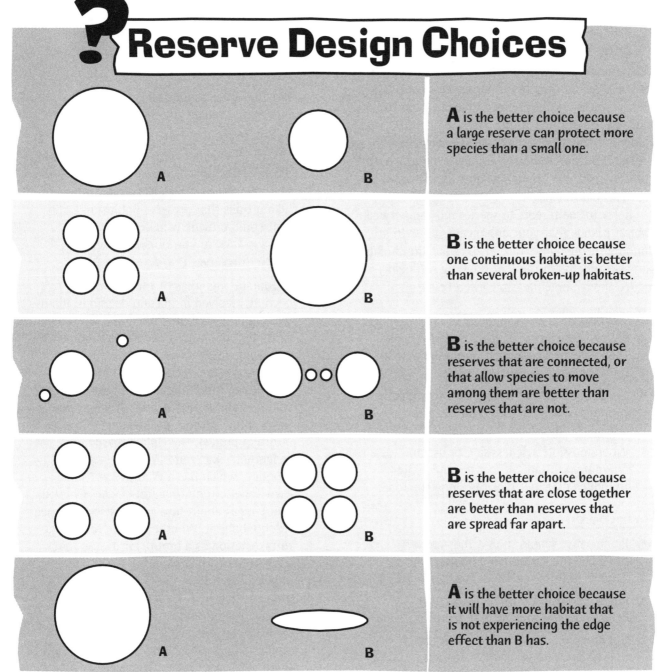

A is the better choice because a large reserve can protect more species than a small one.

B is the better choice because one continuous habitat is better than several broken-up habitats.

B is the better choice because reserves that are connected, or that allow species to move among them are better than reserves that are not.

B is the better choice because reserves that are close together are better than reserves that are spread far apart.

A is the better choice because it will have more habitat that is not experiencing the edge effect than B has.

If you have time, you can test these different designs with more rounds of the game. Use the same rules and see if more species survive with the recommended designs. For a look at some of the other ways scientists design reserves, you might like to try the gap analysis exercise in the activity "Mapping Biodiversity" on page 252.

Tell the students that these designs are often the best way to protect the largest number of species. But they're not always realistic, and they're not always the best plan because they don't consider the needs of specific species. Emphasize that there are

many, many factors to consider when planning reserves. For example, does land need to be purchased? How much will it cost? What else might the land be used for? Does anyone live in the area? What species are using the area? Planners of reserves must weigh the costs and benefits of development and protection of natural areas, which can be a very complex and time-consuming process.

If there are any current plans for the purchase of land for either protection or development in your area, you might discuss them here. How do your students feel about the different options available?

12. Talk about how people can help species in developments.

Most animals probably don't live their whole lives in reserves. Even if they use the reserves most of the time, chances are they'll need to leave them to find food, mates, or other things at some point. Do your students think wildlife has a good chance of surviving outside reserves? Are there things we can do to increase the animals' chances and help preserve biodiversity?

Refer to the threats to species moving between habitat islands that your group came up with in Round 2 of the game. Talk again about the threats and ask students to think of ways we can help reduce them. Some ideas are listed below.

On roads:

- Post wildlife crossing signs to alert drivers that they are likely to encounter an animal on the road.

- Construct highway underpasses so animals can cross roads without the threat of getting hit by a car.

- Where possible, use less salt, or use an alternative to salt, on roads in the winter to reduce damage to plants along the road.

In backyards:

- Plant native plants so that the yard is more like the surrounding habitat and attracts natural species.

- Put up bird feeders for birds traveling between habitat fragments in search of food.

- Cut down on the use of pesticides so that birds and invertebrates can use backyard habitats without the threat of being poisoned.

- Keep pet cats indoors so that they do not prey on the lizards, snakes, birds, squirrels, and other wildlife that may be living in the backyard habitat.

Around the school and other buildings:

- Convert part of your schoolyard or community center to a wildlife habitat.

- Provide water sources such as bird baths, marshy areas, or ponds.

- Put up boxes that birds can use for nesting.

In the community:

- Plan "greenways" such as bike paths and hiking trails that preserve habitat in tracts that could provide wildlife with passageways, or corridors, connecting different reserves that are far apart.

- Encourage members of the community to become involved in making decisions about what land will be developed and how it will be done. Once your students have come up with some ideas, give them the option of taking some type of action—from conducting more research to creating a wildlife habitat area nearby. You may want to try some of their ideas as part of a class or group project. The National Wildlife Federation's Backyard Wildlife Habitat Program, which has a Web site at <www.nwf.org/nwf/habitats>, can give your group ideas about how to create or enhance wildlife habitat. Whether or not you decide to take action as a group, make sure that your students understand that although habitat loss and fragmentation are serious problems for wildlife, making the decision to build roads, homes, or schools doesn't mean we have decided not to protect biodiversity. There are many ways we can share our space with other species. (For more information on taking action, see "Getting Involved" on pages 358–377.)

Find a natural area for your group to work in. You will need an area big enough to hold a 64-sq.-yd. plot with plenty of extra room for the students to move around in. If your schoolyard isn't available, a lawn (not newly planted) or roadside field would work well. You might also try a local park. Because the species-area curve does not work well with small numbers, your students need to be able to find at least twelve different species. You should look over the area first to make sure there are enough species before you send your students out. You should also

make sure there are no poisonous plants in the area. Make copies of "Leaf I.D." (pages 162–163 in the Student Book). Also, see the box on "Endangered Plants" (page 296).

Collect these materials: stakes (pencils or coffee stirrers work well), twine, tape measure or yardstick, easel and paper (optional), marker, graph paper (optional), large clear plastic bags (such as Ziploc™ freezer bags), clear tape, copies of "Leaf I.D.," "Graphing Greens Data Log," and "Java Raptors" (optional) on page162–165 in the Student Book.

What to Do • Part II

Graphing Greens: The Species-Area Curve

In this activity, your students will learn how to make and interpret the species-area curve, one tool scientists use to investigate the level of biodiversity in a habitat. The activity will also give them a firsthand look at the relationship between biodiversity and habitat size because they'll survey plant diversity in a habitat they're familiar with—their schoolyard, community, park, or some other local area. They'll look at it in different-sized chunks to see how the number of species changes as the habitat size gets bigger and bigger. (See "Setting Up Plots" [below] for information on how to set up the plot. For what to do about endangered plants see the box on the next page.)

The species-area curve is a graph that shows the relationship between habitat size and the number of

species in the habitat. In this activity your students will look at the relationship between the size of your selected habitat and the number of plant species in it (they can also look at other kinds of species if you like). Almost every species-area curve has the same general shape: the number of species rises fairly quickly and then levels off (see the species-area curves on pages 299–300). The way that curves differ from this general shape can give scientists important information about the habitat or species they're studying.

This activity would fit well with a unit on graphing. It's designed to get students to think about graphs as tools for looking at the natural world. While this activity does not require a very advanced understanding of graphing, it is probably too advanced to be a good introduction to making and interpreting graphs.

Setting Up Plots

Mark off an 8 yd. by 8 yd. square and divide it into 10 plots using stakes (pencils or coffee stirrers will work) and twine. (You can have students help you do this if you like.)

After the plots are marked off, place a large clear plastic bag (such as a Ziploc™ freezer bag) in each plot and mark the plot number on it.

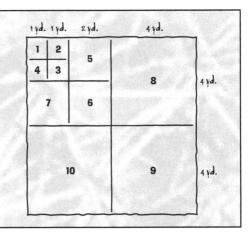

Adapted from "Quantifying Biodiversity" in *Global Environmental Change: Biodiversity* (National Science Teachers Association, 1997).

Endangered Plants

Are there any endangered plants in your area? You might want to find out before you do this activity or have your students do some research on the subject. (They could start by contacting your state fish and game department.) If there is an endangered plant in your area, and there's a chance you could find it in the area where you'll be working, have the students prepare a poster or other display about the plant that includes information about its life history, why it's endangered, and how to identify it. That way, they'll be sure not to harm any endangered plants while they're out collecting samples, and they'll learn about an endangered species that could live right in their own backyard.

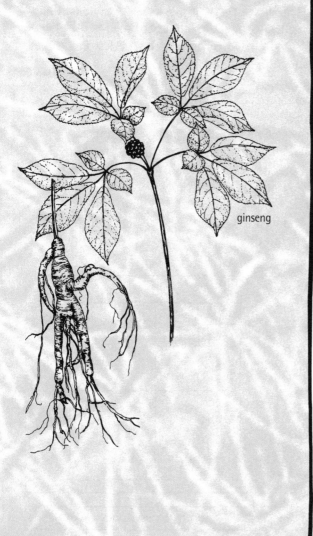

ginseng

DAY 1: Collecting Plants

1. Introduce fragmentation.

If you've already done Part I of this activity, review the concept of fragmentation. Ask students to think about island biogeography and which size islands contain the largest number of species. They should recall that the larger islands hold more species. Tell them that in this activity, they'll try to see if the same is true of larger habitats in their schoolyard or local park. They'll be investigating plant diversity in their schoolyard or local park, and finding out how it changes as they look at larger and larger habitats.

If you haven't done Part I, tell students that you'll be investigating the diversity of plants in their schoolyard. Tell students they'll be trying to see if there are different numbers of species in habitats of different sizes. Explain that many people are concerned about how the size of habitats affects biodiversity because we are breaking up many species' habitats into small chunks by building roads, homes, shopping centers, and other developments—a problem called fragmentation. Many scientists and planners are trying to better understand how fragmentation is affecting biodiversity.

Tell the students that in this activity they'll see one way that fragmentation affects biodiversity. Explain that they'll be counting the number of species in a sample area made up of plots of different sizes, and that each plot represents a different-sized habitat. You might ask for any predictions about how they think a habitat's area will affect the number of species it contains.

2. Explain the collection procedure.

Bring students to the plots (see "Setting Up Plots" on page 295) and explain how they should work in groups to collect their samples. Tell each group to take a leaf from each different species they find in their plot and put it in their plastic bag. Encourage them to be as gentle as possible and not to take more than one leaf if they can avoid it.

Review with students how to tell different plant species apart. Pick leaves of two very different species, and ask the students if they think the two leaves are the same species. They should recognize that the leaves are from two different kinds of plants. Ask them how they know. *(The leaves look different.)* Ask them to be specific about what's different. Refer to the handout "Leaf I.D." for some basic leaf characteristics that students can use to tell one kind of plant from another. Make sure they understand that the names of all the different characteristics of leaves are not important for this activity. What's important is that the students realize that these characteristics, which have been named, are ways that people tell if plants are the same species or not.

Pick leaves of two different species that look more similar and ask students if the leaves are from the same kind of plant. Students should again be specific in telling how the leaves are different. Have a few copies of the "Leaf I.D." handout for students to refer to while they're collecting.

It's best if you don't allow more than 25 students in the plot area at once because it can become too crowded for them to work. You might divide them this way: one student per 1-sq.-yd. plot, three students per 4-sq.-yd. plot, and four students per 16-sq.-yd. plot. If you have more than 25 students, have students who aren't collecting help the students working in the smallest plots identify the plants after they've finished collecting. Or have them trade the job of collecting plants. If you have fewer than 25 students, reduce the number of students in the midsize plots first and in the large plots second.

3. Collect the samples.

Give the students as much time as they need to collect their samples. Times will vary according to the number of species and the number of students. Plan to spend at least 15 minutes collecting.

4. Log the samples.

Bring all of the samples back inside and have the groups empty their bags and sort through the samples in their collection groups. Have them make sure that each group has only one sample of each species. If they have more than one, have them select the leaf that's in the best condition to represent that species.

While the students are sorting through their samples, prepare a data log like the one on the right for them to use to record their data. A piece of butcher paper that you can unroll as you need more space would work well. You can also use easel paper, but you'll probably have to use a few sheets to hold all the samples. **Note:** You may want to prepare the log in advance to save time while doing the activity.

Once you're ready, have the students bring their samples to the data log in order by plot number, beginning with Plot 1. Have the student(s) from Plot 1 tape up a sample of each species in the "species" column and put an "⊗" under the "Plot 1" column next to each species to show that they were first found in Plot 1. Every other time one of these species is found in a plot, simply mark an "X" to show that it is in the plot, but it isn't new to the entire sample of species (see sample log below).

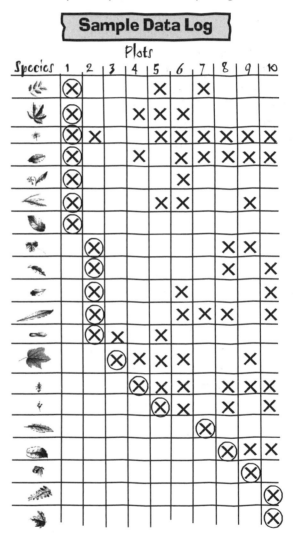

Next have the student(s) from Plot 2 post their samples. You should tape up only samples of new species. If a student has a sample of a plant that is already on the data log, he or she should mark an "X" in that species' row. Any new species should be taped up and an "⊗" should be placed in that species' row under Plot 2 to show that it first appeared in Plot 2. Do the same for the rest of the plots.

Java Raptors: A Down-Time Idea

While you're filling in the data log, there will be a lot of time when most students don't have a task. If you would like to give your group an assignment related to the species-area theme, you can give them the "Java Raptors" handout on page 165 in the Student Book. This handout is designed so that students can work independently while you fill in the data log. It should give the students a good idea of how scientists use graphs both to make sense of data and to learn how different species use space in their habitats. You can have students work independently or in groups on this worksheet while students add their plants to the log.

Alternatively, you could give your students the "Java Raptors" assignment at the end of the activity as a follow-up to get them thinking about other ways of graphing the species-area connection. In that case, you'll need to have some other activity for them to work on while you're making the data log. You might want them to focus on the plant samples they've collected, having the students identify the samples, classify them, or make rubbings of them as a science or an art project.

(Answers to "Java Raptors": area; average number of birds seen per hour; 3—4.5—6.7—8.2; large; these birds are predators and need large areas to hunt for their prey; answers will vary.)

5. Fill in the "Graphing Greens Data Log."

Once you have finished the data log, have the students use the information in the log to fill in the data summary table at the bottom of the "Graphing Greens Data Log". This table will help them make the species-area curve (see sample table below). They'll use the "Total Number of Species" row (the cumulative total of species found in the plots) for their y-axis and the "Total Sample Area" row (the cumulative area of plots that make up the sample area) for their x-axis when they make the graph.

Students can make this graph as a group, or they can make the graph as a homework or an in-class assignment if they need the practice. If they're going to make their graphs on their own, have each student fill in a data summary table. If you're going to make the graph as a class, you can summarize the data into one table as a group. On Day 2, you can either go over the graphs that the students made at home, or you can make one group graph.

Data Summary Table

Plot Number	1	2	3	4	5	6	7	8	9	10
New Species (first seen in sample area; ⊗s in this plot)	7	5	1	1	1	0	1	1	1	2
Total Number of Species (all ⊗s up to now)	7	12	13	14	15	15	16	17	18	20
Plot Area (sq. yd.)	1	1	1	1	4	4	4	16	16	16
Total Sample Area (total of plot areas in sq. yd.)	1	2	3	4	8	12	16	32	48	64

DAY 2: Plotting the Species-Area Curve

6. Graph the results.

The graph should have "Total Area" on the x-axis and "Number of Species" on the y-axis (see sample graph below). Data for the x-axis will come from the "Total Sample Area" row of the data summary table. Data for the y-axis will come from the "Total Number of Species" row of the data table. The number of data points will equal the number of plots (10 in this example). The graph below is based on the sample data on page 298. Your students' graphs should look similar.

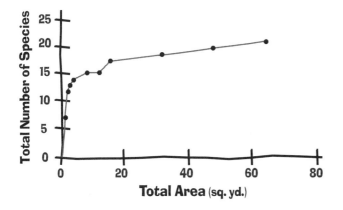

7. Interpret the graph.

Before you can talk about all the different uses of the species-area curve, your students have to interpret it. Did your graph have the same general shape as the sample curve? If it did, tell the students that the curve they made based on their schoolyard habitat or local park is a lot like curves made from samples in other kinds of habitats. Most species-area curves have this general shape. If your curve is very different, show them the more common type of curve so they get a better sense of how species-area curves usually look.

Ask the students why species-area curves look this way. *(In general, most species in North America are commonly found throughout their habitat. In other words, you would probably see in a 10-sq.-yd. plot of forest most of the trees, birds, or mammals that live in a 50-sq.-yd. plot of the same forest. But as you looked at larger and larger plots, your chances of finding rare species or species that require special resources would increase. [Species that are top predators, such as big cats and birds of*

prey, may be rare in a habitat because they need a lot of space to find food. They might only be in a plot if their home or nest is there or if they are passing through in search of food. Species can also be rare if they have more specialized needs than other species and depend on a certain type of soil or food. These species would only be found where the resources they need are found.] As you look at bigger and bigger chunks of a habitat, the chances increase that you'll find these rare species, but you won't find them at the same rate that you found the more widespread species. So the curve will usually rise sharply at the small plot areas, then more slowly as the area increases.)

8. Discuss how scientists use the species-area curve.

Graphs are important tools of scientists. The graphs help scientists make sense of a lot of data by putting it in a form that allows the scientists to see quickly what the numbers mean. Talk with your students about how the curve might allow them to see the connection between species and habitat area better than the log would. What kinds of things do they think they could use the species-area curve to find out?

One way scientists use the curve is to figure out how much of a habitat they have to look at in order to find most of the species living in it. If they want to take samples of most of the species in a habitat, they probably won't have time to go over every square inch of it in search of all the species. But by using the species-area curve, they can look on the graph at the plot area where the graph levels off, or where very few new species are added, and they can look at a plot of that size and feel confident that they will get most of the species in the habitat. Based on the sample curve, for example, a scientist would look at a 20-sq.-yd. plot to find most of the plant species in the habitat. Ask the students what size plot they would have to make to find most of the plants in the area they surveyed.

Another way the species-area curve can be used is to compare different habitats. Although almost all species-area curves have an initial rise in the number of species followed by a more level curve, different habitats can have curves with different shapes. The steepness of the rise, the point where the curve levels off, and how quickly it levels off can be different for different habitats.

Draw a species-area curve that is less steep and that levels off at a lower number of species than the one you made as a class. (See the example below. You may want to change the numbers on the y-axis to be more like the numbers of species your group found.)

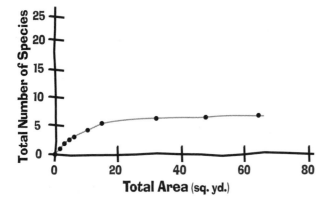

Ask the students what the different shape tells us about how the species in this habitat compare to ones in the habitat they explored. *(There are fewer kinds of species in this habitat than in the one your group explored.)*

Draw a species-area curve that is very steep and levels off quickly at a high number of species.

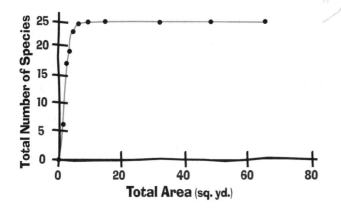

Ask the students how this compares to their schoolyard or habitat sample. *(This habitat is very high in biodiversity, and the species seem to be tightly packed. A very small plot would show almost every species in the habitat. In the students' study area, species are most likely more spread out, and there are probably fewer species.)*

Species-area curves can also be used to look at one habitat over a long period of time to see how it changes. The students could look at their schoolyard at different times of year or after major disturbances such as big storms, insect population explosions, or pesticide applications to see if these events changed the species-area connection.

9. Discuss the species-area curve and conservation.

What does the species-area curve tell us about the problem of fragmentation? Since many curves level off at relatively small plot areas, does it mean that small habitat fragments will still contain most of the species that were in the larger habitat? Unfortunately not. Ask students if they can think of any reasons this isn't true. *(Remind students that they looked at plots that were part of a larger habitat. They weren't looking at habitat fragments. If they've played "Island Hopping," they should remember that, in fragments, the edge effect will affect the number of species. It can change the habitat drastically and can cause a loss of more of the original species from the larger habitat. Also, have them think about the species that are lost at the large habitat sizes. In species-area curves, a few new species usually appear once the curve has leveled off. If we cut a habitat's size to a point where only a few of these species are lost, we may be losing some important species. We would probably lose species that require large areas, such as top predators, and these kinds of species often play important roles in habitats. Without them the habitat and species in it could change. Encourage the students to come up with other reasons there might be fewer species in a habitat fragment than the species-area curve shows there would be in a plot of the same size within a larger habitat.)*

WRAPPING IT UP

Assessment

Distribute a sheet of graph paper (with 1/2-in. to 1-in. squares) to each student. The students are to be "developers" in charge of developing the land represented by the graph paper. One-fourth of the space will be used for housing, one-eighth of the space will be used for roads/parking, one-eighth of the space will be used for commercial development, and one-fourth of the space will be used for industrial development. The remaining space (one-fourth of the total area) will be natural or landscaped area. Have the students design their development. They should label or color-code the design and, on the edges, explain why they used the land in the way they did.

> **Unsatisfactory—**The student's design does not incorporate any of the concepts regarding fragmentation and species diversity.

> **Satisfactory—**The student's design uses basic ideas to avoid fragmentation.

> **Excellent—**The student's design uses ideas both to avoid fragmentation and to increase species diversity.

Portfolio

Part I has no portfolio documentation. For Part II, use the tables and graphs in the portfolio.

Writing Idea

Have the students write an interview between a journalist for *BioTimes* magazine (or some other title they come up with), and an animal or plant whose habitat has gotten smaller and smaller because of development. The interview could include questions such as these: *"So, why are you leaving home?" "Where do you think your travels will take you?" "What are your special habitat needs?" "How could people have reduced the damage this development has caused?"* (You might want to have the students do some research in advance to find out about the specific needs of their species.) Afterward, students can share their interviews by taking turns playing the roles of journalists and species being interviewed.

Extensions

- Stage an in-class debate about a current development issue in your area. Have half the students in favor of developing the land and the other half against it. Those in favor of development should be able to cite some of the potential social and economic benefits of the proposed project, and those opposed should cite some of the project's potential environmental consequences, especially its potential effect on biodiversity. Can the two sides agree on a compromise?

- Look for fragments of the same type of habitat in your community. You might find fragments of pine forests, beach-dune systems, or grasslands. Then take your students out on a field trip to investigate some different-sized fragments. Have them think of ways they could investigate the level of biodiversity in the fragments, and then compare the fragments. You might ask a local park ranger, a naturalist, or some other expert to help you organize the trip.

Resources

Global Environmental Change: Biodiversity
(National Science Teachers Association, 1997). (800) 722-NSTA.

The Song of the Dodo: Island Biogeography in an Age of Extinctions
by David Quammen (Simon and Schuster, 1996).

The Theory of Island Biogeography by Robert H. MacArthur and
Edward O. Wilson (Princeton University Press, 1967).

Visit National Wildlife Federation's Backyard Wildlife Habitat Program's
Web site at <www.nwf.org/nwf/habitats>.

28 | Dollars and Sense

World Wildlife Fund

SUBJECTS

mathematics, social studies (economics), language arts

SKILLS

gathering (reading comprehension), organizing (plotting data, graphing, sequencing), analyzing (calculating, identifying components and relationships among components), interpreting (identifying cause and effect)

FRAMEWORK LINKS

30, 31, 32, 65

VOCABULARY

demand, **externalities** *(external costs), supply, market economy,* **natural resource***, nonrenewable resource, price, product, renewable resource, value*

TIME

Part I—*one session*
Part II—*one session*
Part III—*one session*

MATERIALS

See "Before You Begin" Part I, Part II, and Part III

CONNECTIONS

For students without an economic background, this could be a challenging activity. Both "Putting a Price Tag on Biodiversity" (pages 266–269) and "The Many Sides of Cotton" (pages 278–285) directly relate to different aspects of economics and biodiversity. However, if you want to build a unit that looks at economics and the environment, check out some of the resources listed at the end of this activity that go into more depth on the relationships between economics and environmental issues.

AT A GLANCE

Using chewing gum as an example, explore some of the connections between economics and biodiversity.

OBJECTIVES

Explain the role of consumers and producers in a market economy, and how supply, demand, and price are interrelated. Describe the connection between what we buy and biodiversity and give examples of how culture, biodiversity, and economics are linked.

In our society we are bombarded with advertisements designed to get us to buy certain products. Yet few of us have an understanding of the way our economic system works and how what we buy affects biodiversity and the global economy. This activity is designed to introduce your group to some of the connections between economics and biodiversity, as well as to some of the basic principles of our economic system, including how a free-market works.

This multi-part activity focuses on how our economic system can affect biodiversity. Our *economic system* determines what things are produced in our society and how they are produced. In turn, what *resources* we use depends on our economic system. At the heart of an economic system are *producers* (who supply goods and services) and *consumers* (who demand goods and services). Producers and consumers come together in *markets.* A market may be concentrated in one place (such as a seaside fish market or the Chicago Mercantile Exchange) or spread all over a country (such as the market for lawyers) or the world (such as the market for Boeing 777 airplanes). (See box on page 305 called "Demand, Supply, and Price—What's the Scoop?")

To see how the economic system affects biodiversity, it helps to look at the nature of resources and what influences producers and consumers. Resources include human labor and businesses; assets available in nature such as sunlight, fish, metal ores, and trees; and human-produced aids to production such as machinery, roads, power lines, and buildings. Economists call resources "factors of production," and often group them under four general headings—natural resources, labor, capital, and management.

The economic system depends on resources being used to satisfy human *wants* and *needs* . Since resources are limited and wants are unlimited, resources in a sense are always relatively *scarce.* Clean water and breathable air are examples of resources that have become more and more scarce in some of the world's cities because of the increasing wants of society. Even though these resources were once plentiful, as people demand more products and services, the pollution coming from production of these products and services degrades the limited resources of clean air and water.

Using chewing gum as an example, your students will be introduced to supply (by producers), demand (by consumers and other producers), and price. Because our economic system is very complex, we have simplified the economics presented to make the activity understandable for this age group. For example, there are many influences that act on producers and consumers apart from price. These influences include advertising, government subsidies, trade restrictions, price controls, taxes, and the activities of labor unions. Also there are political, social, and historical influences. Some of these factors are mentioned in the activity, but the focus remains on the relationship between producers, who supply, and consumers, who demand goods and services.

Like many educators, we believe that it's important for students to start thinking about economic issues in middle and high school and to learn how individual consumer choices can influence our economic system. But we don't think it's important for them to get bogged down with too many details. We also think it's helpful for them to discover that our economic system is global and heavily influenced by the social norms of individual countries and regions, which help determine what is produced and how it is valued.

Before You Begin • Part I

Make copies of "What Would You Pay?," "What's the Demand?," "What's the Supply?," and "What's the Price?" (pages 166–169 in the Student Book) for each student. Purchase several packs of chewing gum to use as props.

What to Do • Part I

Supply and Demand

1. Ask students to think about the value of a pack of chewing gum.

Begin by holding up a large pack of chewing gum. Ask your students how much they'd be willing to pay for the gum. First ask how many of them would pay 10 cents. (You might need to give them some direction about how much they would normally pay for gum.) Record the number on the board. Next ask how many would pay 20 cents. Continue until you reach 2 dollars. Have the students fill in the "What Would You Pay?" chart on page 166 in the Student Book as you write the numbers on the board.

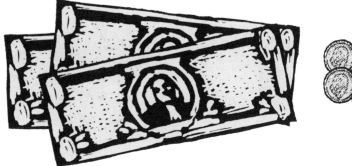

Renewable and Nonrenewable Resources

When thinking about natural resources and the economy, people often group our natural resources into the following two categories:

1. **Renewable Resources:** *This category includes living organisms (plants, animals, and micro-organisms) that can reproduce and grow, and nonliving things (such as air, water, soil, and solar radiation) that are replenished and maintained by natural cycles and energy flows. Some renewable resources, such as soil, can take centuries to be replenished, while other renewable resources, such as sunlight, are replenished daily–at the speed of light!*

2. **Nonrenewable Resources:** *This category includes energy sources (oil, gas, other fossil fuels, and radioactive elements), minerals, and other resources that if used are gone, at least within human time frames. (For example, fossil fuels could develop again–but it would take millions of years and certain climatic and physical conditions.) The depletion of some nonrenewable resources, such as aluminum, can be slowed through conservation measures, but eventually they too will become unavailable to us.*

The use of both renewable and non-renewable resources creates waste (solid waste, air pollution, heat waste, toxic runoff, and other types of pollution) that can affect the health and well-being of living things.

2. Introduce the three worksheets about demand, supply, and price.

Although the next four topics (demand, cost, supply, and price) are closely interrelated, it may be best to introduce them to your students in several steps.

a. Introduce demand.

Pass out copies of "What's the Demand?" (page 167 in the Student Book), and have the students graph a demand curve based on the numbers supplied above the graph. These numbers were determined by looking at the same kind of information your students just provided in the "What Would You Pay?" chart earlier that showed how many students would be willing to buy a pack of gum at each of the prices. The only difference here is that the results from numerous groups have been compiled. To get the information that will be graphed, the number of people willing to buy gum at each of the prices was multiplied by the number of packs of gum each one would purchase. This amount is called the *total quantity demanded* and can be graphed using what is called a *demand curve.* Notice that, as in the group example of "What Would You Pay?" the total quantity demanded goes down as the price goes up. (See answers on page 313.)

b. Introduce cost.

As the students are thinking about demand, introduce the concept of cost. Ask the group why stores don't charge 10 cents for gum instead of 50 cents, because so many more people would be willing to buy gum at 10 cents. *(It's because there are costs involved in producing and selling anything, and all the people involved in the production and sale of a product need to recoup their costs and make a profit.)* Ask your students to think about some of the costs involved in making and selling chewing gum, such as growing

sugar cane, refining sugar, transporting sugar, producing the gum base, producing and adding flavoring, producing the packaging, transporting cases of gum to retail stores, advertising, employing retail workers, and paying other retail costs.

c. Introduce supply.

Pass out copies of "What's the Supply?" (page 168 in the Student Book). Explain that producers cannot stay in business if they sell their gum for less than it costs them to make it. And producers are able to supply more when the price is higher. By graphing the price (on the vertical axis) and the number of packs of gum producers will supply (on the horizontal axis), you get a supply curve.

(The "What's the Supply?" worksheet has a sample supply curve and several questions for your students to answer. Point out that as the price rises suppliers will supply more packs of gum.) (See answers on page 313.)

d. Introduce price.

Pass out copies of "What's the Price?" (page 169 in the Student Book), and have the students draw both the supply curve and the demand curve on the same graph. The place where the curves meet indicates the price (on the vertical axis). (Have the students draw a dotted line from the point where the demand and supply curve lines meet to the vertical axis.) (See answers on page 314.) The indicated price

Demand, Supply, and Price—What's the Scoop?

Demand is the amount of a product or a service that consumers are willing and able to buy at a given price. As price goes up, consumers usually demand less of a product (see demand curve on the right). This demand curve shows the relationship between possible prices and the quantity demanded (and assumes that other factors, such as fashion or the consumer's income, don't change).

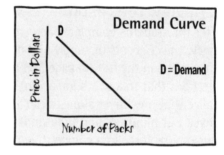

Supply is the amount of a product or a service that suppliers are willing to offer for sale at a given price. As price goes up, producers supply more of a product, creating a *supply curve*. The illustration on the right shows the relationship between prices and quantity supplied (and assumes that other factors, including costs such as taxes, don't change). Manufacturers incur costs to make gum, which affect the price they can accept. In our hypothetical example, manufacturers cannot offer gum for less that 50 cents a pack because if they did they couldn't cover the *cost of production*. Some of the costs of gum production include machinery (capital costs), workers (labor), and ingredients such as sugar and flavorings (natural resources).

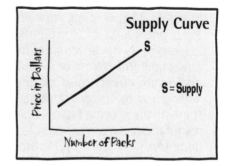

Market price is the price at which the quantity supplied is equal to the quantity demanded (see the illustration on the right). The price is determined by all the interactions between the potential buyers (demand) and the potential sellers (supply). And the market is not one place or store—it is represented by all the suppliers that sell gum and all the customers who buy it.

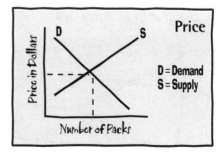

level occurs where the consumers want to buy the same amount of a product or a service as the suppliers are willing to offer for sale. (See "Demand, Supply, and Price—What's the Scoop?" on page 305 for more about price.) Ask the students what the "going price" is for a pack of gum, using the data from "What Would You Pay?" *(The "going price" depends on how much the students we surveyed were willing to pay for gum compared with the amount of gum the producers are willing to supply at a certain price.)* Ask the students if the price looks reasonable. Is it more or less than the price of gum in the community? Why would it be different? *(One reason is that the demand curve we provided is based on the results of a small survey and may not reflect the general public's demand. Another reason is that the supply curve might not accurately reflect the behavior of actual suppliers.)*

3. Review the results.

After the students complete each worksheet, review the concepts of supply and demand, using the information in the box on page 305 and the discussion questions that follow. It's important for students to realize that one buyer alone doesn't have much power. But hundreds and thousands of buyers are a tremendous force in a competitive market economy.

a. Name or describe some specific factors that could make people more or less willing to buy gum. *(Answers will vary. Price, flavor, types and source of ingredients, social responsibility of the manufacturer, environmental effects of manufacturing, and health effects could all affect people's willingness to buy gum.)* Name or describe some of the specific factors that could make manufacturers more or less willing to produce gum. *(Answers will vary. Profitability is the main factor that influences manufacturers' willingness to produce gum. However, many businesses also consider the social impacts along with profitability.)*

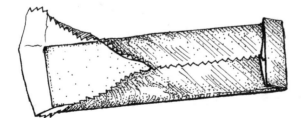

b. What would happen to the supply of gum if the price of sugar, one of the main ingredients in gum, rose sharply? *(The costs of supplying the gum would go up, so gum manufacturers would be willing to supply fewer packs of gum at any given price than before. This would shift the supply curve to the left, from S to S_1. The quantity demanded would drop off as the price rises, and a new, higher market price would be created. See graph #1.)*

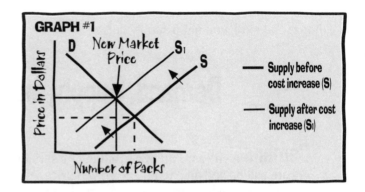

c. Suppose that a national association of dentists runs a huge national campaign to encourage kids to stop chewing gum to prevent cavities. How could this campaign influence demand? What would the new demand curve look like? What would happen to the demand curve for gum if several companies started national campaigns saying that their sugarless gum is better—that their gum doesn't cause cavities or hurt the environment? *(If the dentists' national campaign is successful, fewer kids would want to buy gum, so the demand curve would shift to the left. [See graph #2.] This would create a new market price that's lower than before. If several companies were able to produce and market a new sugarless gum, demand for gum could rise back to the levels before the dentists' campaign and could even be so high that it encourages more manufacturers to produce more sugarless gum.)*

d. Can consumer power influence whether or how something is produced? Give examples. *(Yes. Public outcry over the way things are produced can lead to changes in how manufacturers produce them. For example, public outcry over the plastic foam*

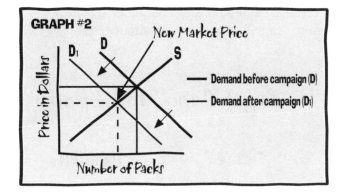

GRAPH #2

containers that McDonald's ™ used for sandwiches eventually led McDonald's to switch to paper packaging. Public support for "positive" products is often as important and effective as boycotts of "negative" products. For example, by campaigning against tuna that hadn't been caught in a "dolphin safe" way and by buying only tuna that had been caught in this way, people pushed tuna packagers to package only "dolphin safe" tuna.)*

Before finishing the activity, write down any questions that students may have and collect their worksheets for the second day of the activity. Remind the students that the economy is extremely complex (as they'll see in Part II) and that supply and demand are influenced by many factors.

Before You Begin • Part II

Make copies of Parts 1 and 2 of the "Stuck on Gum" worksheet (pages 170–171 in the Student Book) for each student. Also make copies of "Tackling Trade-Offs" (page 172 in the Student Book) for each student.

What to Do • Part II

Biodiversity and Gum

What is gum made of? How do supply, demand, and price relate to biodiversity and the environment? In Part II of this activity your students will take a closer look at what goes into making and selling gum and how the production of gum (and other products) connects to the environment.

1. Brainstorm about gum's manufacturing process.

Conduct a brief brainstorming session to get your students to think more about how gum is made and what some of the costs are in producing gum. First make a list of the types of ingredients that the students think go into gum, such as gum base to make gum chewy, flavorings to give gum taste, sugar to make gum sweet, and chemicals to make gum soft. (If the students are not able to think of any

ingredients right away, have them look at the ingredients listed on a pack of gum.) Then ask them to think about the costs involved in manufacturing and distributing gum to the public, such as people involved in all stages of gum production—from growing sugar cane, to transporting gum to stores, to selling gum to customers. (You might want to refer to the list the students made in step 2 of Part I.) Finally ask your students if they think there is any connection between gum and biodiversity. *(Gum uses natural resources such as sugar.)*

Now tell the students that you will be reading them a story about gum that will answer some of the questions they have about how it is made. Leave the brainstorming lists on the board so they can think about what they said as they listen to the story.

2. Read "Stuck on Gum."

Pass out copies of Part 1 and Part 2 of the "Stuck on Gum" worksheet (pages 170–171 in the Student Book). Ask the students to listen carefully as you read Part 1 of the "Stuck on Gum" story (page 312) aloud to the group. (Part 1 of the story will get your kids thinking about how we depend on natural systems for the products we buy. Part 2 focuses on the ingredients and processes used to make gum today.)

After reading Part 1 of the story, divide your students into teams and have each team work together to complete the two worksheets. (You can either read the second part of the story or pass it out and have one person in each team read it aloud to the other team members.) You might want to pass out copies of both Part 1 and Part 2 of the story so the teams can use them to answer their worksheets.

3. Discuss the worksheets.

After the students have completed the worksheets, go over their answers, calling on teams to explain how they came up with their answers. (See pages 315 and 316 for examples of answers.)

4. Introduce the concepts of external factors or hidden costs.

For older or more advanced students, you might want to introduce the concept of external factors (costs to society that are not factored into the price of a product) after you've discussed the worksheets. (This can be done by asking the students to brainstorm a list of external factors.) The exercise will help your students realize that the products we buy have "hidden costs" that are very difficult to measure and quantify but that can decrease the quality of life in our society nonetheless. For example, a stick of gum is more than 50 percent sugar. The cost of labor and machinery to produce, ship, and pack the sugar is accounted for in our system and the costs get passed on to retailers and

consumers in the price. However, some of the costs of producing the sugar are not accounted for. These hidden costs include the loss of forest cleared for sugar plantations (which affects the carbon dioxide recycling potential of forests, the species that live in forests, and so on), the water diverted from rivers for irrigation, pollution resulting from the use of fertilizers and pesticides, and smoke created when sugar cane fields are burned after harvest.

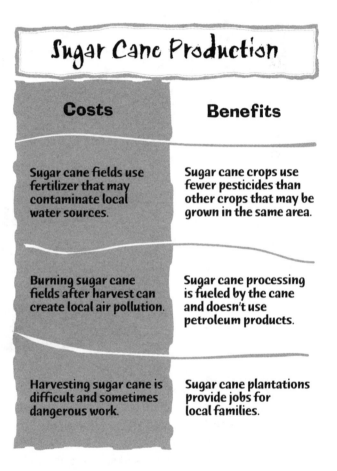

Sugar Cane Production

Costs	Benefits
Sugar cane fields use fertilizer that may contaminate local water sources.	Sugar cane crops use fewer pesticides than other crops that may be grown in the same area.
Burning sugar cane fields after harvest can create local air pollution.	Sugar cane processing is fueled by the cane and doesn't use petroleum products.
Harvesting sugar cane is difficult and sometimes dangerous work.	Sugar cane plantations provide jobs for local families.

Many economists believe that we need to factor in the cost of natural resource depletion, runoff pollution, and other environmental and social factors when we make decisions about how to produce a product—and even *if* we should produce a product. But others argue that many of these externalities involve complex trade-offs and that it's almost impossible to take into account all the possible environmental side effects. And in some cases one type of environmental problem could be preferable to another. For example, sugar cane may be less damaging to the environment than other crops that would be grown on the same plot of land because it might use fewer pesticides and less fuel and labor. (See some examples of the costs and benefits of sugar production in the chart on page 308.)

Your students may have listed some of these hidden costs as they thought about how gum affects biodiversity. They might have also mentioned how gum production has affected the people in the regions where chicle is produced. For example, the chiclero lifestyle (highlighted in Part 1 of the "Stuck on Gum" story) developed as the demand for natural chicle increased. But as synthetic latex and other non-natural ingredients replaced chicle, the culture in many "chicle communities" changed and people had to find new ways to make a living.

In terms of supply and demand, the chicle situation can be diagrammed as below:

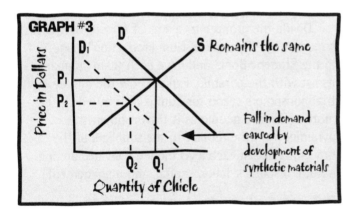

GRAPH #3

Price in Dollars

D_1 D S Remains the same

P_1
P_2

Fall in demand caused by development of synthetic materials

Q_2 Q_1

Quantity of Chicle

Before the development of synthetics, chicle communities could sell a certain amount (for example, Q_1) of chicle a high price at (P_1) per unit. The total income gained can be figured out by multiplying the quantity sold by the price at which it was sold. (For example, total income would have been P_1 x Q_1.) After natural chicle was replaced with synthetic materials in the gum manufacturing process, the communities could not sell as much chicle because the demand dropped, and that, in turn, caused the price to drop. (Note that the supply

curve for chicle is unchanged even though the quantity actually supplied fell.) For the chicle communities the drop in demand was a disaster that they had no control over.

5. Discuss trade-offs.

Ask your students to think about some of the trade-offs that are involved in producing, distributing, and selling products, and how their buying habits affect the environment. For example, you might want to hand out copies of "Tackling Trade-Offs" (page 172 in the Student Book) and have your students work in small groups to answer the questions on the worksheet. (See page 317 for possible answers. Also see the extension on page 320 that looks at production cycles of other products to find out more about the connection between what we buy and the environment.)

You might also want to explain that economic issues are very complex and often involve policy decisions that try to balance the needs of the environment, workers, consumers, and producers. For example, we don't really know exactly what impact not producing gum would have on the global environment or economy. Yet even though the issues are complicated, it's important that your students start to see that people can influence our economic system—from voting for certain candidates to becoming educated consumers to choosing a career that helps make our economic system more environmentally or socially responsible. It's also important for them to realize that every product affects the environment in some way, that every product also affects other aspects of society (health, jobs, and so on), and that consumers have to weigh the environmental pros and cons against many factors.

To help them understand these relationships, you might want to bring in outside speakers who can explain the role of different sectors (i.e. government, consumers, producers, retailers) in trying to balance the sometimes competing demands. The aim is to get kids thinking about the trade-offs in producing or not producing a product, how products are connected to the environment, and the role consumers play in a market economy.

Make copies of "Introducing the Factors of Production" (page 173 in the Student Book) for each student. Make copies of the "Pyramid of Gum Production" (page 174 in the Student Book) for each team. Also provide large sheets of paper (one per team), tape or glue, and scissors.

Looking Deeper

Pine trees, offshore oil rigs, Central American rain forests, and Midwestern cornfields are all involved in producing a stick of chewing gum. In this part of the activity your students will gain a better understanding of the links between the products they buy and activities all over the world. They will also see that our economic system depends on the Earth's biological wealth—including the free services that we get from ecosystems.

1. Introduce the "factors of production."

To produce products, our economic system relies on four main components, which economists call the "factors of production." These factors are as follows:

- **Natural Resources:** the raw materials, such as timber, minerals, and topsoil, needed to produce a product, as well as water, clean air, and so on

- **Labor:** the human effort that goes into creating and producing a product, including muscle power and brain power

- **Capital:** the equipment and structures needed to turn raw materials and labor into finished products

- **Management:** the process that organizes and manages the labor, raw materials, and capital to produce and sell finished products

Write the four factors of production on the board and go over the definitions. See if your students can come up with some examples of the factors on their own and then discuss how economists have grouped them into these four categories.

2. Review gum production.

Review all the steps in the gum production process the students brainstormed in Parts I and II. Do they want to add anything to the list?

Divide the group into teams of four or five. Then hand out the "Pyramid of Gum Production" (page 174 in the Student Book) and have each team compare its list with the pyramid. Explain how the money that consumers spend for gum is distributed to more and more people as it travels down the pyramid, from the retailer to the suppliers of the raw materials. At each level money is divided among natural resources, labor, capital, and management.

3. Take a close look at the factors of production.

Hand out copies of page 173 in the Student Book, "Introducing the Factors of Production," to each student. Give each team a large sheet of paper. Have each team cut out the "pyramid" and glue it to the center of the large sheet of paper. Also have the team members cut apart each of the activities in "Introducing the Factors of Production." Then have them try to match each activity with the appropriate factor of production. For example, someone ringing up a cash register would represent labor. Tell the students that each activity can be matched to more than one factor of production. Finally have the students glue each activity to the appropriate place on the production pyramid. For suggested connections see the answer sheets on pages 318–319.

Explain that when someone buys a pack of gum, the money is divided among the many people who have worked to produce the gum and to get it on the shelves of the store where the consumer bought it—from the farm workers cutting cane to the artist who designed the gum wrapper to the clerk who sells it at the corner store.

4. Explore the biodiversity link.

Finally ask the students to look at the biomes at the bottom of the pyramid that support gum production. Explain that many industries, environmental groups, and other concerned organizations and citizens are working on ways to better protect and value our natural ecosystems within our current economic system. By placing value on such things as clean air and flood control, and by understanding their importance, many people are hoping that we can develop more efficient methods of production that help protect these natural systems and allow us to use them sustainably. There are many economists who think that putting a price on these more intangible things would be impossible. At the same time many people believe that if the United States added hidden costs to the price of goods and services, while other countries didn't, we would be less competitive in a global economy.

To wrap up, ask the students if they think that, if consumers knew more about the economic background of different products, their wants and needs as consumers might change. Why or why not?

STUCK ON GUM STORY—PART 1

JUNGLE CHEW

In 1871 a brand new substance appeared on the shelves of a drugstore in Hoboken, New Jersey. A box labeled "Adams New York Gum" was filled with dozens of small flavorless balls that were meant to be chewed, not swallowed. Chewing gum had made its first appearance in the United States.

But what was this stuff? The balls were made from the sap of a tropical tree called the sapote (sah-POH-tey), which grows in the tropical forests of southern Mexico and northern Central America. (The scientific name of the tree is *Manilkara zapota*.)

Like many other rain forest plants, sapote trees couldn't be grown on plantations, so between late summer and December, people ventured into the jungle to collect sap, or chicle (CHEE-clay), from wild trees. They cut zigzagging slits down the trunk and collected the sticky white sap in buckets. They boiled the sap to remove excess water, shaped it into 20-lb. loaves, left the loaves to cool, and finally shipped the loaves off to be made into chewing gum. The trees and forests remained healthy, and each year the local people could return to collect more chicle.

The people who collected chicle—the chicleros (chee-CLARE-ohs)—lived on the edge of the forest. In addition to harvesting chicle, they were able to hunt and collect other things they needed to live in the forest. They knew many medicinal and edible plants. By harvesting chicle, the people were able to make a living and the forest continued to grow.

Although gum was flavorless at first, a businessman named Wrigley combined sugar and flavorings with the sap, and gum grew in popularity. By 1930 the United States was importing over 15 million pounds of chicle per year. Eventually chemists invented cheaper synthetic materials that could extend or mimic chicle's chewiness. Today these chemicals are used in most gums, and there is only a small demand for natural chicle.

STUCK ON GUM STORY—PART 2

GUM TODAY

So what's in today's gum, if we no longer use boiled down sap? There are several main ingredients in a modern stick of gum.

Sweeteners—Sugar and other sweeteners usually make up between 50 and 80 percent of the gum by weight.

Gum Base—This makes the gum chewy. Natural chicle is still used in some gums, but today it is extended by a wide variety of different materials, including synthetic latex, resins, and wax. Gum base accounts for about 20 to 30 percent of the gum by weight.

Softeners—These usually constitute between 1 and 25 percent of the gum. In sugar-sweetened gum, corn syrup is the most commonly used softener. Other softeners may include honey, molasses, palm syrup, and fruit juice concentrates.

Flavorings—Only about 1 to 3 percent of gum is flavorings. These may include any combination of fruit essences, essential oils (peppermint, spearmint, cinnamon, wintergreen), and artificial flavors.

How do manufacturers put all these ingredients together? They begin with softening the hard gum base by stirring in corn syrup and other softeners. At this point flavorings may be added so that they will take a longer time to be released when the gum is chewed. The last things to be added are the sweeteners. When you chew a stick of gum, the sweeteners quickly dissolve in your mouth, giving a burst of flavor. Flavorings such as peppermint oil can last longer, but most gum loses its sweetness within just a few minutes.

Answers
WHAT'S THE DEMAND?–Worksheet 1

Price and Number of Packs Demanded			
$2.00 = 0 packs	$1.50 = 5 packs	$1.00 = 10 packs	$0.50 = 15 packs
$1.90 = 1 pack	$1.40 = 6 packs	$0.90 = 11 packs	$0.40 = 16 packs
$1.80 = 2 packs	$1.30 = 7 packs	$0.80 = 12 packs	$0.30 = 17 packs
$1.70 = 3 packs	$1.20 = 8 packs	$0.70 = 13 packs	$0.20 = 18 packs
$1.60 = 4 packs	$1.10 = 9 packs	$0.60 = 14 packs	$0.10 = 19 packs

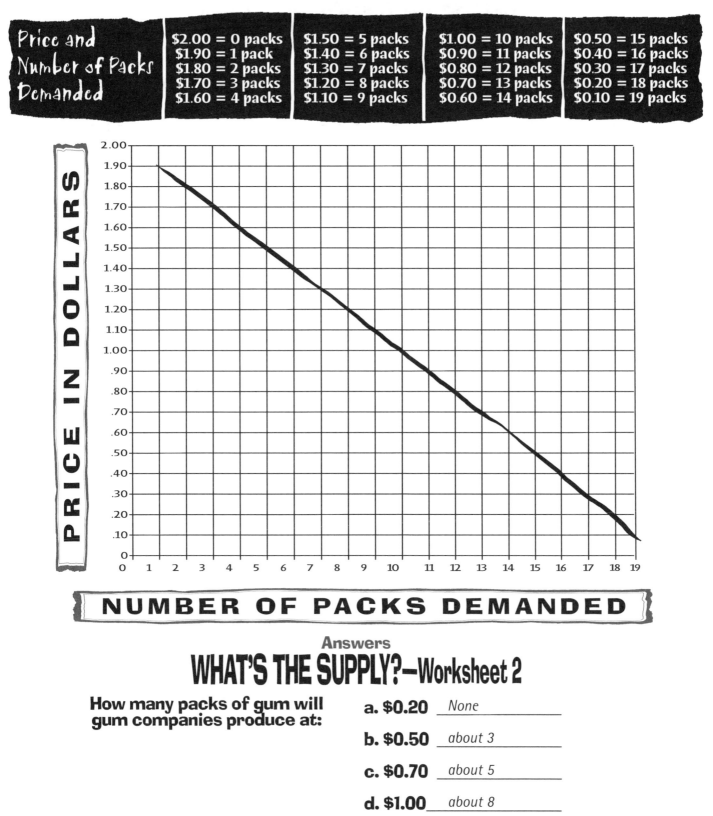

PRICE IN DOLLARS

NUMBER OF PACKS DEMANDED

Answers
WHAT'S THE SUPPLY?–Worksheet 2

How many packs of gum will gum companies produce at:

a. $0.20 *None*

b. $0.50 *about 3*

c. $0.70 *about 5*

d. $1.00 *about 8*

Answers
WHAT'S THE PRICE?–Worksheet 3

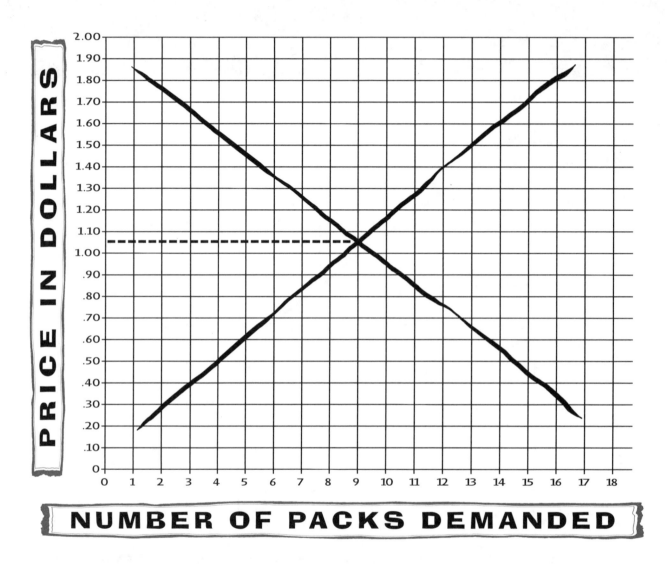

PRICE IN DOLLARS

} **NUMBER OF PACKS DEMANDED** **{**

According to your data, what is the price of a pack of gum? *$1.05*

Is the price more or less than the price in real life? Why? *More. One reason is because the data used in this exercise don't necessarily represent the whole group of consumers within a society. For example, some students might be willing to pay more for a pack of gum because there might not be machines that sell gum within the school; that makes gum harder to find and, therefore, causes those students to be willing to pay more for it. Another reason is that the supply curve might not accurately reflect the behavior of actual suppliers.*

Answers
"STUCK ON GUM" WORKSHEET
Part 1

1. At one time a large amount of gum, or chicle, was gathered in the rain forests of Central America. On the table below, name three positive and three negative effects this harvest might have had on the people and the environment in the area where it was gathered.

Answers will vary. Here are some possibilities:

POSITIVE EFFECTS	NEGATIVE EFFECTS
1 *The sale of chicle provided income for the chicleros.*	**1** *Cutting the sapote tree's bark could expose it to disease.*
2 *The chicleros developed an intimate knowledge of the plants and animals of the forest and how they could be used for medicines and food.*	**2** *The presence of too many people in the forest could harm the plants and animals that live there.*
3 *People did not have to create plantations of sapote trees, which would have required that they cut down the native forest.*	**3** *Boiling the sap caused more trees to be cut for firewood.*

2. When chicle was replaced by synthetic latex, what might have happened to the people who had gathered chicle? How could the change have affected the culture of the communities that relied on the sale of chicle?

The chicleros would have had to find another way to earn a living to maintain their standard of living. In some cases finding new jobs required that people leave the chiclero communities to find work. However, in areas where the synthetic latex is produced, jobs were created for other people in the new factories.

The main point to emphasize is that the introduction of new technologies has positive AND negative effects on the economy, the environment, and the people (and cultures) involved. Changing from natural to synthetic ingredients and products has many complex effects, both good and bad, that must be taken into account when discussing what occurs as a result of different manufacturing processes and different ingredients used to make the products. (For example, the price of gum for most consumers dropped with the discovery of synthetics, but at the same time, many chicleros lost their jobs.)

Answers

"STUCK ON GUM" WORKSHEET
Part 2
Breaking It Down

1. The drawing below shows an average stick of gum. Write which ingredient each part represents.

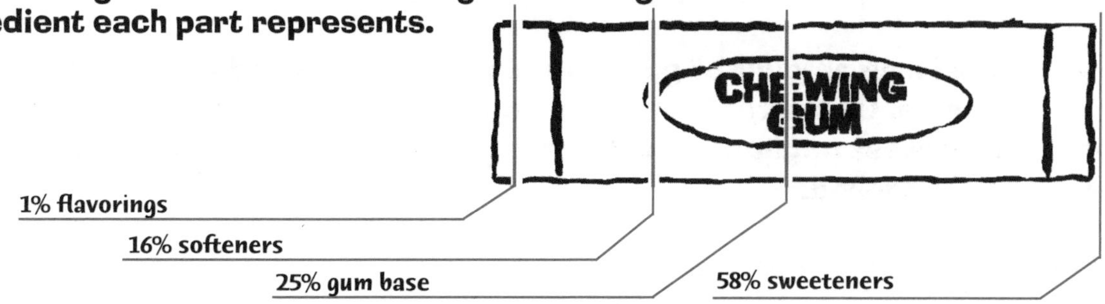

CHEWING GUM

1% flavorings

16% softeners

25% gum base

58% sweeteners

2. Today gum is made from a variety of sugars, plant oils, and other chemicals. Name some of the costs involved in making gum. *Costs can be grouped according to the factors of production (see page 310). Capital costs include the equipment and structures needed to make the gum; labor costs include the salaries and benefits of the people needed to make gum; the costs of natural resources include the purchase of sugar, flavorings, and other raw materials needed in the production process; and the management costs include the people who organize the labor, raw materials, and capital to produce and sell the products.*

3. Name some of the positive and negative effects today's gum production can have on people and the environment. What other information do you need to answer the question? *Other information that might be needed includes details on the waste that comes from the sugar production and refining process, how the flavorings are made, where the synthetic gum base comes from, and so on. Answers regarding the positive and negative effects will vary. Here are some possibilities:*

POSITIVE EFFECTS	NEGATIVE EFFECTS
1 *Gum production is no longer dependent on tropical forests for gum base.*	**1** *Some people that used to gather chicle had to find other jobs, and some have probably lost their intimate knowledge of the forest.*
2 *People no longer upset the forest ecosystem looking for chicle.*	**2** *Air pollution and water pollution are generated from the machinery used to produce and transport gum.*
3 *Less firewood is cut because boiling the chicle is no longer required.*	**3** *Some ingredients, such as sugar, come from plantations that use a lot of pesticides.*

4. Although gum base is no longer made from chicle, most of the other things that make up a stick of gum are made from products that come from plants. List five plants that add flavor or sweetness to a stick of gum. *sugar cane, sugar beets, cinnamon, spearmint, peppermint, vanilla, licorice (anise), varieties of fruits*

Answers

TACKLING TRADE-OFFS

Every product you buy is connected to the environment in some way. All products either use natural resources as a raw ingredient or cause some type of pollution as they are produced, used, or thrown away. People around the world—from business owners to consumer groups—are looking at how to make products that have fewer negative impacts on the environment. At the same time they're looking at the role consumers play in shaping what people buy and sell.

The following questions will get you thinking about some of these trade-offs:

1. If producers stopped making gum, name some of the positive and negative effects this might have on people, the environment, and the economy.
<u>Positive Effects:</u> *1) Because less sugar would be consumed, less sugar cane would be harvested. The fields then would be burned less often and less air pollution would result. 2) Kids might have fewer cavities. 3) Fewer pesticides might be used. 4) Money originally spent on buying gum might be spent on other products that, in turn, might create new jobs in other industries.* <u>**Negative Effects:**</u> *1) A lot of people would lose their jobs—from the sugar cane harvesters to gum factory workers, truck drivers, and distributors. 2) People would no longer have the choice of consuming gum. 3) Investors in the gum/candy industry would lose money. 4) New products and jobs would affect the environment in other ways.*

2. Think about the products you and your family buy every week. What are several ways your buying habits affect people and environments around the world? How could you find out more about the impact of your consumer choices? *All products affect the environment in some way. For example, students might talk about where the raw materials come from to make particular products, where and how those products are produced, transported, and disposed of. The point here is just to get students thinking about how complex and globally connected our economic system is. Consumers can find out more by contacting organizations like Green Seal, Consumer Reports, local consumer groups, and environmental organizations.*

3. What types of careers are involved in making sure that producers, consumers, and the environment are protected in our society? *engineers, city planners, natural resource economists, scientists (chemists, soil scientists, etc.), agricultural experts, government regulators, educators, environmental attorneys, and so on.*

Answers
INTRODUCING THE FACTORS OF PRODUCTION
What does it take to make something?

Natural Resources	**Labor**	**Capital**	**Management**
The raw materials needed to produce a product (e.g., timber, minerals, water)	The human effort that goes into producing a product, including muscle power and brain power	The equipment and structures needed to turn raw materials and labor into finished products	The process that organizes and manages the labor, raw materials, and capital to produce and sell finished products

Production Factor 1

Julie, a local high school student, runs the cash register at the corner store.

(Labor) (Pyramid Level: Retailer)

Production Factor 2

Gordon, a citizen of Jamaica, comes to the United States every year to harvest sugar cane and earn extra money for his family.

(Labor) (Pyramid Level: Supplier of Raw Products)

Production Factor 3

Celeste has been a beekeeper since she was a little girl. She looks after about 20 hives that produce wax and honey.

(Labor and Resources) (Pyramid Level: Supplier of Raw Products)

Production Factor 4

Giant metal mixers stir the sweetener, gum base, softener and flavorings together.

(Capital) (Pyramid Level: Gum Manufacturer)

Production Factor 5

Rafael is an artist. He designed the flashy new gum wrapper for the super-sized gum packs.

(Labor) (Pyramid Level: Gum Manufacturer)

Production Factor 6

Peppermint grows wild in many parts of the state. It is also grown commercially.

(Resources) (Pyramid Level: Supplier of Raw Products or Refined Products Manufacturer)

Production Factor 7

Bob owns and operates a farm that grows corn.

(Management) (Pyramid Level: Supplier of Raw Products)

Production Factor 8

John manages a manufacturing plant that makes gum base.

(Management) (Pyramid Level: Gum Manufacturer)

Production Factor 9

At the paper processing plant, timber is converted into thick sheets used to wrap paper products.

(Resources) (Pyramid Level: Refined Products Manufacturer)

Production Factor 10

Selim drives a truck to transport essential plant oils to different factories where the oils go into soaps, perfumes, and food.

(Labor) (Pyramid Level: Supplier of Raw Products)

Production Factor 11

Powdered sugar coats each stick of gum and keeps it from sticking to others.

(Resources) (Pyramid Level: Refined Products Manufacturer)

Production Factor 12

Computers automatically track the gum factory's production output, sales, and inventory.

(Capital) (Pyramid Level: Gum Manufacturer)

Production Factor 13

Prairies have fertile soil that is perfect for growing corn and sugar beets.

(Resources) (Pyramid Level: Biome That Provided Natural Capital)

Production Factor 14

A machete is used to cut slits in the side of the sapote tree to collect sap.

(Capital and Resources) (Pyramid Level: Biome That Provided Natural Capital)

Answers
PYRAMID OF GUM PRODUCTION

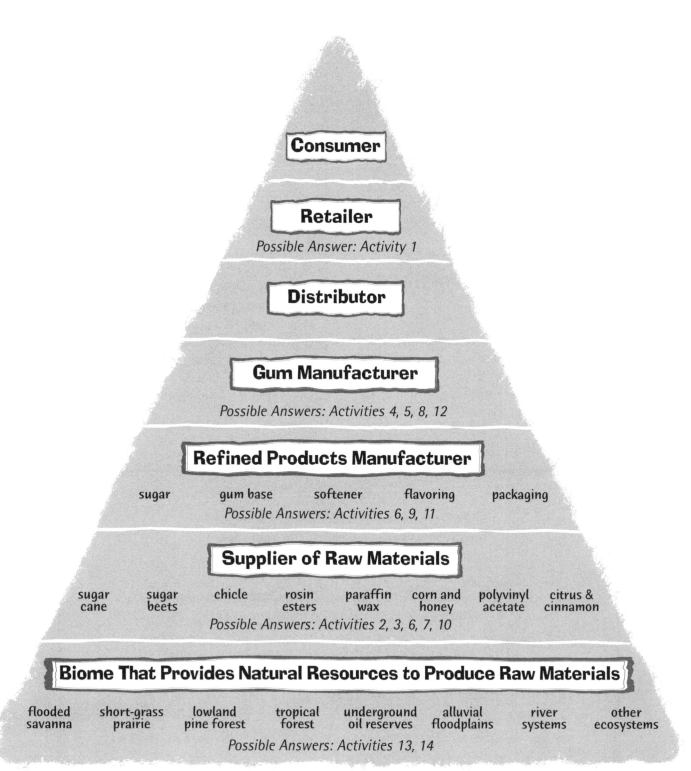

Consumer

Retailer

Possible Answer: Activity 1

Distributor

Gum Manufacturer

Possible Answers: Activities 4, 5, 8, 12

Refined Products Manufacturer

sugar   gum base   softener   flavoring   packaging

Possible Answers: Activities 6, 9, 11

Supplier of Raw Materials

sugar   sugar   chicle   rosin   paraffin   corn and   polyvinyl   citrus &
cane   beets   esters   wax   honey   acetate   cinnamon

Possible Answers: Activities 2, 3, 6, 7, 10

Biome That Provides Natural Resources to Produce Raw Materials

flooded   short-grass   lowland   tropical   underground   alluvial   river   other
savanna   prairie   pine forest   forest   oil reserves   floodplains   systems   ecosystems

Possible Answers: Activities 13, 14

WRAPPING IT UP

Assessment

The worksheets can be used for assessment.

Unsatisfactory—The student is not able to complete the worksheets.

Satisfactory—The student completes the worksheets in a manner that demonstrates a basic understanding of the economic and environmental principles highlighted in this activity.

Excellent—The student reveals creative and practical thought in the completion of the worksheets and goes beyond basic comprehension by applying the information to more complex situations.

Portfolio

The worksheets can be used for portfolio evidence.

Writing Idea

Have students keep a log of everything they purchase for one week. For each product have them identify and list the major ingredients. Finally have students write a short paragraph on how three of the most commonly identified ingredients are related to biodiversity.

Extension

After the students have explored the many ways that chewing gum is connected to ecosystems, people, and the economy, have them choose another common product and follow it through its production cycle—from raw materials to final product to eventual disposal. As they do this, have them look at the benefits of the current system of production (What are the benefits of the product? Who gets what from the production process and the final products?) and at some of the costs (Are natural resources or people harmed in the process? Are there some long-term effects that are not included in the costs of the products?). To help your students think about this, you might want to pass out the sugar cane production cost/benefit chart (page 308) as an example. Remind them that we all use numerous goods and services every day and that all products have some impact on the environment—from using energy to extract natural resources to generating waste throughout the production cycle. The challenge is to find ways to develop products that are good for people and the environment.

Resources

Economics and the Environment by Curt L. Anderson
(National Council on Economic Education, 1996).

Economics for Environmental and Ecological Management
by C. A. Tisdell (Elsevier Science, 1993).

Eco Sense—An Economic Environmental Learning Kit by Debbie Johnston
(Business Economics Education Foundation, 1992).

Environomics: Exploring the Links between the Economy and the Environment
by Gary Rabbior (Canadian Foundation for Economic Education, 1996).

A Framework for Teaching Basic Economic Concepts with Scope and Sequence Guidelines/K–12
edited by Phillip Saunders and June Gilliard (National Council on Economic Education, 1995).

Visit the National Council on Economic Education's Web site at <www.nationalcouncil.org>.

Visit the Wisconsin Council on Economic Education's Web site at <www.uwsp.edu/WisEcon>.

*"Show me a person who understands
the wise use of resources—capital, labor, and
the environment—and I will show you a
business leader of the future."*

–Samuel C. Johnson, businessperson

How Can We Prot

The activities in this section investigate how people around the world are working to protect biodiversity and create more sustainable societies—through education, research, and futures studies. For background information, see pages 50-61.

ect Biodiversity?

©David Wall/Lonely Planet Images

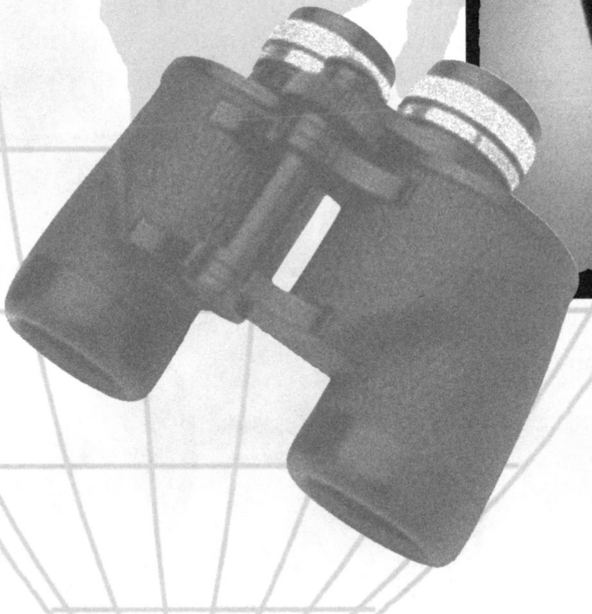

"Creating sustainable
communities, economies,
and lifestyles . . . is the
challenge of our times."

−Daniel D. Chiras, scientist

29 | Future Worlds

 AT A GLANCE

Build a pyramid to reflect personal priorities for the future, and investigate ways people are working to protect the environment and improve the quality of life on Earth.

OBJECTIVES

Create a personal vision for the future, especially as it relates to biodiversity. Reach group consensus using negotiation and conflict resolution skills. Brainstorm about ways of arriving at the envisioned future. Analyze various approaches people are taking to arrive at those futures.

Some people look into our future and see a gloomy, inhospitable picture: a world that is less healthy, less safe, less diverse, more crowded, and more polluted. While such portraits of doom may scare some people into action, they often have the opposite effect. Negative forecasts can become self-fulfilling prophecies, making people—especially young people—resigned and hopeless. This activity doesn't take a doom and gloom approach. Instead, it is designed to get your students to begin envisioning the future they *want* to inhabit and to learn about some real-life examples of how people are working to make the future brighter.

©Susannah Low/Masterfile

Before You Begin

Option #1
Make a copy of "Future Blocks," "Priority Pyramid," and "Making It Happen" (pages 177–182 in the Student Book) for each student and for each group. Also provide scissors and glue or tape.

Option #2
Make a copy of the "Priority Pyramid" and "Making It Happen" (pages 178–182 in the Student Book) for each student and for each group. Also provide scissors and glue or tape.

What to Do

Option #1

1. Make personal pyramids.
Give each student a pair of scissors and a copy of both the "Future Blocks" and the "Priority Pyramid." Explain to the students that the blocks list 15 different possible conditions or components of their future world. Have the students read through all the conditions. Make sure they understand all the words in the blocks. Then ask them to think about which of these conditions they most want to have as part of their world when they're, say, 50 years old. Clean air? Tigers? Less crime?

Have the students rank all the components from the most important to the least important. Tell them to cut out the squares and arrange them in the pyramid. The most important component should be placed in the top box of the pyramid, the next two components on the next tier, and so on. Once the students have arranged all the blocks, have them mark each block with the number that reflects the priority rating they gave it. (The priority rating numbers are one through five and correspond to the levels of the pyramid.)

2. Create group pyramids.
Arrange the group into teams of four or five students. Have the students work together to come up with a single pyramid that represents the team's priorities for the future. Everyone on the team should record the priority ratings each member of the group gave each block and compare them with his or her own personal pyramid.

3. Suggest methods of reaching consensus.
It might not be easy for teams to reach consensus regarding their group pyramid. The students may discuss their reasons for various choices, prioritize options, or try other methods of reaching a group decision. But if they get stuck or seem to be struggling to achieve a fair process, you may want to interrupt and share the following negotiation suggestions:

a. Make a list of all the possibilities for the top spot. Are there more than three? If not, would everyone accept having his or her first choice listed among the top three? If that's the case, students may have an easier time resolving the question of order.

b. Give everyone a chance to present his or her choice for the top square of the pyramid and explain why it was chosen. The other students should listen closely to these explanations. A student who lists bees as a top priority, for example, may explain that she's concerned with how bee pollination is critical for food production. Then other students who have pushed to place "enough food for all people" in the top spot may realize that they share the same concern as the "bees" proponent. This strategy may help narrow choices or change the nature of the discussion.

c. If the students are still struggling to reach a resolution, have them take a time-out to reflect on the process. Are certain views being overlooked because some students are quieter or less stubborn than others? Is the group uncomfortable with anyone's way of working out the problem? By reflecting on the dynamics of their discussion, the students may be able to isolate the obstacles to group consensus.

d. Encourage each student to offer a solution that involves concessions on all sides. Afterward, the students should vote on which compromise package they prefer. If there are ties, hold a tie-breaking vote.

4. Discuss the consensus process.

Bring the class back together and ask a representative from each team to present the team's top three priorities to the class. Students can also summarize some of the conflicts the group experienced, as well as how the conflicts were resolved.

After each team has presented, ask your students to reflect on the process of reaching consensus. Were they surprised by the disagreement among the members of the group about future visions? Did any of the students change their own views by talking with other group members? How do they think their team's decision making process might reflect some of the challenges that communities and societies face in working toward a positive future? (Make sure the students think about how hard it is to make positive changes for the world, especially if people have different ideas about what they want and what they feel is important or right.) How much harder would it have been if the students had only enough time or money to ensure that their top two or three priorities would be achieved? (Again, you might want to point out the difficulties encountered by governments and organizations that are struggling to improve the world when their resources are limited.)

5. Discuss solutions.

Ask the students if they have ever thought much about how they want the world to look in 50 years. Is simply thinking about dreams for the future enough to make them come true? If not, is it valuable? Why or why not?

Ask students to look over the blocks in their pyramids. Can they think of anything they or other people are doing to ensure that these things will happen in the future? Are there other things they would add to the pyramid?

6. Hand out "Making It Happen" sheets.

Tell the students that they have a two-part assignment. First, they should compare their personal pyramid to the pyramid their group created. How do they explain the similarities and differences? How do they feel about the process they went through? How do they feel about the result?

Next, hand out copies of "Making It Happen" (pages 179–182 in the Student Book). Ask the students to read about each of the projects and choose three approaches that they think best achieved the top three priorities set by their team's pyramid. (See "Making It Happen" Answer Sheet.) They should write one or two sentences explaining how each of the projects they picked is working to achieve the specified objectives, and what they think each project's strongest and weakest aspects are.

7. Have a wrap-up discussion.

After students have completed the assignment, make a list on the board of every priority square that made it onto a group's top-three list. Then ask the students to describe the projects they matched to those squares. Did people pick different projects to achieve the same goal? Why? Do they think any of the actions address more than one objective? Discuss possible strengths and weaknesses of the approaches listed. Do students think local projects are more effective than passing national legislation? Do they prefer preventive approaches (like education) or fix-ups (like pollution-cleaning technology)? Point out that there are a lot of ways to effect change in the world—such as educating people, passing legislation, and developing innovative technologies or strategies.

Option #2

1. Brainstorm about conditions for the future.

Ask the students to think about what they want the future to be like. What things or conditions do they want when they're adults or when their own children are grown up? List their ideas on the chalkboard.

2. Make personal pyramids.

Give each student a copy of the "Priority Pyramid." Explain to the students that they need to fill in the pyramid with the conditions they want to see in the future, ranking the conditions from the most to the least important. The most important condition should be placed in the top box of their pyramid, the next two conditions on the next tier, and so on. The students can use the ideas the group generated and add their own ideas.

3. Share personal pyramids and create group pyramids.

Arrange the class into groups of four or five students. Have the students share their pyramids with their team. Then have them work together to come up with a single pyramid that represents the team's priorities for the future. Everyone on the team should record the arrangement the group comes up with.

4. Suggest methods of reaching consensus.

See step 3 in Option #1.

5. Discuss the class process.

Bring the group back together and ask a representative of each team to present the team's top three priorities to the class. Students can also summarize some of their group's conflicts and how the conflicts were resolved.

As the groups present their pyramids, discuss with them the implications of each of their top three priorities. What kinds of things are entailed in each of their conditions? For example, if they chose world peace as a condition of the future, they should see that world peace is more likely when people everywhere have the food, water, and other resources they need for a satisfying and productive life. As the students discuss each condition, they should come to realize that everything they selected can be traced, at least in part, to a healthy environment.

When all the teams have presented, ask your students to reflect on the process of reaching consensus. See step 4 in Option #1.

6. Discuss solutions, hand out "Making It Happen" sheets, and have a wrap-up discussion.

See steps 5 through 7 in Option #1. Adapt the discussion to fit the different groups' pyramids.

"MAKING IT HAPPEN"—Answers

Each of the pyramid "blocks" can be matched with at least one of the "Making It Happen" examples. Here is a quick reference sheet for your use, but your students might make other connections that also work.

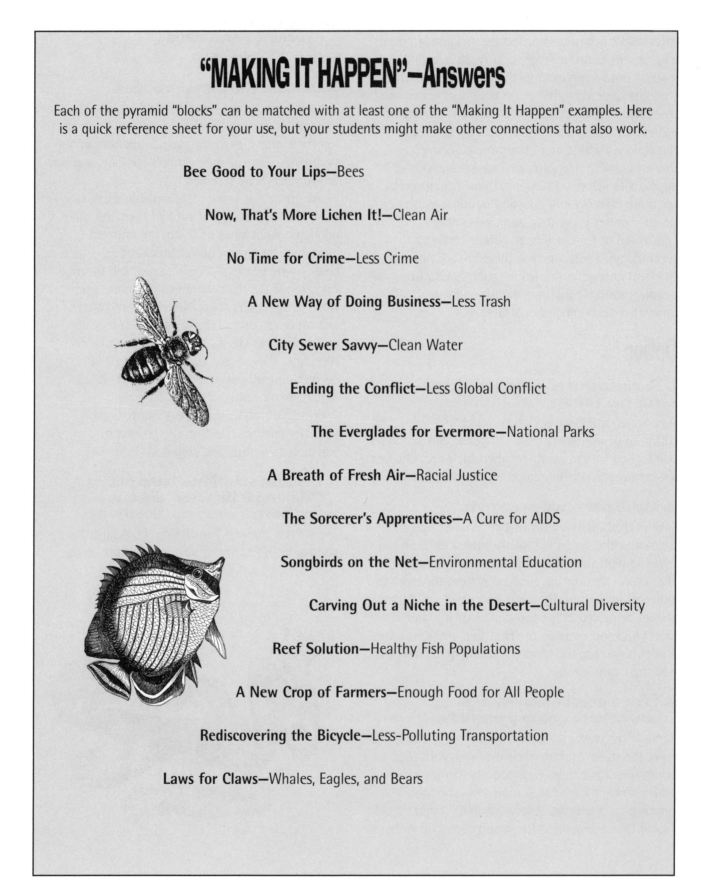

Bee Good to Your Lips—Bees

Now, That's More Lichen It!—Clean Air

No Time for Crime—Less Crime

A New Way of Doing Business—Less Trash

City Sewer Savvy—Clean Water

Ending the Conflict—Less Global Conflict

The Everglades for Evermore—National Parks

A Breath of Fresh Air—Racial Justice

The Sorcerer's Apprentices—A Cure for AIDS

Songbirds on the Net—Environmental Education

Carving Out a Niche in the Desert—Cultural Diversity

Reef Solution—Healthy Fish Populations

A New Crop of Farmers—Enough Food for All People

Rediscovering the Bicycle—Less-Polluting Transportation

Laws for Claws—Whales, Eagles, and Bears

WRAPPING IT UP

Assessment

Use both worksheets for the assessment. Inform the students that they will be assessed on these assignments.

Unsatisfactory—The student fails to complete the assignment or does cursory work.

Satisfactory—The student compares individual and group priorities, discusses the group process for collaboration, and reveals the thought process behind his or her selection of alternatives.

Excellent—The student shows careful consideration in the comparison of individual to group priorities and in the process of the group decision making. The student reveals the ability to identify the consensus process in the group. The student has a very clear explanation that supports his or her individual choice.

Portfolio

The personal "Priority Pyramid" can be used in the portfolio.

Writing Ideas

- Before constructing pyramids, ask your students to write about their top priority for the future and the reasons it is most important to them. Use the following journal starter: "When I imagine a sustainable future, I think of . . ."

- Have each student write a newspaper article on an imaginary, but realistic, event that addresses the number one concern from his or her personal pyramid.

Extension

Have the students keep a "future log" for a day. Explain that the point of the log is to focus on how their activities, behaviors, and even their thoughts can affect the future. In the log they should simply write down what they do, think, say, and so forth—just as they would in a diary. In the log, however, they should use "bullet style" instead of paragraphs. At the end of the day, they should think about and write down how each "bullet" affects the world around them, either positively or negatively, and what the ramifications could be for the future. For example, if a bullet reads, "Had a PBJ sandwich for lunch, with a bag of chips and ginger ale," the student might write down the fact that he or she recycled the soda can—a positive action for the future that saves natural resources and landfill space. But the student might also mention throwing away his or her lunch bag instead of using it again, or instead of bringing lunch to school in a heavy-duty container that can be used over and over. Have the students add ways that they can do more to create a positive future by changing their daily actions.

recycled

recyclable

Resources

The Diversity of Life by Edward O. Wilson (Harvard University Press, 1992).

Earthwater Stencils (supplier of storm drain stencils) 4425 140th Ave. SW, Rochester, WA 98579, (360) 956-3774. <earthwater-stencils.com>.

Eco-Solutions: It's in Your Hands by Oliver Owen (Abdo and Daughters, 1993).

The Value of Life: Biological Diversity and Human Society by Stephen Kellert (Island Press, 1996).

SUBJECTS

language arts, social studies

SKILLS

gathering (reading comprehension), organizing (sequencing), analyzing (identifying components and relationships among components), interpreting (relating), presenting (illustrating)

FRAMEWORK LINKS

12, 29, 38, 40, 41, 44, 50

VOCABULARY

archaeology, cannibalism, habitable, paleontology, pollen, quarry, radiocarbon, sedges, sediment, strata (most of these words appear only in the article and not in the comic strip)

TIME

two sessions

MATERIALS

copies of "Easter's End" article (pages 183–188 in the Student Book) or comic strip (pages 189–192 in the Student Book), paper, rulers, colored pencils or markers

CONNECTIONS

This activity works well after "Thinking About Tomorrow" (pages 348–357). "Future Worlds" (pages 324–329) and "Food for Thought" (pages 270–277) are good follow-up activities that will help you explore the challenges of sustainable resource use.

AT A GLANCE

Read an article or a comic strip about Easter Island to get a historical perspective on the connections between biodiversity and human survival.

OBJECTIVES

Create a time line that shows changes in Easter Island's biodiversity and human society. Describe the conditions that some people think led to the decline of the human civilization and biodiversity on Easter Island, and discuss the relevance of the story to people today.

When early explorers came to Easter Island in the 1700s, they discovered enormous stone statues lining the coast. But the origin of the statues was a mystery. How could the 80-ton statues have been moved and erected when the island had no tall trees for making hoists and other machines, no strong ropes, no draft animals, no wheels, and no source of power except human muscle?

Now, with the help of modern science, we are much closer to understanding Easter Island's history. Not only do many scientists believe that they can explain how the statues were erected, but they've also uncovered incredible information about the decline of Easter Island's biodiversity and human civilization. Today, the story of Easter Island presents a fascinating case study of resource extraction and the slow decline of an island ecosystem.

Easter Island

Before You Begin

Depending on the reading level of your students, make one copy of the article "Easter's End" (pages 183–188 in the Student Book) or the comic strip version (pages 189–192 in the Student Book) for each student. Collect paper, pencils, rulers, and colored pencils or markers for students to use in creating their time lines.

What to Do

Day 1

1. Introduce the topic of Easter Island.

Ask students if they have ever heard of Easter Island. Ask them what they know about it. They may know, for example, about the giant statues found on the island and the questions surrounding their origin. Explain that the "lost civilization" of Easter Island presented a great mystery to people for many years, but modern science is now helping to solve the puzzle.

2. Hand out copies of "Easter's End" in article or comic strip form.

Depending on the reading level of your group, hand out either the "Easter's End" article (advanced) or the comic strip (intermediate). Tell the students to read the handout for homework. While they read, they should pay particular attention to changes in plant and animal populations on the island and to the corresponding changes in human activities. You may also want to suggest that the students underline every date mentioned in the text.

Day 2

3. Create time lines for Easter Island.

Ask students if they have any questions or general comments about what they read. Then tell them they're going to work in groups of three or four to construct time lines that convey the changes that took place on Easter Island. You may want to provide an example of what a time line might look like, but students should feel free to construct the time lines however they choose (see page 332 for examples). Encourage students to include pictures as well as written descriptions of each time period on the island.

Explain to the students that they should include as much information on their time lines as they can find in the story regarding the following issues:

1. Which species were known to be living on the island during different time periods?

2. Which species were known to have become extinct on the island during different time periods?

3. What were people doing at different time periods that may have affected other species? (Consider construction, diet, wars, and other factors.)

Be sure to tell students that there are not always exact dates mentioned for all the events that occurred. For example, Easter Island's story begins 1,500 years ago, so students will have to subtract from the present date to begin the time line at about the year 500. Exact dates for the extinction of some species aren't always given, but students should have enough information to paint a picture of the changing conditions on Easter Island.

4. Discuss the time lines.

When the students have completed their time lines, ask each group to share theirs with the class. Then hold a wrap-up discussion. In addition to soliciting general comments, you may want to ask the following specific questions:

1. What was the relationship between the human inhabitants of Easter Island and the island's biodiversity?

Note: In this activity, we present the theory of Dr. Jared Diamond, a physicist and island ecologist who is currently working on a book about island extinction.

2. Do you think the Islanders noticed the decline in the forests and island animals? Why or why not? Why didn't they do anything to stop the decline?

3. Do you think anything like this could happen again? Why or why not? (You might point out that some people think Earth should be considered one large "island.")

4. Jared Diamond (the author of this story) thinks that mistakes like those made by the inhabitants of Easter Island can be prevented today because we are better able to record changes in our environment over time and to learn lessons from past mistakes. Do you agree with him? Why or why not?

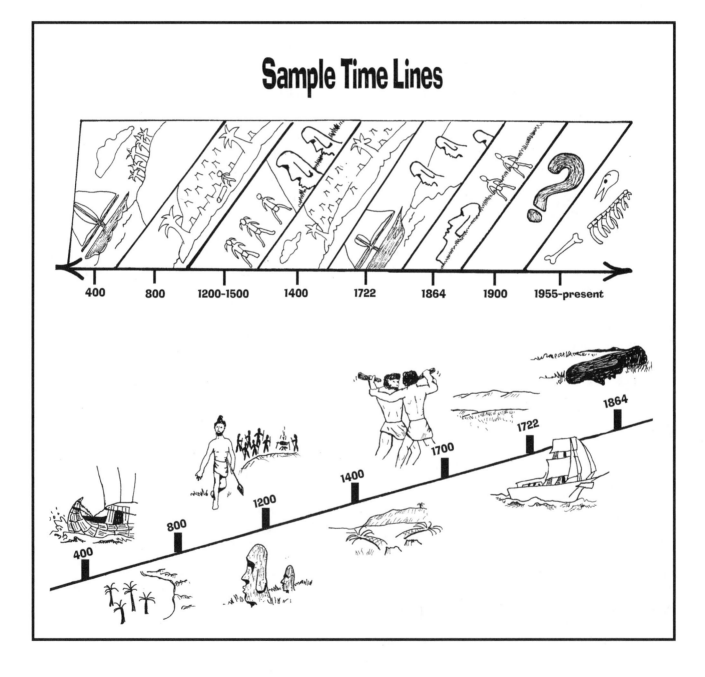

Sample Time Lines

400   800   1200-1500   1400   1722   1864   1900   1955-present

400   800   1200   1400   1700   1722   1864

WRAPPING IT UP

Assessment

Have students write a fictional story about what could have changed Easter Island's history to make the society more sustainable.

Unsatisfactory—Either the student does not complete the activity, or the story does not relate to sustainability or address the concept of loss of biodiversity within the story.

Satisfactory—The student creates a story that contains a description of how Easter Island maintained biodiversity and used the resources in a more sustainable manner.

Excellent—The student creates a story that encompasses cultural aspects as well as varied uses of resources to create a more sustainable society.

Portfolio

The time line is documentation for this lesson. The story can also be used in a portfolio.

Writing Ideas

- Use the assessment activity as a writing exercise.

- Have the students write a first person account of the fall of Easter Island.

- Have the students pretend to be reporters who go back in time. They can write a newspaper article about the end of Easter Island.

Extensions

- Have students decorate their classroom as Easter Island might have been if its resources were managed responsibly. They can hang island birds from the ceiling and decorate the walls with pictures of the toromiro trees and Easter Island palms that supported the Islanders. They may also want to make some statues of their own.

- Students could use their decorations as props for a play depicting the history of Easter Island and the lessons we've learned from its story. Have the students write the script and play the parts of all of the major participants in Easter Island's history.

- Have students read other articles by Jared Diamond and write a report on his thinking about biodiversity.

Resources

The Cartoon Guide to the Environment by L. Gonick and A. Outwater (Harper Perennial, 1996).

"Easter's End" by Jared Diamond. *Discover*, August 1995. (Article reprinted with permission.)

Visit the Easter Island Home Page Web site at <www.netaxs.com/~trance/rapanui.html>.

31 | The Biodiversity Campaign

SUBJECTS

art, language arts

SKILLS

gathering (reading comprehension), analyzing (identifying components and relationships among components), applying (designing, creating), evaluating (critiquing), presenting (writing, illustrating, persuading), citizenship (evaluating the need for action, planning and taking action)

FRAMEWORK LINKS

62, 72

VOCABULARY

public service announcement (PSA)

TIME

one to two sessions

MATERIALS

copies of "Biodiversity Ads" (pages 193–197 in the Student Book), poster-sized paper, colored markers, and other drawing materials

CONNECTIONS

For more on what people are doing to help protect biodiversity in their personal and professional lives, try "Career Moves" (pages 338–343) or "Future Worlds" (pages 324–329) before or after this activity.

AT A GLANCE

Review educational ads focusing on biodiversity, then create your own biodiversity ads to educate others.

OBJECTIVES

Discuss and analyze a variety of public service announcements about biodiversity. Design and create ads to educate others about biodiversity. Apply concepts of good advertising to messages about biodiversity.

Although most of us think of advertising as someone trying to sell us a product (shoes, hamburgers, or cars), advertising is also a great way to provide information about concepts or ideas. Educating the public about a complex and sometimes misunderstood subject like biodiversity can be a real challenge. But organizations such as Green Team, an advertising firm in New York, and many conservation groups are tackling this challenge head on by creating biodiversity advertisements designed to increase the public's awareness of the world around them. By presenting biodiversity in a way that people can relate to, the Green Team and others are trying to help everybody become aware of biodiversity's value.

If you carefully watch a commercial for a particular car, you will notice that the advertisement is trying to sell you the car by selling you a concept. Whether you see a person wearing a suit in a luxury car or a person wearing beachwear in a convertible, the advertisers want you to think that you could be that person. Because most effective advertisements deal with concepts that affect your life (satisfying your hunger, organizing your life, or looking good in a new shirt), advertisements can capture your attention and persuade you to buy or to do things. Consequently, they are a great way to get people interested in protecting the environment or some other social cause.

In this activity your students will take a shot at creating their own biodiversity advertisements for display. It's a good way for them not only to reinforce their understanding of biodiversity, but also to introduce the concept to others while developing their creativity.

Make copies of the "Biodiversity Ads" on pages 193–197 in the Student Book. (The number you copy will depend on whether you want the students to work alone or in small groups.) Collect poster-sized paper, colored markers, and other drawing materials.

1. Hand out and discuss the ads.

Explain the advertisements and their use in protecting biodiversity. (See the introduction to the activity.) Then give the students time to look at and read the ads. Afterward, ask how the ads are similar to and different from more typical ads, such as those found in most magazines. *(Like other ads, these ads try to promote a concept by grabbing the reader's attention with catchy language and images.)*

Explain that ads like those included in this activity, which focus on information and education rather than products, are often called public service announcements or PSAs. Public service announcements are generally made by noncommercial groups to support education, safety, health, and other issues that contribute to the public good. But they are also made by commercial groups to get a point across.

2. Present discussion questions.

Write the following questions where everyone can see them:

- Do the ads grab your attention? Why or why not?
- What is the key message of each ad? Is it clear?
- Is the message something that most people can relate to and understand? Why or why not?
- Are the ads successful in helping people understand and value biodiversity? Why or why not?

Before you lead a group discussion, have the students write their responses to the questions. You might also want to have them discuss the questions in small groups.

He may seem
big and ugly
to you,
but to someone
with heart disease,
he's a prince.

The Houston Toad produces alkaloids,
which can help prevent heart attacks.

Biodiversity
It's bigger than you think.

© 1999 Green Team Advertising, NY

How
Ethiopia
helped feed
the U.S.

A barley plant from Ethiopia was found to have a gene that now
protects the $160 million barley crop in California.
This is just another example of how biodiversity affects us all.

Biodiversity

It's bigger than you think.

3. Explain the ad project.

Challenge your students to design their own biodiversity ads. Any aspect of biodiversity can be the subject of the ads, from the phenomenal number and variety of species or ecosystems on Earth to the value of genetic diversity. Like the Green Team ads and other PSAs, the students' ads can focus on foods and medicines that benefit people. But encourage the students to explore other aspects of biodiversity as well, such as recreational or artistic benefits or the amazing relationships and connections that link species to one another, to their ecosystems, and to humans. The ads could even be from the point of view of a nonhuman species—for example, of a frog asking you to save its habitat.

Tell your students that before designing their ads, they should decide who their audience is. For example, they could "target" a particular age range (peers, parents, young kids, senior citizens) or a particular interest group (consumers, sports fans, music lovers). They should also decide what their message is and whether they need examples to illustrate the message. Explain that by doing these things first they'll find it easier to come up with an interesting design.

Before the students begin, have them brainstorm about characteristics that make ads effective. Among other things, a good advertisement

- presents a clear message
- relates the message to people's lives
- is attractive, clever, and interesting enough to grab people's attention
- connects to people's values and what they care about most

4. Draft ads before finalizing them and displaying them.

Have students draft their ads before working in final form. Then provide plenty of art supplies for their final ads. Hang the finished works in a library, mall, town hall, or other place where the public can see them.

WRAPPING IT UP

Assessment

Compare the advertisement to the criteria for "good" biodiversity advertisements from step 3.

Unsatisfactory—The student does not demonstrate an understanding of effective advertising.

Satisfactory—The student demonstrates an understanding of effective advertising.

Excellent—The student demonstrates a clear understanding of effective advertising by presenting a creative ad with a strong and accurate message.

Portfolio

The student's advertisement can be used in the portfolio.

Writing Ideas

- Have the students write a constructive critique of one of the ads made in class. (Remember to remove the names of the ad designers from the ads before critiquing.)

- Let the students use the following as a journal starter: "One of the ads really got my attention because . . . " or "I really liked (or didn't like) a certain ad because . . ."

Extension

Have the students design an advertising campaign around a local issue such as protecting a nearby park or natural area. Have them place the ads in school and local newspapers, on school bulletin boards, at parents' workplaces, and around area stores. Follow up the campaign by interviewing people in the community to see if they have seen the advertisements and become aware of the issue.

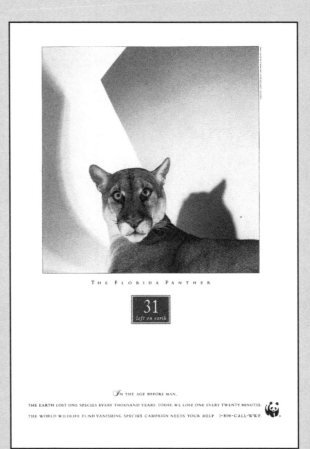

THE FLORIDA PANTHER

31 left on earth

IN THE AGE BEFORE MAN,
THE EARTH LOST ONE SPECIES EVERY THOUSAND YEARS. TODAY, WE LOSE ONE EVERY TWENTY MINUTES.
THE WORLD WILDLIFE FUND VANISHING SPECIES CAMPAIGN NEEDS YOUR HELP. 1-800-CALL-WWF.

Resources

Blaze Communications <www.blazegroup.com>.

Co-op America's Green Pages Online <www.coopamerica.org/gp>.

Green Team Advertising, Inc. <www.greenteamusa.com>.

Huey/Paprocki, Ltd., 865 Adair Ave., NE, Atlanta, GA 30306. (404) 872-2747.

Karen Gedney Communications, 375 3rd St., #1, Brooklyn, NY 11215-2801. (505)466-2866.

SUBJECTS

science, social studies, language arts

SKILLS

gathering (reading comprehension, interviewing), analyzing (questioning), interpreting (summarizing), presenting (public speaking)

FRAMEWORK LINKS

6, 57, 62, 72

VOCABULARY

See individual career profiles for possible unfamiliar words.

TIME

variable (much of the activity is done outside of class time)

MATERIALS

copies of "Tips and Tricks of Interviewing" (page 205 in the Student Book), "Sample Interview Questions" (page 341), and "Career Profiles" (pages 198–204 in the Student Book)

CONNECTIONS

To help your students think about the future and to introduce them to some of the actions that people are taking to create a more sustainable society, try "Future Worlds" (pages 324–329) before or after this activity.

AT A GLANCE

Read profiles of people who work in biodiversity-related professions, then interview people in your community with similar occupations.

OBJECTIVES

Describe various careers related to biodiversity. Identify and discuss biodiversity-related careers held by members of the community. Conduct a successful interview.

Protecting biodiversity can take many different forms. Some people, for example, focus on minimizing their own impact on the environment by shopping carefully, recycling as much as possible, avoiding the use of harsh chemicals on their gardens and in their homes, installing low-flow shower heads, and so on. Other people take more of a community-action approach, getting involved in local projects designed to protect or enhance the environment. And some people devote much of their lives to biodiversity and its protection through their choice of careers.

Here's a way to introduce your students to people whose careers are linked to biodiversity issues. Your students will find out what people are doing, who in the local community works in a biodiversity-related field, and what kind of education is needed for particular careers. They might also discover some ways they can use their own interests and skills to get involved with biodiversity issues.

photographer Gerry Ellis

Gerry Ellis/ENP Images

Make one copy of the "Career Profiles" (pages 198–204 in the Student Book) and cut out each profile. Make one copy of "Tips and Tricks of Interviewing" (page 205 in the Student Book) for each student. (You can also make copies of the "Sample Interview Questions" on page 341.)

1. Divide the group into pairs and introduce the activity.

Discuss with your students some of the ways people make contributions to issues they care about (volunteering, trying to live in a way that's consistent with their values, helping others to become involved, and so on). Point out that one way some people contribute is by choosing a career related to an issue that's important to them. Then explain that they'll be conducting interviews to learn about people whose careers are related to environmental issues, including biodiversity.

2. Pass out the "Career Profiles" (one profile per pair), and explain the project.

Give the students time to read their profiles. Then tell them that each pair of students will be working together to locate and interview someone in the community who has an occupation similar to the one in the profile they read.* The work that the community member does may differ from the work of the person profiled, but the occupation should be the same or similar. For example, the environmental attorney in the "Making Waves" profile focuses on ocean life. But the students may find a local environmental attorney who focuses on development issues within the community. (If the students are unable to find someone with an occupation similar to the one in their profile, they can choose another occupation from the list on page 342.)

Explain that the students can conduct their interviews either over the phone, by e-mail, or in person. (Be sure to allow time for coordinating the students' use of a phone.) Afterward, the students will be using the information they've gathered to develop a creative presentation.

3. Discuss ways to locate people to interview.

Your students may know people in the community whom they can interview, or they might have ideas about where to start looking. Write their ideas where everyone can see them, then add the following suggestions:

- state departments of wildlife or fish and game
- local park authorities or departments of parks and recreation
- local offices of federal agencies and departments (such as the U.S. Environmental Protection Agency and the U.S. Fish and Wildlife Service)
- local zoos, aquariums, or natural history museums
- nature centers
- local or regional environmental organizations
- universities and colleges
- community newspapers
- businesses with a strong environmental focus
- companies or stores that provide alternatives to environmentally harmful products (such as supermarkets that specialize in environmentally sensitive products)
- organizations that have successfully integrated environmentally sound programs into their operations (such as office-wide recycling)

* An alternative to having pairs of students setting up and conducting the interviews is to invite a professional, or a panel of professionals, to the class. The class as a whole could come up with interview questions in advance, and students could take turns asking the questions.

Have each team write down a list of the people or the organizations they'll contact in their search for a person to interview. Tell them to order the list so that they contact their most likely resources first. (You may want to check over their lists to make sure they're on the right track and that several groups aren't planning to contact the same organization.)

4. Review how to set up interviews.

Ask your students to describe or act out what they would say on the telephone if they called an organization to get the name of somebody to interview. The following is an example of one approach they might take:

> *"Hello, my name is _____. I'm a student at _____ and I'm doing a class project that involves interviewing people about their careers. I'd like to interview a(n) _____. Do you know of somebody I might be able to interview? If not, is there anyone else available who might be able to suggest someone? May I ask your name? Thank you very much."*

Next ask your students to describe or act out how they would ask a particular individual for an interview. Here's one approach:

> *"Hello, my name is _____. I got your name from _____. I'm a student at _____, and I'm doing a class project that involves interviewing people about careers related to biodiversity and the environment. Would you be willing to be interviewed? When would be a good time? Thanks very much for agreeing to talk with me."*

Remind the students that they'll need to discuss a suitable time and place if the person agrees to be interviewed. They should also ask whether he or she would like to have a copy of the questions in advance. If the person does not wish to be interviewed, remind your students to thank him or her for speaking with them.

5. Develop interview questions and review "Tips and Tricks of Interviewing."

Write on the board or hand out the "Sample Interview Questions" (page 341). Have your students brainstorm to come up with a list of general questions they think would be appropriate to ask during their interviews. Write their ideas on the board, then help the class organize and edit the list. Encourage the students to come up with open-ended questions rather than yes or no questions.

The students can use the basic list as a guide when conducting their interviews. Each pair should also come up with additional questions that are specifically related to the career of the person they're interviewing.

Next hand out copies of "Tips and Tricks of Interviewing" (page 205 in the Student Book). Carefully explain each tip while the students follow along.

If you'd like to give your students practice conducting interviews, form groups of four by bringing two pairs together. Then have each pair develop five questions to ask the other pair. You might want to have one person in each pair ask the questions while the other takes notes, or you might have the two alternate so each student can practice asking questions and taking notes. You might also encourage your students to practice by interviewing their parents or neighbors.

6. Conduct the interviews.

Help your students figure out the best way to conduct their interviews. In some cases, the interviewees might be able to come to the school. In other cases, they might be able to meet with the students after school at their place of work or another convenient location. And in some (perhaps most) cases, interviews might take place over the telephone or by e-mail.

7. Briefly discuss interviews and career options.

After the students have completed their interviews, give them time to discuss the process and share their experience with the group. To get them started, ask them how the interviews went. Was the process easier or harder than they expected? Why? Which careers didn't they know about before? Which careers sounded the most interesting, and why? What did they learn that surprised them?

SAMPLE INTERVIEW QUESTIONS

**Here are examples of some of the kinds of questions
you might want to ask during your interview.**

✎ What's a typical day on the job like for you ?

✎ What's your educational background?

✎ What skills are especially useful in your profession?

✎ Does your job involve protecting biodiversity? If so, how?

✎ What's the most challenging aspect of your job?

✎ What's the most rewarding aspect of your job?

✎ How did you become interested in your field?

✎ Did you always want to be involved in a career like the one you're in?

✎ Do you have any advice for students who may be considering a career like yours?

Some students might have been surprised to find out that a person doesn't have to have a career directly related to the environment or to biodiversity to be able to make a difference. Point out that more and more careers include aspects of environmental protection these days. And many corporations are hiring people with environmental backgrounds. For example, 20 years ago it would have been hard to find an ecologist working for an electric power company, but today many ecologists are helping such companies make sure they aren't harming the environment.

8. Develop and give presentations.

Have each pair put together a brief presentation focusing on the career of the person they interviewed. Encourage the students to create posters and other visual aids to explain and illustrate how the career they're highlighting relates to biodiversity. Make sure that both students in each pair are involved in the presentation. (You might also want the students to write up their interviews as an article. See the writing idea at the end of this activity for ideas on how to do this.)

Once the presentations are finished, ask your students for their reactions to the careers they learned about.

CAREER CORNER

Botanist

Community Activist

Conservation Biologist

Ecologist

Ecotourism Trip Leader

Environmental Educator

Environmental Journalist

Environmental Specialist at a Foundation

Fisheries Biologist

Forester

Hydrologist

Integrated Pest Management (IPM) Specialist

Land-Use Planner

Curator at a Natural History Museum

Naturalist/Interpreter in a Park or a Wildlife Sanctuary

Nature Photographer

Natural Resource Economist

Organic Farmer

Population Expert

Preserve/Refuge Manager

Recycling Coordinator

Soil Conservation Specialist

Solid Waste Manager

Toxicologist

Zoo Keeper

Zoologist

WRAPPING IT UP

Assessment

Collect the students' interview notes. Compare the information gathered in the interview with the information included in the presentation.

Unsatisfactory—The interview questions were not well thought out, the answers were incomplete in notation, the presentation did not reflect the interview, and/or the interview and the presentation did not tie the profession in with biodiversity.

Satisfactory—The questions were reasonable and the answers were accurately recorded. The presentation content is evident from the interview notes and the tie to biodiversity is clear in the presentation.

Excellent—The interview questions were well developed and insightful, and notes from the interview were well documented. The presentation directly reflects information from the interview and makes clear the career's tie to biodiversity.

Portfolio

Interview notes can be used in the portfolio.

Writing Ideas

- Have the students write up their notes as an interview-style article. (For ideas, you might want to have students read the article "Ask Dr. B!" in *WOW!—A Biodiversity Primer.*)

- Let the students use the following as a journal starter: "The jobs I found most interesting are . . . because . . ."

Extension

Have the students research other careers that are related to biodiversity conservation or environmental protection. (See the list entitled "Career Corner" for ideas.) They can use their research as a launching pad to write career profiles of people in the community, conduct more interviews, create displays, or write reports.

Resources

Blacks in Science: Astrophysicist to Zoologist by Hattie Carwell (Exposition Press of Florida, 1977).

Eco-Careers: A Guide to Jobs in the Environmental Field by John Hamilton and Stuart Kallen (Abdo and Daughters, 1993).

Scientists on Biodiversity edited by Linda Koebner, Jane E. S. Sokolow, Francesca T. Grifo, and Sharon Simpson (American Museum of Natural History, 1998).

Stepping into the Future: Hispanics in Science and Engineering by Estrella M. Triana, Anne Abbruzzese, and Marsha Lakes Matyas (American Association for the Advancement of Science, 1992).

Women Life Scientists: Past, Present, and Future by Marsha Lakes Matyas (American Physiological Society, 1997).

WOW!—A Biodiversity Primer (World Wildlife Fund, 1994).

33 | What's a Zoo to Do?

SUBJECTS

science, social studies (ethics), language arts

SKILLS

gathering (reading comprehension), analyzing (identifying components and relationships among components), interpreting (defining problems, summarizing), applying (problem solving, proposing solutions), evaluating (identifying bias), presenting (articulating), citizenship (debating, evaluating, seeking consensus, taking and defending a position, working in a group)

FRAMEWORK LINKS

6, 40, 60, 67

VOCABULARY

bias, botanical garden, boycott, captive breeding, curator, **genetic diversity**, implantation, predation, **reintroduce**

TIME

two to three sessions

MATERIALS

paper and pencils, copies of the dilemmas (pages 206–223 in the Student Book)

CONNECTIONS

To introduce your students to a variety of viewpoints about biodiversity protection, try "The Spice of Life" (pages 206–213) before this activity. For other activities dealing with challenging issues and personal opinions, see "The Many Sides of Cotton" (pages 278–285), "Food for Thought" (pages 270–277), and "Future Worlds" (pages 324–329).

AT A GLANCE

After exploring various perspectives on controversial issues, come up with ideas to deal with some of the difficult dilemmas that zoos, aquariums, botanical gardens, and other community institutions face.

OBJECTIVES

Explore the ethical issues and challenges surrounding biodiversity dilemmas faced by institutions such as zoos, aquariums, and botanical gardens. Discuss differing viewpoints and articulate personal views regarding how to address these dilemmas. Make recommendations to resolve a particular dilemma.

Zoos, aquariums, botanical gardens, and similar institutions are important storehouses of biodiversity. They often have endangered species in their collections and support captive breeding or propagation programs designed to boost declining populations. Some also conduct and support field research and conservation focusing on endangered species in their native habitats. But the efforts of these institutions to protect biodiversity are often hampered by difficult dilemmas that require the institutions to balance the viewpoints of many people, the welfare of individual animals, and the welfare of species and ecosystems.

This activity gives students insight into some of the efforts that zoos and similar institutions undertake to protect biodiversity, as well as some of the hard choices they often face. It also challenges students to clarify their own views regarding actions an institution can take.

Make copies of the dilemmas (pages 206–223 in the Student Book). You'll need one dilemma (with accompanying information) for each group of students. If you'd like more background information on dealing with issues, beliefs, problem solving, and detecting bias, see "The Components of an Environmental Issue" (page 280) in "The Many Sides of Cotton."

What to Do

1. Divide into groups and assign dilemmas.

Divide the class into groups of three to five and provide each group with one of the dilemmas. Explain that each group's dilemma represents a problem that zoos, botanical gardens, and other institutions sometimes face. Be sure the students understand that the particular animals and circumstances discussed in each dilemma, while fictitious, are based on real animals and events.

A Few Things to Think About...

This activity is designed to stimulate thinking and get your students to understand the tough choices that zoos, aquariums, botanical gardens, and other conservation and education institutions often have to make. It's important for your students to know that the actual dilemmas are fictional, but are based on real animals, plants, and events. We also think it's important for them to realize that these issues are very complex and that there are no right or wrong answers. It's the process of thinking and discussing that's important. At the same time, we hope your students will see the role attitudes, values, and beliefs play in decision making as they try to come up with strategies for solving the problems.

Although we've simplified the scenarios for this activity, we believe they can help students understand the complexity of issues faced by community institutions. Before the students get into small groups, you might want to mention that it's important not to stereotype individuals, organizations, or groups of people. For example, not all drug companies or zoo officials think or act alike. But for the purpose of this activity, they will need to discuss the information they've been given. You might also want to talk about stereotypes at the end of the presentations so that your group can reflect on how they dealt with the issue during their discussions. We hope that this activity will help your students see that most issues are not black and white and that people have different ways of looking at the same issue.

In this activity, Dilemma 1 focuses on a real species, the lowland gorilla. Dilemma 2 describes one real species (the thick-billed parrot) and one fictitious species (the orange-necked parrot). In Dilemma 3, the rusty antelope, eland, and okapi are real species while the Brooks antelope is fictitious.

You might want to have the students conduct additional research on gorillas in captivity, breeding programs at zoos, biodiversity prospecting, or any of the other issues highlighted in the scenarios. You could also have them try the extension on page 347 to investigate a real dilemma.

2. Work in groups to develop recommendations.

Explain to the students that their job will be to use all of the information provided to develop a recommendation addressing their group's dilemma. It is important to inform your students that the following dilemmas are based on situations that either have happened or could happen at zoos, aquariums, botanical gardens, and other institutions.

Give the students time to read all their information, then have them discuss the various viewpoints presented as well as their own thoughts and feelings about the issue. Next, have the students talk about some ways the issue might be resolved. On the basis of its discussion, each group should work to develop a recommendation for how they think the institution should deal with its dilemma. Have each group write down its recommendation and the reasons for it, as well as what some of the limitations or long-term effects might be. If more information is needed to come up with a sound recommendation, the group should state what additional information is necessary and where it could be found.

3. Prepare presentations.

Give each group time to develop a presentation to share with the rest of the class. Explain that each presentation should include (1) a description of the group's dilemma, (2) an overview of the different viewpoints regarding what should be done, (3) the students' recommendation and reasoning, (4) the potential drawbacks of the group's idea, and (5) any additional information that would have helped the group develop its recommendation. Ask the students to develop visual aids such as charts and illustrations for their presentations. Also ask the groups to be sure each person has a role in his or her group's presentation.

4. Share dilemmas and recommendations.

Leave time after each presentation for questions and further discussion. You might also want to give the students in the "audience" some specific things to think about as the group presents.

Alternative Format:

Have all of the groups examine the same dilemma(s), and ask each group to present its recommendation. After the presentations, have the class discuss the different ideas presented and, as a group, develop a final recommendation.

Resources

For information on wildlife issues as well as links to zoos around the world,
visit the American Zoological Association Web site at <www.aza.org>.

Animal Kingdoms: Wildlife Sanctuaries of the World by Patrick R. Booz, Graham Pizzey,
and Tom Melham (National Geographic Society, 1996).

The Company We Keep: America's Endangered Species by Douglas H. Chadwick, Joel Sartore,
and the National Geographic Society (National Geographic Society, 1996).

Ethics on the Ark: Zoos, Animal Welfare, and Wildlife Conservation
edited by Bryan G. Norton, Michael Hutchins, Elizabeth F. Stevens, and David Ehrenfeld
(Smithsonian Institution Press, 1996).

WRAPPING IT UP

Assessment

If your students are familiar with peer assessment, the following can easily be adapted to peer grading, especially by presentation teams. After completing the activity, ask your students to describe in writing how the presentations helped them understand the dilemmas that institutions face. Have each student rate his or her own participation in the group based on the concepts below:

1. Did you participate fully in your group's presentation?
2. Did your group evenly distribute the roles?
3. How well did you function as a team member?
4. Did the group use your strengths?
5. Did you and your group cover all the required components of the presentation, including the discussion?
6. Were your ideas for the group well thought out?
7. Did you share your ideas clearly with your group?
8. Did you clearly communicate your part of the presentation?

> **Unsatisfactory**—The student does not participate in the group presentation and/or fails to fairly evaluate his or her role.
>
> **Satisfactory**—The student considers fairly his or her role in the group presentation and presents a good evaluation of his or her learning.
>
> **Excellent**—The student has a well-thought-out evaluation of his or her learning and his or her role in the group presentation, including strengths and areas for improvement.

Portfolio

Each student can outline or write a short paragraph about his or her team's presentation. Include descriptions of the dilemma, the different points of view, recommendations, potential drawbacks, and additional information needed.

Writing Ideas

- Have the students write position papers that describe their opinions about the dilemma their group discussed. The paper must include a thesis, supporting information, and a conclusion.
- Have the students write a few paragraphs that describe what they learned through this activity about the dilemmas that zoos, aquariums, and other community institutions face.

Extension

Have the students research a real-life controversy or dilemma faced by a zoo, aquarium, or botanical garden in the United States. Look for newspaper and magazine articles or Web sites that discuss the situation. Call or e-mail the public relations department at the zoo or garden to see if they have issued a statement on the controversy.

From the real-life story, have the students identify the dilemma, the different points of view, the parties involved, and, if the situation has been resolved, the final outcome. If the situation has not yet been resolved, have the students come up with their own recommendations for the situation based on their group discussions and presentations in this activity.

Web sites you might want to check for this kind of information include: the American Zoological Association, <www.aza.org>; the Brookfield Zoo, <www.brookfield-zoo.mus.il.us>; and the Monterey Bay Aquarium, <www.mbayaq.org>.

SUBJECTS

science, social studies (geography)

SKILLS

gathering (simulating, reading comprehension), organizing (sequencing), interpreting (reasoning), applying (planning, problem solving, proposing solutions), presenting (explaining), citizenship (working in a group, evaluating the need for citizen action, planning action, compromising)

FRAMEWORK LINKS

29, 38, 50

VOCABULARY

desalination, drip irrigation, groundwater, **natural resource**, nonrenewable, reclaimed water, renewable, **sustainable**

TIME

Part I—one session
Part II—two sessions

MATERIALS

Part I—small cups, spoons, measuring cup, one-pound bag of beans (or jelly beans), watch or timer that displays seconds, flip chart paper or chalkboard, shallow tray to hold beans, copies of "Condition Cards" (page 353)
Part II—copies of "Water Works" and "The Options" (pages 224–226 in the Student Book)

CONNECTIONS

This activity provides an introduction to resource use. To further explore the problems and challenges of resource use, try "Food for Thought" (pages 270–277), "Future Worlds" (pages 324–329), or "Easter's End" (pages 330–333).

 AT A GLANCE

Take part in a demonstration of unsustainable use of a natural resource, and use a case study that focuses on drought as you develop a plan to manage a scarce resource.

OBJECTIVES

Define sustainable use of natural resources. Describe several consequences of unsustainable use of natural resources for people and for other species. Recognize the difficulty in identifying sustainable ways to use resources when demand and supplies fluctuate. Develop ideas for using water in more sustainable ways.

Many scientists would agree that we are using our natural resources faster than they can be replenished. Since we rely on renewable natural resources for our survival, we can neither use them all up nor stop using them completely. Between the two extremes, however, is "sustainable use." Sustainably using a natural resource means using the resource in a way that allows people to get what they need while ensuring that future generations and other species can also get what they need. Although no one knows for sure if the "sustainable use of natural resources" is really feasible for the long haul, most would argue that sustainability is what we need to strive for.

Figuring out how to sustainably use a natural resource is not easy. For instance, entirely different techniques are needed to sustainably manage a temperate forest than to sustainably manage a tropical forest. And because the speed with which we consume natural resources is always changing, it is sometimes difficult to tell whether the resource is being used sustainably at any particular moment. Sustainable management also depends on whether a resource is "renewable" or "nonrenewable." Petroleum, for instance, is a nonrenewable resource because it takes millions of years for it to form. Sunlight, however, is a renewable resource because we'll have a steady supply of it, no matter how much we use. Scientists think the sun has enough material to radiate for another five billion years.

Whether a resource can be considered to be renewable or nonrenewable is a matter of perspective. A resource may replenish itself over time, but if it takes millions of years, it has little relevance to us. For a resource to be considered truly renewable, it must replenish itself in a couple of generations or less. For example, trees may be a renewable resource because they can be grown relatively quickly, but a mature, ancient forest is not renewable because it may take hundreds of years for all of the interacting parts to be reestablished, if they can be reestablished at all.

In Part I of this activity, your students will learn about the concepts of sustainability and managing renewable and nonrenewable resources. Then, in Part II, they will look more closely at a natural resource that is critical to all life on Earth—water. They'll learn that our choices about water use—whether we deplete limited supplies, pipe it in from elsewhere, build reservoirs to contain it, or use it sparingly—affect both people and biodiversity. Your students will practice making some of these decisions using a recent drought in Santa Barbara, California, as a case study.

Before You Begin • Part I

Make sure you have plenty of room to do the simulations. Collect twenty small cups, four spoons, a measuring cup, one pound of beans, a watch or a timer that displays seconds, a flip chart or a chalkboard, and one shallow tray. Pour the beans in a tray for the class and place the tray on a table or a desk in an open area of the room. You will be dividing the class into four equal-sized groups or "families." Put a spoon and a stack of five cups for each family next to each side of the tray.

Do a trial run of the simulations on your own to figure out how many beans you'll need for each group. Find out how many beans you can scoop in 10 seconds. Repeat this for the number of people you'll have in each family (usually four or five). Count the total number of beans you've scooped. If you have four families, you'll need four times as many beans in the tray to cover the entire class. Playing the game with too many beans won't illustrate the concepts as clearly. The average student should be able to scoop about 55 beans in a turn, provided there are enough beans in the tray. Have extra beans available to add to the tray according to the instructions on the condition cards.

Across the top of the flip chart paper or the chalkboard, write Simulation #1, and the family names that each group decides on. The students can then record their data in columns. Make a similar chart labeled Simulation #2 and Simulation #3. (See sample chart on the following page.) Photocopy and cut out the "Condition Cards" (page 353).

Sample Charts

Simulation #1

	Family Name	Family Name	Family Name
Great-great-grandparent	# of beans	# of beans	# of beans
Great-grandparent	# of beans	# of beans	# of beans
Grandparent	# of beans	# of beans	# of beans
Parent	# of beans	# of beans	# of beans
Children	# of beans	# of beans	# of beans
Total # of beans remaining			

Simulation #2

	Family Name	Family Name	Family Name
Great-great-grandparent	# of beans	# of beans	# of beans
Great-grandparent	# of beans	# of beans	# of beans
Grandparent	# of beans	# of beans	# of beans
Parent	# of beans	# of beans	# of beans
Children	# of beans	# of beans	# of beans
Total # of beans remaining			

Simulation #3   Effects of drought, average yield, contamination, surplus, and flood

Condition		Family Name	Family Name	Family Name
	Great-great-grandparent	# of beans	# of beans	# of beans
	Great-grandparent	# of beans	# of beans	# of beans
	Grandparent	# of beans	# of beans	# of beans
	Parent	# of beans	# of beans	# of beans
	Children	# of beans	# of beans	# of beans
Total # of beans remaining				

World Wildlife Fund

What to Do • Part I

1. Divide students into groups.

Divide the students into four equal-sized groups (there can be four or five students in each group, but the groups need to be equal in size). The remaining students can be timekeepers or data recorders. Tell the students that each group represents a family and each person represents one generation of that family. Have each group select a family name, as having a name will help the students fill out the chart and remember which family they belong to. One group member will represent a great-great-grandparent, one will represent a great-grandparent, one will represent a grandparent, one a parent, and one a child. Make sure each student remembers the family and the generation that he or she represents. That information is important in order to fill out the chart accurately. Each family should form a line at a different side of the tray of beans.

2. Explain the first simulation.

Tell the students that the tray of beans represents a *nonrenewable* natural resource (petroleum, minerals, or wood from ancient forests, for example). You may want to ask students if they can think of other nonrenewable natural resources people use. Emphasize that their "families" are part of a community that relies on the nonrenewable natural resource in the tray. Each family member will have the opportunity to "extract" some of the natural resource (beans) for 10 seconds using the spoons to scoop beans into the cup. The family member can use only one hand to scoop. The cup must remain on the table and cannot be lifted to the tray. Beans that have been spilled on the floor cannot be scooped. They represent wasted resources. Then students will record their results on the flip chart paper or the chalkboard.

3. Begin Simulation #1.

Have the first family member (great-great-grandparent) from each family pick up a spoon and explain that when the timekeeper says "go," the

great-great-grandparents can scoop as many beans as they *care* to into the cup. They must use the spoon to scoop the beans into the cup. The cup must remain on the desk and the student may scoop with one hand only. When the 10 seconds are up, the students will then go to the flip chart or chalkboard, count their beans, and record the results. After the first family member (a great-great-grandparent) has had a chance to "extract," allow the second family member (great-grandparent) from each family to do the same. Repeat the process until each generation has had 10 seconds to collect beans. The next generation of family members can begin collecting beans while the members of the previous generation count and record their results. Measure how many beans are left at the end of the simulation. The amount of beans remaining represents the amount of natural resource left for future generations.

4. Begin Simulation #2.

Place the beans back into the tray. Have the beans represent the same nonrenewable natural resource as in the previous simulation. Inform the students that they *need only 35 beans* worth of the natural resource to survive. Repeat the procedure as in Simulation #1, having the members of each generation draw as many beans as they *care* to from the tray and then recording their results on a different chart paper labeled Simulation #2. After all the family members have extracted from the tray, measure how many beans are left. The amount of beans remaining represents the amount of natural resource left for future generations. Ask how many students were able to extract at least 35 beans. (They represent those family members who were able to get enough of the natural resource to survive.) Ask how many students were not able to extract at least 35 beans. (They represent family members who could not get enough of the natural resource to survive.) Ask how many students extracted 36 beans or more. (They represent family members that used more of the natural resource than needed.)

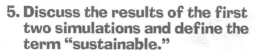

5. Discuss the results of the first two simulations and define the term "sustainable."

Using the following questions, have the class discuss their observations about the simulations:

- For both simulations, did each successive generation scoop more or fewer beans than the previous generation? *(If your students are at all competitive, the last generations found fewer and fewer beans to scoop up.)*

- Compare the number of beans left over after the first simulation to the number left over after the second simulation. Did one simulation have more left than another? Why? *(If the students were taking only enough beans to survive, all students in the second simulation should have had enough beans, and more beans should have been left over than in the first simulation.)*

- Why might a particular generation consume as many natural resources as possible? *(People may not realize they are depleting their natural resources, or they may think they really need to use them to survive. In some cases, people are just greedy.)*

- Are there any reasons a particular generation would want to conserve its natural resources? *(People may feel it is important to save enough resources for their children or grandchildren to use. They may also feel it's necessary to conserve natural resources for the survival of the community and culture, the future health of the environment, or for the sake of other species.)* Explain to students that using natural resources in a way that protects them for future generations of people and other species is called "sustainable" use.

- Who did not survive, and how did it feel? Who got too much of a resource? Did it affect whether the next generation in your family got enough of the resource?

- Did knowing how much of a natural resource you need to survive help you decide how much to scoop out of the tray? *(Students who want to make sure all of their family members get enough beans might make sure to take only as much as they need to survive.)*

- Did either simulation produce a lot of wasted resources? What could have been done to reduce the amount of wasted resources? *(In rushing to fill their cups, the students will drop many beans on the floor. Designing a community strategy for effectively filling up the cups could decrease the amount of resource wasted.)*

6. Explain Simulation #3.

Simulations #1 and #2 dealt with a nonrenewable natural resource. Tell the students that the tray of beans now represents a *renewable* natural resource. Ask students if they can think of some *renewable* natural resources people use (water, food crops, solar energy, and so forth). Each person still represents a member of a family that is part of a community. Each member will still have 10 seconds to extract some of the natural resource, but the natural resource will be renewed by different conditions between each round. Explain that most renewable resources have limits to their rate of replenishment. Physical, environmental, and human conditions can change how much a resource is renewed per generation. This simulation will illustrate sustainability under different conditions.

7. Begin Simulation #3.

Place the beans back in the tray. Follow the same procedure for scooping beans as in Simulations #1 and #2; in this simulation, however, after each student takes a turn, the resource (beans) will be adjusted according to five conditions: drought, flood, surplus, average yield, or contamination by disease. Two conditions increase the number of beans in the tray: average yield (+1 cup)—only an average amount of the resource was replenished; surplus (+1½ cups)—more of the resource was replenished than expected. Three conditions remove beans from the tray: drought (-½ cup)—lack of available water caused a decrease in the resource; flood (-½ cup), excess water wiped out some of the resource; and contamination by disease (-⅔ cup)—some of the resource was infected to unusable levels. All the conditions represent situations that can affect natural resources. A blank card has been included so that the students can add their own situation. Stack the condition cards in random order. When the 10 seconds are up, someone will draw a card and ask each group to add or subtract beans to or from its cup accordingly. (In each round of Simulation 3, all groups should be following the same condition card that was read aloud.) After each round, the cards should be replaced in the deck and the deck should be reshuffled. Meanwhile, the students will count their beans, go to the flip chart or the chalkboard, and record the results. Repeat the process until each generation has had 10 seconds to collect beans, drawing a condition card between each turn. Family members can begin scooping beans while those of the previous generation count and record their results. (Remind the students that they'll still need 35 beans to survive.)

8. Discuss the results of the simulations.

Using the following questions, have the students discuss their observations about the simulations:

- What will happen if we use our natural resources faster than they can be replaced? *(They will eventually run out.)* Can you think of examples where this has already occurred? *(Animals such as the passenger pigeon, the New Zealand moa, the great auk, and the heath hen were driven to extinction because*

they were overhunted. Some northeast-coast fisheries have been greatly reduced in size because of overharvesting. Forests around the world have been wiped out because trees were cut without being replaced. Grasslands have been converted to desert because they were overgrazed. Groundwater levels in many parts of the world have been reduced because of overconsumption and poor management of water resources.)

- Did the condition cards affect the number of beans you scooped? *(Some students may have decided not to take so many beans if a "drought" or a "contamination" card was drawn. Some may have decided to scoop more beans after a surplus card was drawn.)*

- What are the similarities in strategies for sustainably managing a renewable resource and a nonrenewable resource? *(Both involve conservation and identification of alternate resources. Even though there are often renewable alternatives to nonrenewable resources, a renewable resource can be depleted to the point where it can no longer renew itself.)*

CONDITION CARDS

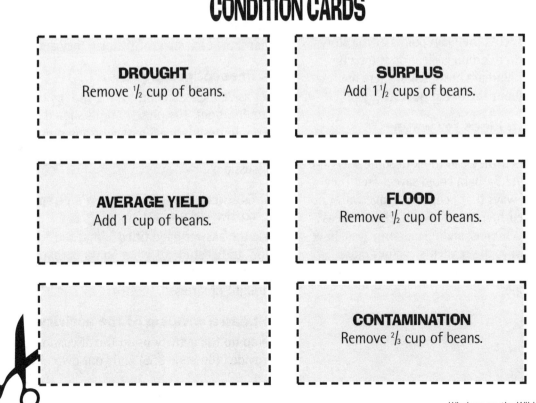

DROUGHT
Remove ½ cup of beans.

SURPLUS
Add 1½ cups of beans.

AVERAGE YIELD
Add 1 cup of beans.

FLOOD
Remove ½ cup of beans.

CONTAMINATION
Remove ⅔ cup of beans.

Before You Begin • Part II

Make one copy of "Water Works" (page 224 in the Student Book) and one copy of the "The Options" (pages 225–226 in the Student Book) for each group of four or five students.

What to Do • Part II

1. Introduce the assignment.

Tell the students they will be developing a plan for a city confronted with a natural resource crisis. In this case, they will be studying the drought experienced by Santa Barbara, California, in the late 1980s and early 1990s. First they'll read some background information, and then they'll brainstorm about water conservation measures that could be used to save water in Santa Barbara. Next, they'll receive a list of options considered by Santa Barbara. Using their brainstormed list and the Santa Barbara options list, they'll create a three-stage plan to deal with a worsening drought crisis and prepare a presentation for the rest of the class.

2. Divide students into groups and hand out background information.

Divide students into groups of four or five and give each group a copy of "Water Works." As a class, read the background information. When the students have finished, discuss the main points of the story to ensure that every group is clear on the facts about the Santa Barbara crisis. Ask if there are any questions about the background.

3. Brainstorm ideas for water conservation measures.

Ask the students if they can think of some ways the residents of Santa Barbara could save water. They should think of ways that people can save water not only around their homes but also in "shared areas" such as parks, fountains, and restaurants. Also, have them think of ways the residents can get more water. Keep track of their ideas on a chalkboard or on flip chart paper.

4. Hand out one copy of "The Options" to each group.

Hand out a copy of "The Options" to each group and give the students time to read the list. These options represent a list that the city of Santa Barbara came up with to deal with the drought. The options vary in how restrictive they are on water use. Some of the options require that all water use stop for certain activities while other options ask for reduced use of water. Tell students that they can use one or two class periods to work in their groups to develop their plan and presentation. Or, you can have students meet outside of class.

Each group's plan should be made up of three stages. Each stage will deal with a worsening drought crisis. For instance, water use may not be very restrictive during Stage I, but may become increasingly restrictive during Stage II and Stage III if the drought gets worse. You might suggest that the students prepare posters or handouts for their presentation that show each stage of the plan they develop.

5. Present group plans.

Set aside a class period for the groups to give their presentations. The presentations should be five to ten minutes long. After each group presents, make sure to leave a few minutes for questions and dissussion.

6. Discuss Santa Barbara's response to the drought.

Use the answer page titled "What Did Santa Barbara Do?" to tell students what Santa Barbara included during each stage of its drought plan and how the drought progressed.

7. Lead a wrap-up of the activity.

Wrap up the activity using the discussion questions provided (on page 356) and your own questions.

WHAT DID SANTA BARBARA DO?

Santa Barbara started with Stage I in March 1989, Stage II in January 1990, and Stage III in February 1990. There was enough rain in 1991 and 1992 to end the drought emergency in 1992.

Stage I

During Stage I, water officials asked the residents of Santa Barbara to voluntarily cut back their water use by ten percent. The city encouraged residents to let their lawns become even drier before watering, to turn off faucets while brushing their teeth, to check faucets and toilets for leaks, and to install water-saving toilets and shower heads.

Stage II

Stage II was seen as a time to test new regulations. Stage II regulations ended up being similar to Stage III regulations, but officials were more flexible about making modifications during this period. It lasted only six weeks.

Stage III

This stage included

- a rate increase with a temporary exception for large households

- a desalination plant for converting seawater to drinking water (Construction of the plant began in 1991 and was completed in 1992 [an unusually short time for such a project]. The plant was used for only three months before the drought ended. Since producing water this way is expensive, the desalination plant will be used only during future water emergencies. Fortunately, there has not been a drought as severe as the one that ended in 1992.)

- regulations such as

   – a ban on cleaning pavement with water

   – a requirement that fountains reuse water

   – a ban on adding water to swimming pools (followed by an exemption for pools with covers)

   – a requirement that restaurants serve water only when a customer asked for it

   – a requirement that hotel/motel owners post drought notices in every room

   – a ban on watering lawns (Trees and shrubs could be watered only if it was done with drip irrigation.)

   – a ban on car washing, except at car washes that reuse water or at home with a bucket or automatic shut-off hose

- hiring two "water cops" to enforce the new regulations (People who broke the laws were fined up to $250.)
- encouraging voluntary water conservation (For example, low-flow shower heads were given away free, and special rebates were given to people who replaced their old toilets with water-saving toilets.)
- a water reclamation project (This project was already underway before the drought hit. It had been part of the long-range plan for conserving water in Santa Barbara. Because the first phase of the project was completed in 1989 and the second phase was on its way to completion, it was very useful during the drought.)
- public education and information about the regulations and new water projects (The information was disseminated through special flyers that came with water bills; ads in newspapers, radio, and television; a "Drought Awareness Fair" that taught conservation techniques; and a newsletter.)

WHAT DID SANTA BARBARA DO? (Cont'd.)

Results

People living in single-family homes used 56 percent less water during Stage III. Over 20,000 regular toilets were replaced with low-flow toilets, more than 41,000 low-flow shower heads were installed, and an unknown number of people fixed leaking faucets.

Leftover Problems and Issues

Once the drought was over, water officials were left with a predicament. People were demanding that water rates be lowered again. But if that were done and people continued using less water, the city would not be able to collect enough money to run the current water system or work on new water projects. Ironically, full reservoirs could cause problems, also. If people continued to conserve water once the reservoirs started to fill up, the reservoirs could overflow and water would be wasted.

Discussion Questions

- What problems did you have designing a plan that would save water but would not cause hardship for the residents of Santa Barbara? *(Answers will vary.)*

- How could Santa Barbara have gotten the extra water it needed during the drought? What effect would your ideas have had on other species? *(Water could have been brought in from other areas such as aquifers, rivers, and lakes, but that may have affected the animals and plants that use the water. A new reservoir could have been built, but reservoirs can permanently flood habitat for animals and plants. People could get water from the ground if it were available, but it would also need to be carefully used to keep it from running out.)*

- Do you think Santa Barbara's water plan would have continued to work if the drought went on for another five years? *(Santa Barbara might have been OK because they could have relied on the desalination plant.)*

- If Santa Barbara doubles in size by the next drought, do you think Santa Barbara's plan would still work? How about your plan? *(Santa Barbara would probably have to find new sources of water or expand the desalination plant. The city could also expand its water conservation efforts, but water conservation may not be able to keep up with the new demand.)*

- Is Santa Barbara using water sustainably? *(It's hard to say. It's possible that Santa Barbara's water supply could have been stretched for a while, especially with the desalination plant being available. If the population grew or if the drought lasted a lot longer, however, Santa Barbara residents may have needed to make more adjustments.)*

- Do you think people can tell if they are using a natural resource—such as water, forests, or fisheries—sustainably? Does sustainable use vary with time and under different circumstances? *(It is hard to tell when a natural resource is being sustainably used because we don't always have a complete understanding of how the natural world works and because conditions are always changing. It is perhaps easier to tell when we are overusing a natural resource. Even so, we may choose to ignore the signs.)*

Note: It is important to point out that the concept of sustainability is still fairly new and that people around the world are trying to understand what sustainable use means. But many people argue that using resources in a sustainable way is a goal we should be striving for.

WRAPPING IT UP

Assessment

Have each student choose a local "resource" such as a stream, a desert area, or a park. The student should write a "plan" (between one and three pages) for ensuring sustainable use of the resource. The plan should address the concepts of a fluctuating resource versus a constant resource, sustainable use for people, sustainable use for other species, and threats to the resource similar to the threats on the condition cards in the class activity.

Unsatisfactory—The student does not address one or more of the ideas that he or she was asked to address in the plan.

Satisfactory—The student addresses each of the stated components in the plan.

Excellent—The student goes beyond the stated components to examine complex issues related to sustainability of the resource identified and is able to use competing needs of different species and habitats in the discussion.

Portfolio

The student's group plan can be used for the portfolio.

Writing Idea

Let the students use the following as a journal starter: "The images that come to mind when I think of a sustainable future include . . . "

Extensions

- Have students investigate a local water-use issue. Some regions of the country may not be susceptible to droughts, but there are often other water concerns such as water quality, unsustainable use of groundwater, and environmentally damaging water projects.

- Have students investigate the use of another natural resource and discuss whether they think the resource is being sustainably used. Some possible topics include forests, fisheries, wildlife, and agricultural lands.

Resources

Demand Reduction in Response to Drought: The City of Santa Barbara Experience
by Bill Ferguson and Alison Whitney (City of Santa Barbara Public Works Department, 1993).

Endangered Ecosystems: A Status Report on America's Vanishing Habitat and Wildlife
by Reed Noss and Robert Peters (Defenders of Wildlife, 1995).

Santa Barbara's Temporary Emergency Desalination Project: After the Drought is Over by Stephen F. Mack
(Presented at the Desalination Technical Transfer Seminar, City of Santa Barbara Public Works Department, 1993).

Water Rates and Revenue Impacts of Severe Drought Response, City of Santa Barbara,
1990–1993 by Stephen F. Mack and Bill Ferguson (City of Santa Barbara Public Works Department, 1993).

Getting Involved!

"The future

is not what's

going to happen,

it's what we're

going to do."

—Jorge Luis Borges,
writer

ncouraging students to take a community-level approach to biodiversity issues helps them make connections between what they've learned about biodiversity and how they can directly help protect it. Now that your students have tried a variety of biodiversity activities, they're ready to apply the skills and knowledge they've gained to projects in their own community. Students can participate with individuals and organizations in many towns and cities that have started restoration projects and other activities related to biodiversity. And many nonformal educators at nature centers, zoos, museums, and parks have teamed up with teachers to work with students on action projects. Your students can also take part at the local level in a variety of regional or national efforts that are being made on behalf of biodiversity. Or your students may want to come up with a project of their own.

However your students choose to get involved, community activities can make a big difference—not only to the environment but also to the students. By contributing to their community, your students can feel a sense of accomplishment and satisfaction. Getting involved in community projects can extend the understanding and knowledge your students have already gained and allow them to experience an increased sense of worth and competence. This section provides information on a number of national efforts that your students can become involved with locally, as well as guidelines for facilitating projects that your students can create themselves.

EXPLORING COMMUNITY SERVICE AND OTHER PROJECTS

There are dozens of types of projects that your students might want to tackle that will help them learn more about biodiversity and at the same time

address a local or national problem. Some projects might focus on teaching others about environmental issues. Some might help to physically improve the environment, such as building a trail or cleaning up a park. Other projects might focus on political, consumer, or legislative action, or on research or monitoring in the community. Taking part in a restoration or monitoring project, for example, is one of the best ways for your students to get involved in biodiversity-related community service. Many groups are helping to restore damaged ecosystems (like wetlands and prairies) or are monitoring species and populations (everything from birds to frogs to butterflies to bats) and the habitats they live in. Your class may choose to get involved with state fish and wildlife agencies, conservation organizations, universities, or gardening clubs that have ongoing restoration and monitoring projects. Many national monitoring projects welcome student participation. (See "Biodiversity Monitoring Projects" on page 371.) Whether you are working with students in your classroom, after school, or as part of a nature center, zoo, aquarium, or youth program, this section will assist you in facilitating all stages of your project.

PLANNING A PROJECT

The following are some basic steps that will help your students get their projects off the ground. Please note that planning and implementing a project is not necessarily a linear process. You're likely to evaluate and rethink the steps in a course that will send you back through some of the steps again. With that in mind, adapt the steps to fit your needs.

What a Nose!—Polar bears have a great sense of smell. A hungry bear may pick up the scent of a dead whale or seal and then walk and swim for up to 100 miles until it reaches its meal.

1. Explore project topics and community issues.

Your students can become informed about projects by collecting information from newspapers and magazines, interviewing community members and parents, and contacting organizations and government agencies that focus on environmental issues. If you can, arrange for students to get out and see local environmental problems firsthand. Even if students choose to investigate a problem that's occurring thousands of miles from their community, the exposure to concerns in their own backyard will be an important learning experience. Many projects focus on matters such as enhancing a neighborhood environment or maintaining a unique habitat for native species.

2. Create a list of possibilities.

Once the students have highlighted a number of potential topics, have them work in groups to develop a list of topics that they think are the most interesting or worthwhile. Then have them come up with a list of projects that could address all or part of each topic. Have students select the topic they most want to tackle. Next, they can brainstorm specific projects that might help the situation. They can list any additional information they'll need to evaluate the project. (You might want to remind your group about the "rules" of brainstorming to get the most out of your discussions. See the box on the right.) It might be helpful to explain that environmental topics can be very broad and that several project possibilities almost always exist for each topic. For example, if students are concerned about the use of pesticides in their community, they can survey homeowners, launch an education program for pesticide users, work with local resource agencies to explore state or regional pesticide issues, collect data on the effects of pesticides on wildlife, and so on.

3. Narrow the choices.

For each project listed, ask the students specific questions that will help them think about the process they'll use to accomplish certain tasks.

Students may first want to develop a list of criteria for selecting a project: How much time will the project take? How complex is it? What resources are needed? Whom will they need to talk to? For example, if one of the project ideas is to create a city park in a vacant lot or to restore a park, the students might ask the following questions:

- Why is it important for the city to have a park?

- What problems would be addressed with the development of a park?

- How would different people in the community feel about a park?

- Who might support or oppose a project like this?

- Who will benefit from the park? Why?

- What costs and obstacles are in the way?

Tips For Better Brainstorming

- Don't criticize any idea until the brainstorming session is over.

- Write down the ideas using the exact words of the speaker.

- Encourage creativity. Remind the group that there are no right or wrong ideas.

- All ideas belong to the group. Get as many ideas as possible.

- Remember that lots of "off the wall" ideas can lead to a great idea. Record all ideas.

- Encourage the group to use other people's ideas as springboards for new ideas.

- At the end, look for recurring themes or issues. Try to make groupings of like thoughts and build toward consensus.

Adapted with permission from *Learning by Giving: K–8 Service-Learning Curriculum Guide* by Rich Willits Cairn and Theresa Coble (National Youth Leadership Council, 1993).

NARROWING THE CHOICES

How do you go from dozens of ideas to one? Dr. William Stapp, a leader in the field of action research and environmental education, gives the following guidelines to help students narrow their selection of ideas:

Focus—Use the diagram below to select a focus for projects. In general, projects in the smaller circles are easier to tackle. Projects in the larger circles can be more complex and involve more time and more steps, but they can also have the potential to influence many more people.

Interest—Help students think about which projects they would find most interesting to work on. Strong personal interest leads to increased prospects of success.

Time—Is the timeline realistic? Some projects may extend beyond a school or calendar year and take longer than the students have time to give. Remember, projects almost always take longer than you expect.

Complexity—Some projects are just too complicated. Help students understand that working on a smaller chunk of a larger problem might be the most feasible thing to do.

Change—If students want to see immediate changes as a result of an activity, they need to think about that goal when selecting a project. Some projects will bring about more rapid and concrete results than others.

Information and Resources—Ask students if they think they will have access to the information and other resources they'll need to complete an action project. If not, they are setting themselves up for frustration and disappointment.

Help—Some projects might require a great deal of other people's time. First, help students think about who will need to be involved in the project. Second, investigate whether other groups may have already done work on this topic and could be of assistance. Third, you may decide that the project requires too much outside help. If so, students should shift their focus to another project.

In the case of the park, the students might be addressing the problem of a lack of green space in their city. By discussing options, they might decide that creating or restoring a park is not the only solution to the problem. Planting trees and flowers throughout the city or passing a new ordinance to limit development may be other options.

Sometimes it's difficult for students to decide among local, national, and global projects. Although each will provide learning opportunities, an advantage of a local project is that students will learn more about how their own community works. Students will also be more likely to see real results. To begin with, they might want to focus on one or two small-scale projects. Try not to take on projects that go beyond available resources and time. A large-scale project could become too time consuming or discouraging and lose its impact.

4. Learn more about the issues and pick a project.

Have students choose three to five possible projects, develop a list of questions for each one, and work in groups to research the answers. Give them adequate time to do research, and encourage them to use the library, search the Internet, collect newspaper articles, interview experts, conduct surveys, monitor local TV news, and so on.

Afterward, organize presentations or class discussions designed to help students demonstrate their knowledge. Invite experts or resource people to discuss problems, find potential solutions, and help evaluate ideas.

As students approach their final decision about which project to pursue, have each group present a case for one or more of the projects that the group feels strongly about. Then hold a group vote. Or have a large group discussion and try to reach consensus. The important thing is to let students have as much say in the decision-making process as possible. At the end, group members should have chosen a project that they think is both interesting and doable.

5. Create an action plan.

Help your students get started on their action plan by asking, "What do you hope you'll accomplish by doing this project?" Guide them in coming up with a goal or vision for the project and specific, concrete objectives that need to be accomplished along the way. If the students are tackling a community problem, have them reflect on questions such as the following: What is the current status of the problem? What changed to make it become an issue? Do other communities and countries experience the same problem? What do other people think about the problem? Are there many different opinions? Are there any conflicts? The students can also use the questions in "Mapping Your Action Project" (page 364) to help them plan and focus. A large-format task and timeline chart may help the groups keep track of responsibilities and deadlines. If you've decided to let each group take on its own project, you should adapt this process to fit small group work.

As the students work on their action plan, guide them toward realistic objectives. One of the most common problems for students is thinking too big. Help them focus and simplify the project by discussing the responses to the questions and by asking them to carefully consider hard questions. "How will you raise that much money?" "Is it realistic to think you can spend every weekend working on this project?" "How might you tackle a smaller, but more manageable, piece of the problem?"

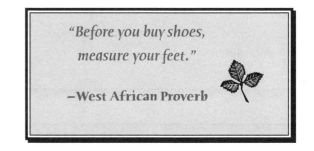

"Before you buy shoes, measure your feet."

–West African Proverb

Missing the Mark?—Over the past 50 years, the amount of pesticides used on crops in the United States has increased by 900 percent. Only 0.003 percent of all pesticides actually reach their target insects each year.

MAPPING YOUR ACTION PROJECT

1. *What environmental issue will your project focus on? What is the current status of the issue?*

2. *Briefly describe the goal of your project and your strategy to accomplish this goal. What would you like to see change as a result of your project?*

3. *What are the specific objectives that will help you reach your overall goal?*

4. *What are the approximate starting and ending dates of your project?*

5. *List the tasks that need to be accomplished to meet each objective. Include a tentative completion date for each task, the names of people responsible for each task, the supplies and equipment required, any funding needed, and ideas about sources for materials and funding.*

6. *Write down the names of people and organizations that may be able to provide you with useful information, specific skills or expertise, or other help.*

7. *List ideas for ways to publicize and generate support for your project.*

8. *Describe criteria for measuring your success.*

6. Put the plan into action.

As students get started, explain that their projects will work best if they keep detailed and clear records. They will need to keep track of who's doing what, what they've done, when they did it, whom they've contacted, and so on.

It's important that students evaluate the project periodically to see if they're on target and to make modifications if necessary. Remind them that it's OK to rethink their goals and objectives and to revise their plan of action in light of new information or unexpected obstacles.

To build support for their projects, have the students publicize their successes through press releases to local newspapers and TV stations, posters or bulletin boards for their school, articles in school newspapers, and information on school or local Web sites.

7. Assess, generalize, and apply.

As the project draws to an end, guide students in assessing the project itself, the process, and their feelings about the experience. A group discussion will allow students to reflect on their feelings and attitudes by providing the chance to listen to what

"Our society will not be unsustainable one day and sustainable the next. Sustainability is a process with a beginning and no end. The challenge will remain with us and our children and their children."

—Stephen Viderman, President, Jesse Smith Noyes Foundation

SOME TIPS TO KEEP IN MIND ABOUT PROJECT PLANNING

Encourage student ownership and initiative. *The more your students are involved in the project the more they'll get out of it. As much as possible, allow them to make their own decisions (what problem to focus on, how to conduct the project, and how to share results).*

Encourage parents and other community members to buy into the project. *Conflict can sometimes surface when students interact with community members and parents who don't agree with a specific activity or who don't feel that action projects are an appropriate educational approach. In many cases, you can diffuse this response by discussing projects with parents and community members beforehand and by explaining how environmental projects enhance the goals of your school or institution.*

Keep your opinions in perspective. *Part of being an effective educator is allowing everyone to have the chance to openly express his or her opinions, no matter how different they may be from your own. Many educators recommend that teachers hold back on sharing their opinions so students have a chance to discuss the issues, research material, and form their own ideas about how they feel.*

Encourage student cooperation, compromise, and understanding. *Have your students work in small groups whenever possible. Besides the well-documented educational benefits of cooperative learning, group work offers a taste of real-life problem solving. Teams of people—scientists, politicians, concerned citizens—often arrive at a plan of action together. Multiple perspectives encourage thoughtful debate, boost critical thinking skills, and allow students to make informed choices.*

Help students evaluate their methods and change their plans if necessary. *Being able to adapt—or even totally change directions if something isn't working—is a plus when it comes to solving problems. Encourage specific methods for evaluating success as the project develops. Students might keep journals and record what they think is going well and what seems to need to change. They can also survey community members to get reactions to the project. At the same time, periodic checkups to gauge reactions from colleagues, teachers, or other students can be valuable.*

Help students appreciate the value of their work. *It's important for students to know that their project, no matter how small, is significant. Even if the students' actions don't seem to have much effect right away, the long-term results can be very important. And often the actions of an individual or of a small group of citizens can inspire others to get involved.*

others say about the project and the experiences they have had. To facilitate the discussion and assess your students' projects, you could ask: "Did your project accomplish its goal and objectives?" "What was the most successful part of your project?" "What was the least successful part?" "Who was influenced or motivated by your actions?" "Whom might those people influence in turn?" "If you repeat the project, what, if anything, would you do differently and why?" "How do you feel about your involvement in the project?" "Have your feelings and opinions about the issue you worked on changed since you began the project? If so, how?" "What did you learn during this project that you'll be able to apply to other situations that you'd like to change?" "Would you get involved in another environmental project? Why or why not?" "What advice would you give to other students who are planning an action project?" "Do you think it's important for citizens to volunteer for community service? Why or why not?"

IDEAS FOR MEASURING SUCCESS

Taking time to evaluate a project helps students understand what they've accomplished and allows them to recognize how their project has facilitated their personal growth. Here are some ways to measure student knowledge and project success:

Assessing Student Knowledge

- Keep a video or photo log of project highlights. After the project is completed, use the videos or photos as a springboard for discussion with students.

- Create a scrapbook in which students can write personal comments and assemble memorabilia.

- Ask students whether their project changed their thinking or behaviors. Have the students write essays describing those changes and the possible reasons for them.

- Have students keep a journal to record feelings about the project, its progress, and its setbacks, and to keep notes about the challenges and rewards of working with others.

- Have students evaluate other members of their group as well as themselves. Before the students do this, give them guidelines on positive, constructive feedback that focuses on specific points such as contributions and efforts.

- Have community members who were involved in the project assess student performance.

Assessing Project Success

- Have students describe how well they think their project accomplished the objectives they outlined at the start.

- Have students conduct surveys or interviews to assess the success of their completed project. What worked? What didn't? Why?

- Evaluate how the students planned for ongoing maintenance and sustainability of the project.

- Have community members and others who were involved in the project assess the project outcome.

FACILITATING ACTION PROJECTS

Among the keys to success for educators who are initiating student projects are enthusiasm and commitment, the ability to create a sense of ownership in students, and a willingness to work with students to overcome barriers and to celebrate success (even when projects don't turn out as expected). The challenge is to facilitate a project so that students have the power to make decisions and learn to rely on themselves for answers or direction. Creating an atmosphere of cooperation and trust will go a long way toward encouraging ownership and responsibility, and toward empowering students to make positive changes in their communities.

Encourage your students to take the following actions:

Be prepared. Arming yourself with facts and background knowledge will serve you well. Researching accurate information is critical to a balanced approach to environmental issues.

Be realistic. Set goals that you feel you can accomplish. Do not take on projects that are beyond your available resources and time.

Stay positive. Be *for* something you care about, rather than *against* something you disagree with.

Be respectful. Ideally, environmental action projects stem from a concern and respect for the health of our world. That same respect should apply to people as well. Lack of respect and the tendency to generalize or stereotype are two of the biggest pitfalls of action projects.

Keep trying. Don't let early setbacks or failures derail your project. Rethink your approach, brainstorm new avenues, and try again.

Welcome change and surprise. Continually reassess your project. How can you do things better? Are you still on track? What can you learn from problems?

PROJECT IDEAS

Ideas for projects can come from many sources, including your own interests and experiences. You and your students may find the perfect action project listed below or you may want to create your own.

Adopt an endangered species. There are a number of programs that enable kids to adopt an animal, plant, and, in some cases, part of an endangered habitat. Most of the programs provide students with information on the species and the threats it faces. (See "Wildlife and Wildlands Adoption Programs" on page 373 for a list.)

Restore a creek. Educating landowners and homeowners in a community about reducing the use of fertilizers and pesticides, planting eroded banks with native grasses and shrubs, and removing trash are just some of the ways to improve the health of local creeks and waterways for all of their inhabitants.

Start an organic garden. By starting an organic garden on your school grounds, students will learn ways to lessen the environmental impact of growing food and get to eat the results.

Start a compost system. Composting is an easy way to turn certain wastes into products that are useful for gardening, landscaping, and soil restoration.

Remove introduced plant species. Removing invasive, introduced plant species gives native species of plants and animals a chance to reclaim their former habitat. Contact your local natural resource department for a listing of native and invasive plant and animal species in your area.

Start a native plant nursery. If students want a hands-on project that can improve wildlife habitat, have them start a community native plant nursery. They can investigate suitable plants, materials, locations, and fundraising options, as well as ways to distribute the plants once they are grown. If they decide to have a plant sale, nature centers and botanical gardens might help with the setup and publicity. Or better yet—have a plant giveaway! Contact the National Gardening Association, 180 Flynn Ave., Burlington, VT 05401, (800) 538-7476, for information on the Native Plant Society in your area.

Make a video about a local biodiversity issue. Making a video about a biodiversity issue is a creative way for students to research an issue, organize their thoughts, and educate others. Explore partnerships with local public TV and cable companies.

Develop a newspaper or newsletter. Creating a newspaper or newsletter on biodiversity issues helps students fine-tune their writing skills and explore and share their thoughts. They might consider distributing their newspaper or newsletter beyond their school. If the students have a school Web site, the newspaper or newsletter could be posted on the Internet.

Survey the community about an issue. How do community members feel about protecting a local threatened or endangered species? What about a nearby water pollution problem? Students can gauge community opinion on an environmental issue by surveying people about their thoughts and feelings. The survey can also lay the groundwork for future action projects by identifying issues that people are most concerned about or issues that are ignored.

Participate in an annual species count. The annual Audubon Christmas Bird Count is a great example of how individual participation can add up. Each year in late December, around Christmas, people survey the types and numbers of birds in their area. The data is used to study trends in bird populations across the country. (See page 371 for information on the Audubon Christmas Bird Count.)

Develop and perform a play for the community. What better way to put your students' talents to work for biodiversity than to have them create a play on an important biodiversity issue? A play can not only educate others, but it can also serve as a fund raiser for other biodiversity projects. Work with local drama teachers, community college theater departments, and other local experts for help in getting your play started.

Paint signs for storm drains. What goes in a storm drain often comes out into nearby streams, rivers, or wetlands. Many people are not aware that hazardous materials entering a storm drain can affect biodiversity somewhere else. Students around the country have been trying to change this "out of sight, out of mind" mentality by painting warnings on storm drains that state where the drains dump their contents. Pre-made storm drain stencils can be ordered by contacting Earthwater Stencils, 4425 140th Ave. SW, Rochester, WA 98579-9703, <www.earthwater-stencils.com>.

Participate in a coastal cleanup.

Students who take part in beach cleanups learn firsthand about marine biodiversity and how people affect it. By participating in a nationwide coastal clean-up program, students can provide data that will help everyone learn about pollution trends across the country. Begin by contacting the Center for Marine Conservation, which coordinates the International Coastal Cleanup. (For contact information, see page 374.)

Start a school energy patrol. Saving

energy conserves natural resources and protects habitats. Students can save energy at school by turning off lights and keeping windows and doors closed when the heat or air conditioning is on. A student energy patrol can remind other students and teachers to conserve energy, promote energy conservation at home, and share energy conservation tips.

Hold a used oil collection day for the

community. At least 61 percent of the car oil that is changed by "do-it-yourselfers" is not disposed of properly. Oil that drains into the street can eventually wash into local streams, rivers, and lakes where fish and other organisms live. But used oil can be recycled. Organizing a used oil collection day will motivate people to recycle used oil. Contact your local recycling centers for more information. A local organization or agency may be willing to help

organize and publicize the event. Also check with environmentally friendly service stations and repair shops to see if they will help.

Develop a carpool plan for parents

and others. Break out the maps and work schedules, and see if students can develop a carpool plan for their parents or other groups in the community. If your students can convince their parents to try carpooling, they will be saving natural resources by conserving energy and cutting down on pollution. Your kids can also work with local businesses, government agencies, and other organizations that offer carpool services to their employees and see if they can help launch a carpool education program.

Buy habitat acres. Individuals and groups

can get directly involved in protecting habitat for animals and plants by buying acres. From rain forests to coral reefs, students can pitch in by raising money to purchase and protect habitat—one acre at a time. (See the list of habitat protection programs on page 373.)

Hold a town meeting on a local

biodiversity issue. Town meetings inform citizens and give people a way to speak out on issues that concern them. A student-organized town meeting may be a catalyst for community change. A student/teacher guide for planning a town meeting is available from Earth Force. (For contact information, see page 455.)

Participate in local government.

Students can learn a lot about the environment and citizenship skills by getting involved in issues faced by the local government. They may need to contact organizations, request information, write letters, conduct surveys, circulate petitions, build coalitions with other schools or groups, and make presentations.

Build a nature trail.

Is biodiversity being overlooked in your school's backyard? Is there a patch of land nearby that could use some sprucing up? Could the local nature center or botanical garden use the students' help to create a trail on their grounds? Creating a nature trail, making a nature study area, or restoring a native habitat are just a few ways to get students and adults involved in learning more about local biodiversity. Also check with local government agencies and planning groups to see if they have ideas for projects.

Design and sell Earth-friendly grocery bags.

Plastic or paper? If you bring canvas bags to the grocery store, you'll never have to answer that question again. You'll also save trees, fossil fuels, and landfill space. Students can encourage the use of reusable grocery bags by designing attractive bags and selling them to parents, neighbors, and others. Bags designed by local students are sure to be a hot seller! And they could be sold in the gift shop of your local nature center, zoo, or museum.

Have an energy-efficient light bulb sale.

A new twist on the candy drive or bake sale is the sale of energy-efficient light bulbs. Although they are more expensive initially, energy-efficient light bulbs use less energy and last much longer than standard light bulbs.

Do a school waste audit.

Students don't need to jump into garbage bins to do a waste audit, but finding out what kind of waste their school tosses away can help them develop a plan to reduce and recycle waste. Conduct ongoing audits of the school's waste and document the results. Present the results to your school administrators, along with ways to reduce waste. Since landfills require land, any garbage that can be recycled leaves more habitat for animals and plants.

Develop and distribute fact sheets.

Rather than having students write another research report, why not have them develop creative fact sheets connected to a local issue of concern? Fact sheets can be easily distributed to other students and adults, and they can help focus attention on specific biodiversity-related issues.

Write letters.

Letter writing can be part of any project. Students will need to study an issue from all perspectives, develop an opinion, put their thoughts into writing, and identify the individuals, organizations, agencies, or businesses that should receive the letters.

Develop a magazine share among households.

Magazines are hard to recycle but easy to share. Students can make a list of magazine subscriptions received in their household and then coordinate an exchange with other students and their families. This project will give more people access to the magazines before they are thrown away, and it could cut down on some subscriptions—thereby saving raw materials, energy, and landfill space.

Hold a community paint exchange.

To keep unused paint out of landfills, hold a community paint exchange or giveaway. The color that didn't work in your kitchen may be perfect for someone else's bathroom.

Two Pea-Eating Aphids—Pea aphids in southern Wisconsin come in two colors: red and green. Although both the red and green aphids feed on the same pea plants, each color attracts a different predator. Ladybugs appear to eat more red aphids than green aphids. Scientists think this is because the red ones stand out against the green pea plants. But parasitic wasps go for the green ones. The wasps lay their eggs inside the green aphids. When the eggs hatch, the wasp larvae eat their way out. Scientists believe that over time, the wasps "learned" that if they laid their eggs in red aphids, they would get eaten by ladybugs. So to protect their young, they pick the green ones. Adapted from "Discover," 1998.

Test your local water. Water testing can be the first step toward discovering local water problems and helping to solve those problems. Test water at several locations, and compare the results to water quality standards. If results don't meet the standards, alert community officials. Nonformal institutions can help by purchasing several water testing kits and lending them to local schools. And they can offer short workshops for teachers on how to use the kits. For more information on this type of activity, contact the Global Rivers Environmental Education Network (GREEN). (For contact information, see page 372.)

Organize a biodiversity day. Sponsoring a community event focused on biodiversity can help raise awareness about local, national, and global issues. Have schools team up with zoos, museums, nature centers, environmental organizations, and others to sponsor a local "Biodiversity Day." You can also hold theme-based festivals, such as a "Tiger Day" at the local zoo to highlight tiger natural history and efforts to protect tigers worldwide. Or hold a "Save a Place for Wildlife Day" at a local nature center to help people learn more about protecting habitat and what they can do locally and globally.

Use the Internet. You can use the Internet to ask for help in generating action ideas, keep in contact with local schools, distribute information, answer questions, publicize events, and more. You can also use the Internet as a collection point for data on anything from rare bird sightings to water testing results to spring blooming dates. (It's important to remember that Internet addresses can become outdated and are subject to change. If an address is no longer valid, try doing a general Internet search for either the organization or a key term.)

RESOURCES FOR ACTION PROJECTS

The following list contains information on biodiversity monitoring projects and "adoption agencies," as well as the names of organizations that might be helpful to you and your students as you conduct your action project.

Biodiversity Monitoring Projects

Amphibians as Bio-Indicators
840 SW 97th St.
Wakarusa, KS 66546
(913) 623-4258, Web site
<www.geocities.com/RainForest/Vines/2986>
Students help collect data on declining amphibian populations. The program emphasizes species recognition, field safety, and environmental ethics.

National Audubon Society (Christmas Bird Count)
700 Broadway
New York, NY 10003-9501
(212) 979-3000, Web site <www.audubon.org>
Audubon Society's Christmas Bird Count began in 1900 and is the oldest continuous wildlife survey in North America. It uses volunteers to monitor bird populations throughout the United States. Also check to see if there's a local Audubon chapter in your community.

Global Learning and Observations to Benefit the Environment (GLOBE)
744 Jackson Pl., NW
Washington, DC 20503
(202) 395-7600
Web site <globe.fsl.noaa.gov>
GLOBE creates international partnerships among students, teachers, and the scientific research community. Students collect data, make observations, and investigate a variety of topics, including the atmosphere, hydrology, soil, and biology. Student observations are linked worldwide through the Internet.

Global Rivers Environmental Education Network (GREEN)

206 South Fifth Ave., Ste. 150
Ann Arbor, MI 48104
(734) 761-8142
Web site <www.econet.apc.org/green/>

GREEN is an international water-quality monitoring program. A number of monitoring guides are available, including one on low-cost monitoring. Groups can enter their findings into a computer database and compare them with results from other parts of the same watershed or from other parts of the world.

Journey North

125 N. First St.
Minneapolis, MN 55401
(612) 339-6959
Web site <www.learner.org/jnorth/>

This organization tracks the migration of birds, butterflies, and whales. Students participate by making their own observations and reporting them on the Internet. Field biologists enter their sightings into a database on the Internet, and each week there is an "Ask the Expert" session. Students can also follow the migration patterns of loggerhead sea turtles, peregrine falcons, bald eagles, caribou, and loons through satellite imagery.

Naturemapping

Defenders of Wildlife, West Coast Office
1637 Laurel St.
Lake Oswego, OR 97034
(503) 697-3222
Web site <1www.defenders.org>

As part of this program, students or other community members make wildlife observations in their local communities, across statewide ecoregions, or in areas where local resource managers are seeking more information. Participation in Naturemapping involves a training program for group leaders. Training programs are currently available in Oregon and Washington only but will be available in other states in the future. There is a similar program in Virginia called "Wildlife Mapping." Contact the Virginia Department of Game and Inland Fisheries at 4010 W. Broad St., Richmond, VA 23230, (804) 367-1000, Web site <dit1.state.va.us/~dgif>.

Plantwatch

Devonian Botanic Garden
University of Alberta
Edmonton, Alberta T6G 2E1
Canada
(403) 987-3054, Web site
<www.biology.ualberta.ca/devonian.hp/pwatch.htm>

This program encourages students and the general public to record flowering dates for plant species in their area. The information received from tracking flowering times helps to determine how trends in climate change and weather variability affect plants.

Project FeederWatch

Cornell Laboratory of Ornithology
P.O. Box 11
Ithaca, NY 14851-0011
(800) 843-BIRD
Web site <www.ornith.cornell.edu >

People collect data on the kinds and numbers of birds at bird feeders during the winter. The information is used to help scientists better understand North American bird populations. Participants receive a research packet, a bird calendar, and a subscription to the quarterly newsletter *Birdscope*. Classroom FeederWatch is a part of this program that includes a curriculum focusing on bird biology and behavior, as well as on the contribution of observational data to scientific analysis.

Project PigeonWatch

Cornell Laboratory of Ornithology
P.O. Box 11
Ithaca, NY 14851-0011
(800) 843-BIRD
Web site <www.ornith.cornell.edu>

Students make simple observations of pigeon behavior and appearance and then send the information to the Cornell Laboratory of Ornithology. Participants receive a research packet and a subscription to the quarterly newsletter *Birdscope*.

Save Our Streams Program

The Izaak Walton League of America
707 Conservation Ln.
Gaithersburg, MD 20878-2983
(301) 548-0150, Web site <www.iwla.org/iwla>
This program teaches volunteers how to monitor, protect, and restore their local watersheds using techniques that can be used by people of all ages.

Wildlife and Wildlands Adoption Programs

Bats

Bat Conservation International
Adopt-a-Bat Program
P.O. Box 162603
Austin, TX 78716
(800) 538-BATS, Web site <www.batcon.org>
Membership includes a bat adoption certificate, photo, magazines, catalogs, and special event invitations for $25. An adoption certificate without membership is $15.

Manatees

Save the Manatee Club
Adopt-a-Manatee Program
500 N. Maitland Ave.
Maitland, FL 32751
(800) 432-JOIN
Web site <www.objectlinks.com/manatee>
A manatee adoption certificate, photo, biography of "your" manatee, handbook, and the *Save the Manatee Club Newsletter* are included with membership ($25). *Manatees: An Educator's Guide* is available free for educators.

Whales

Pacific Whale Foundation
Adopt-a-Whale
101 N. Kihei Rd.
Kihei, Maui, HI 96753
(800) WHALE-11
Web site <www.pacificwhale.org>
Fees are $35 to adopt a named humpback whale. The adoption "kit" includes an adoption certificate, a photograph of the adopted whale, a whale watching guide, a parent I.D. card, and a copy of the *Soundings* newsletter.

International Wildlife Coalition
Whale Adoption Project
70 East Falmouth Hwy.
East Falmouth, MA 02536
(508) 548-8328, Web site <www.iwc.org>
The project allows children to adopt 1 of more than 50 specific whales that are regularly monitored off Cape Cod, Massachusetts.

Reefs

The Nature Conservancy
Rescue the Reef Program
4245 N. Fairfax Dr., Ste. 100
Arlington, VA 22203-1606
(800) 84-ADOPT, Web site <www.tnc.org>
Sponsorship of a square meter of coral reef ($35) helps to support the research and policing of reefs in the Florida Keys, the Dominican Republic, and the Republic of Palau.

Bison

The Nature Conservancy
Adopt-a-Bison Program
23 W. 4th St., Ste. 200
Tulsa, OK 74103
(800) 628-6860
Web site <www.tnc.org/involved.html>
Adoption of a bison on the tallgrass prairie of Oklahoma is $25.

Monarch Butterflies

Monarch Watch
c/o Orley Taylor
Department of Entomology
University of Kansas
Lawrence, KS 66045
(888) TAGGING, Web site
<MonarchWatch.org/migrate/migrate.htm>
Kits cost $25 each. (Kit One includes larvae ready to pupate. Kit Two has younger larvae.) Kids raise larvae in the classroom and then release them with a "tag" to help a monarch research project. Monarch Watch involves areas east of the Rocky Mountains. However, curriculum materials provided on the Web site are pertinent to monarchs in all parts of the country.

Rain Forests

The Nature Conservancy
Adopt-an-Acre
1815 N. Lynn St.
Arlington, VA 22209
(800) 84-ADOPT, Web site <www.tnc.org>
An acre of Costa Rican rain forest is available for adoption for $35.

Save the Rainforest
P.O. Box 16271
Las Cruces, NM 88004
(888) 608-9435
Web site <www.lascruces.com/~saverfn>
Adopt an acre of rain forest in Panama for $35 or buy an acre of land in Ecuador for $70.

Sea Turtles

Caribbean Conservation Corporation
4424 NW 13th St., Ste. A1
Gainesville, FL 32609
(352) 373-6441
Web site <www.cccturtle.org>
Adoption of a tagged, satellite-tracked sea turtle is $25. Sponsorship includes an educator's guide, a turtle adoption certificate, a sea turtle fact sheet, a decal, and a one-year subscription to CCC's membership publication, the *Velador*, which has a special children's section called *Turtle Tides*. Kids can track the location of their and other students' turtles by accessing maps on CCC's Web site.

Coasts

Center for Marine Conservation

Atlantic/Gulf Coast Office	Pacific Coast Office
1725 DeSales St., NW	312 Sutter St., Ste. 606
Washington, DC 20036	San Francisco, CA 94108
(202) 429-5607	(415) 391-6204

Web site <www.cmc-ocean.org>
Ask for a free Adopt-a-Beach kit to get involved in a local marine, lake, or river cleanup.

Service Learning Organizations

The Corporation for National Service

1201 New York Ave., NW
Washington, DC 20525
(202) 606-5000 ext. 123, Web site <www.cns.gov>
This organization assists communities in developing and planning community-based service-learning education programs for kids.

Quest International

1984 Coffman Rd., P.O. Box 4850
Newark, OH 43058-4850
(800) 446-2700, Web site <www.quest.edu>
This program empowers and supports adults around the world to encourage responsibility and caring in young people through service-learning activities.

Campus Outreach Opportunity League (COOL)

1511 K St., NW, Ste. 307
Washington, DC 20005
(202) 637-7004, Web site <www.cool2serve.org>
This program assists students and communities in developing student-run, high-quality service programs.

Close-Up Foundation

44 Canal Center Plaza
Alexandria, VA 22314-1592
(800) CLOSEUP, Web site <www.closeup.org>
The foundation supports service-learning educators by providing training workshops, program development seminars, curriculum-writing institutes, "training of trainers" sessions and materials, youth service summits, student guidebooks, and teacher manuals to help integrate service learning with curriculum.

National Service Resource Center (NSRC)

P.O. Box 1830
Santa Cruz, CA 95061-1830
(800) 860-2684
Web site <www.etr-associates.org/NSRC>

The organization acts as a central point for sharing training and technical assistance information and resources with AmeriCorps Programs, State Commissions on National Service, other AmeriCorps training and technical assistance providers, and The Corporation for National Service.

Youth Serve America

1101 15th St., NW, Ste. 200
Washington, DC 20005
(202) 296-2990, Web site <www.SERVEnet.org>

This program seeks to build strong communities through fostering citizenship, knowledge, and personal development by building a powerful network of service opportunities for young Americans.

National Service Learning Cooperative Clearinghouse

University of Minnesota, Department of Work, Community, and Family Education
1954 Buford Ave., Rm. R-460
St. Paul, MN 55108
(800) 808-SERV
Web site <www.nicsl.coled.umn.edu>
This program assists "K–12 Learn and Serve America" programs, educators, and community agencies to develop and expand service-learning opportunities for all youth.

Other Organizations

Agriculture on School Grounds

Ag in the Classroom
U.S. Department of Agriculture
14th St. and Independence Ave., SW
4307 S. Bldg.
Washington, DC 20250-0991
(202) 720-7925, Web site
<www.reeusda.gov/serd/hep/agclass.htm>

This program provides a free resource guide and monthly newsletter, as well as state-specific assistance.

Project Food, Land, and People
1990 N. Alma School Rd., #136
Chandler, AZ 85224
(602) 963-7959

This group provides supplemental curriculum materials to promote a better understanding of the interrelationships among agriculture, the environment, and people of the world.

Marine Conservation

The American Cetacean Society
P.O. Box 1391
San Pedro, CA 90733-1391
(213) 548-6279, Web site <www.acsonline.org>

This society works to protect marine mammals through education, conservation, and research. ACS publishes cetacean fact sheets and a newsletter as well as other educational materials.

BIOFACT

Tons of Termites—For every 100 pounds of people in the world, there are about 1,000 pounds of termites.

The Cousteau Society
870 Greenbrier Circle, Ste. 402
Chesapeake, VA 23320
(804) 523-9335, Web site
<www.scubaworld.com/cousteau/member.htm>
This society is dedicated to the protection and
improvement of marine life. It produces films,
lectures, books, and other publications.

Marine Conservation Biology Institute
15806 NE 47th Ct.
Redmond, WA 98502-5208
(425) 883-8914, Web site <www.mcbi.org>
This organization is dedicated to safeguarding
life in the sea by advancing science.

Acid Rain

Acid Rain Foundation, Inc.
1410 Varsity Dr.
Raleigh, NC 27606
This organization provides resources on acid
rain, air pollution, and global climate change.

Energy Conservation

Energy Education Project
State University of New York at Albany
1400 Washington Ave.
Albany, NY 12222
The group publishes several energy-education
curriculum guides.

National Energy Education Development
(NEED) Project
102 Elden St., Ste. 15
Herndon, VA 20710
(703) 471-6263, Web site <www.tipro.com/need>
The program sponsors an annual project of national
energy education for grades K–12 and publishes a
booklet of educational activities.

National Energy Foundation
5225 Wiley Post Way, Ste. 170
Salt Lake City, UT 84116
(801) 539-1406
Web site <www.xmission.com:80/~nef/>
The foundation publishes materials for grades
K–12 and ideas for energy projects.

National Energy Information Center
Rm. 1F-048, Forrestal Bldg.
1000 Independence Ave., SW
Washington, DC 20585
(202) 586-8800, Web site
<www.sandia.gov/ESTEEM/home.html>
The center publishes K–12 curriculum guides about
electricity, water, and natural gas.

U.S. Environmental Protection Agency
401 M St., SW
Washington, DC 20460
(202) 382-2090, Web site <www.epa.gov>
The agency has information on a variety of
environmental issues.

Environmental Lobbyists

Americans for the Environment
1400 16th St., NW
Washington, DC 20036
(202) 797-6665
Web site <www.afore.org>
This group provides information on the environment
and the electoral process.

League of Conservation Voters
1707 L St., NW, Ste., 550
Washington, DC 20036
(202) 785-8683
Web site <www.lcv.org>
The LCV publishes information on the environmental
voting records of members of the U.S. Congress.

Outdoor Habitats

American Forestry Association
Global ReLeaf Program
P.O. Box 2000
1516 P St., NW
Washington, DC 20013
(800) 368-5748, Web site <www.natural-
connection.com/institutes/forestry.html>
The association promotes reforestation and distributes
a teacher's packet.

The Garden Club of America
598 Madison Ave.
New York, NY 10022
(212) 753-8287, Web site <www.gcamerica.org>
The organization works on natural resource conservation issues and distributes an educational packet.

Life Lab Science Program
1156 High St.
Santa Cruz, CA 95064
(408) 459-2001, Web site
<www.ed.gov/pubs/EPTW/eptw7/eptw7j.html>
The organization publishes an elementary curriculum that integrates gardening activities with science, nutrition, and environment. It also offers a newsletter and educator workshops.

National Wildlife Federation
8925 Leesburg Pike
Vienna, VA 22184-0011
(703) 790-4000, Web site <www.nwf.org>
The federation provides a certificate to recognize outdoor school site development through its Schoolyard Habitat program. The program also offers workshops, grants, and activity ideas for improving and using outdoor school sites.

The Xerces Society
10 SW Ash St.
Portland, OR 97204
(503) 222-2788, Web site <www.xerces.org>
The society provides information on butterfly gardening.

Solid Waste
Citizens Clearinghouse for Hazardous Wastes, Inc.
P.O. Box 926
Arlington, VA 22216
(703) 276-7070
Web site <www.essential.org/cchw/>
The clearinghouse provides publications on recycling and other solid waste issues.

Environmental Action Foundation
6930 Carroll Ave, Ste. 600
Takoma Park, MD 20912
(301) 891-1100
The foundation provides recycling information.

Water Quality
American Rivers
1025 Vermont Ave., NW, Ste. 720
Washington, DC 20005
(202) 547-6900, Web site <www.amrivers.org>
This organization promotes river preservation.

Soil and Water Conservation Society
7515 NE Ankeny Rd.
Ankeny, IA 50021-9764
(515) 289-2331, Web site <www.swcs.org>
The society promotes sustainable land and water use worldwide.

"Getting Involved!" was adapted from
Taking Action: An Educator's Guide to Involving Students in Environmental Action Projects
by Project WILD and World Wildlife Fund
(Council for Environmental Education, 1995).

Putting the Pieces

"*The most*

important task,

if we are to save

the Earth, is

to educate."

—Peter Scott,
conservationist, artist

World Wildlife Fund

Together

Building Creative Biodiversity Units

This section provides ideas about how to link activities to build effective biodiversity units. Whether you're a science teacher, a language arts teacher, a social studies teacher, or a nonformal educator, we heartily encourage you to adapt the activities, combine them with other resources, and devise organizing themes that will best meet your particular educational objectives. To that end, we have created several sample units (see pages 384–408) and listed some of the ideas we took into consideration in developing these units. You can pick up and use any one of these units to present biodiversity to your students. But we know that no one can create better units of study for your students than you can—whether you teach in a school, a zoo, a museum, or a community center. We encourage you to tailor the sample units we've provided to your particular needs, or—better yet—to develop your own biodiversity units that draw on local issues and content relevant and engaging to your students. Below is an overview of the thinking we put into the sample unit plans we developed. For more about these topics, see the resources beginning on page 452.

The Thinking Behind the Sample Units

1. Framing the Unit

For the *Biodiversity Basics* module, we've developed a content framework organized around four major themes: *What is biodiversity? Why is biodiversity important? What is the status of biodiversity? How can we protect biodiversity?* Specific learning objectives accompany each of these themes, and the activities tie to specific concepts in the framework on page 418.

In addition, we've developed a framework of skills adapted from the North American Association for Environmental Education's (NAAEE) *Environmental Education Guidelines for Excellence: What School-Age Learners Should Know and Be Able to Do,* and the thinking and process skills developed by the Association for Supervision and Curriculum Development and the American Association for the Advancement of Science. We've also incorporated skills from the Middle School Association and the national standards work in civics, science, math, reading, language arts, and geography. Our skills framework focuses on thinking, process, and citizen action skills that we believe lead to informed and engaged citizens.

Together, the content and skills frameworks guided the development of the activities in this module and the units that follow.

2. Starting with a Storyline

We've organized each of our sample units around a storyline. Storylines are organizing questions or attention-grabbing issues that help focus a course of inquiry. For example, the sample unit "Biodiversity Inquiry" on pages 391–393 uses scientific inquiry itself as the storyline. The question, "How do scientists figure things out?" becomes the unifying thread for understanding biodiversity issues. Another sample unit asks the question, "Should we be concerned about biodiversity loss?" as a way to get right to the heart of the biodiversity crisis.

You can use storylines at the lesson level to help students understand individual concepts, or at the unit level to give sets of activities an added dimension of relevance or interest. Either way, storylines can function as effectively in the classroom as they do in a good mystery: enhancing curiosity and making students eager to find out more.

3. Building on Concepts

In the sample units starting on page 384, we've organized content in a way that builds on previous material. That may sound simple, but it's not always easy to do with a topic as multidimensional as

A HEAVY MATTER—If you weighed all the animals in an acre of Brazilian rain forest, over 90 percent of the total weight would be invertebrates and one-third of that would be termites and ants.

biodiversity. In developing your own units, you'll need to create your own sequence of activities. To help in that process, the "Connections" section at the beginning of each activity lists other activities that are good to do before or after that activity. But in some cases, you will need to insert activities from other sources to fill in some of the "holes." And in other cases, your students will need prior knowledge to get the most out of the activity.

4. A Focus on Learning Styles and Multiple Intelligences

One of our goals in developing this module was to make biodiversity come alive for all students— regardless of how they learn best. To do that, we've tried to cater to a variety of learning styles and intelligences. You'll find a mix of activities and teaching strategies—from hands-on discovery to small group discussions—and you'll see these different learning styles highlighted in different ways throughout the sample units. For example, in the "Biodiversity Inquiry" unit, conducting family culture-nature surveys is likely to appeal most to dynamic learners who like to interact with others. And making ads for a biodiversity campaign will probably be most inspiring for imaginative learners who thrive on creative expression.

In addition to recognizing different learning styles, we have also tried to emphasize strategies that promote the variety of strengths your students already have and help them develop new strengths. (For more about multiple intelligences and learning styles, see the resources on page 458.)

5. Thinking, Questioning, and Creating

We developed *Biodiversity Basics* to get students to think, question, and be creative. Throughout this module you will see a variety of activities that encourage questioning, problem solving, issue investigation, creative writing, and other critical and creative thinking skills. And each of our sample units incorporates these activities. We encourage you to

build on what we've developed to help your students learn to be critical thinkers and to use their creativity to come up with new ways of seeing the world. If you're new to teaching, there are many resources that can give you ideas about how to develop more effective questions, inspire creativity, encourage effective writing, and encourage your students to think at a higher level. For example, writing down questions before you start a unit allows you to think more carefully about the types of questions you use and can help you get your kids to think at a higher level. Encouraging students to brainstorm, without passing judgment on other people's ideas, can help students feel more uninhibited in coming up with creative ideas. We also encourage you to use the writing suggestions at the end of each activity as a way to emphasize different types of writing and to let kids know that you think writing is important—regardless of the subject you teach.

6. Curriculum Connections

The sample units on the following pages provide a variety of ideas for integrating biodiversity across the curriculum. Although we know it's difficult in many schools to link science, math, language arts, and social studies, we also know that many educators around the country are doing just that.

No matter what or where you teach, there are creative ways to bring biodiversity into your teaching. For example, you might already do a unit on classification. Or perhaps you conduct outdoor walks for students in local parks. In these and other cases, the activities in this module can help you add a new twist to your teaching by tying your activities to biodiversity issues. We've provided a sample classification unit that shows one way to do this. However, you may want to incorporate some activities from *Biodiversity Basics* into a unit you already do. You can also use the activities in this module to teach specific thinking and process skills (which are highlighted on pages 446–450 and at the beginning of each activity) or to improve writing skills (ideas are highlighted at the end of each activity).

7. Language Learners and Biodiversity

We know that typical middle school classrooms are made up of students with a variety of language backgrounds. The same holds true of young people taking part in nature clubs, zoo and aquarium courses, and museum classes. In developing this module, we tried to incorporate suggestions from linguistic experts to help address the needs of students who speak English as a second language and students reading below grade level. And on pages 430–431, we've listed some of the tips from linguists and educators about how best to think about language issues as you plan your units. We've grouped the tips into listening and speaking skills, vocabulary, reading, and writing.

8. An Experiential Design

We relied on several experiential design models to develop the activities and units in this module. We know that students and adults learn better when they are actively engaged in their learning and have a chance to take part in varied experiences. In many cases, one activity does not include all the steps of the experiential learning cycle. So in our sample units we combined and sequenced activities in a way that allows students to experience an exciting introduction (which can take the shape of a question, an experiment, a quiz, or some other activity that gets them engaged), a chance to take part in some experience (an activity, an experiment, an investigation, and so on), a chance to process what they did (often a small or large group discussion), a chance to reflect and generalize (what did they learn, what insights did they gain that they can apply in other situations, what did others think and feel?), and finally, a chance to apply and practice (an opportunity to show what they learned, preferably

through a real project or activity). We encourage you to explore the many models that exist for designing effective units. The diagrams below show two experiential learning models.

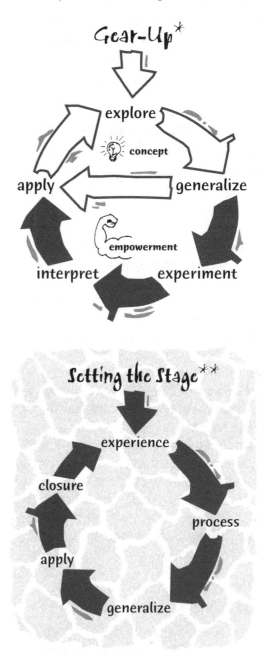

*Adapted from Alaska Science Consortium
**Adapted from Training Resources Group (TRG)

BIOFACT

A Whale of a Whale—The largest of the toothed whales is the sperm whale. A male may grow to be as long as two school buses parked end to end, and its heart alone weighs 400 pounds. Its brain is the largest of any animal that has ever lived.

9. Working Together

Solving the complex problems of the world will take people with diverse ideas coming together to forge creative solutions. We believe that one of the most important skills we can teach young people is how to work effectively in groups. Cooperative learning helps students practice active listening, discussing, questioning, consensus building, defining personal positions, and more.

Working together also helps students learn to respect other's opinions, beliefs, and values. At the same time, they learn how to agree to disagree and work together to understand controversial situations. Each of our sample units includes group experiences to help students practice the teamwork skills that they will use for a lifetime.

> *"More than five hundred research studies now report that students learn better when they work cooperatively."*
>
> **–David W. Johnson and Roger T. Johnson**

10. Wrapping It Up

You'll see that almost every activity has a suggestion for how to assess student learning. (Those that don't are activities designed to introduce concepts or generate interest.) In our sample unit plans, we often use one of the final activities as an assessment to pull the unit together and provide closure for your students. We believe that the best wrap-ups (to an activity, a unit, or a course) help students see how much they have learned, how all this knowledge fits together, and how well they can apply it to a final product or project. We also believe that the best assessments go beyond multiple choice questions and show how students have mastered knowledge or skills by completing something real.

11. Action Strategies

Two of the units (see units on page 384 and page 391) include strategies for facilitating action projects with your students. These units give examples of how your students can get involved in community projects that relate to biodiversity. We believe that it's important for your students to have a grounding in biodiversity before tackling local, national, or global biodiversity action projects. However, you can add an action component to any of the units, including the ones you develop on your own. See "Getting Involved!" on page 358 for more about how to facilitate environmental action projects with your students.

12. Going Beyond Basics

We used a variety of resources to develop *Biodiversity Basics*. And we list many of them in each activity. As you plan your units, we encourage you to bring in other resources—from books to speakers to all sorts of videos, CD-ROMs, and other multimedia programs. See page 452 for a list of additional resources you may want to use. We also encourage you to use newspapers, magazines, and Web sites to bring current events into your teaching.

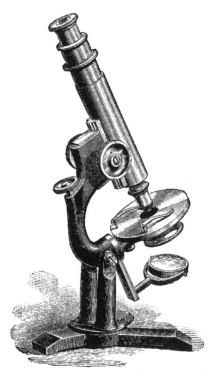

Introducing Biodiversity
version 1 (about 30 sessions)

This unit is designed to give students an overview of what biodiversity is, why it's important, and what's being done to protect it. By leading students through a logical sequence of questions and activities, it challenges students not only to learn biodiversity facts, but also to internalize how biodiversity issues affect them. Using students' natural interest in endangered species as a starting point, the unit begins by focusing on species diversity. As it progresses, it highlights genetic diversity, ecosystem diversity, and the importance of each. Next it encourages students to think about the keys to protecting biodiversity, as well as to examine their own beliefs about biodiversity's importance. The culmination of this unit is an action project that enables students to apply their new knowledge to local issues. (For another approach to an introductory unit on biodiversity, see page 388.)

Session 1: "Endangered Species Gallery Walk"

This endangered species project will pique students' interest as they choose a species to research and incorporate into a poster of their own creation. Make sure that no two students choose the same species.

Sessions 2 and 3: "The Case of the Florida Panther"

At the end of session 2, assign "The Hippo Dilemma" in *WOW!—A Biodiversity Primer* as homework reading.

If the discussion in this activity (step 4) goes too smoothly, you may want to play devil's advocate and pose the question, "Given the fact that we can find other panthers (usually called cougars or mountain lions) in lots of other places, and the fact that panthers in certain areas (such as Colorado) have been known to attack people, should we be going to all this trouble to save the Florida panther?" Let students start sharing their ideas, but don't expect resolution at this time.

Session 4: Work period

Give the students time to work on their endangered species poster projects. Make sure they can go to the library or computer lab, or provide resources for their research. (You might also want to have magazines available for pictures.)

Session 5: "Ten-Minute Mysteries," "Endangered Species Gallery Walk"

"Ten-Minute Mysteries" should illustrate that the worth of a species can be measured in many different ways. After completing the activity, collect and evaluate the students' endangered species posters. Hang them in a prominent place, and design the endangered species scavenger hunt worksheet.

Session 6: "Endangered Species Gallery Walk"

Encourage students to get around to see as many of the posters as they can and to fill out as much of their scavenger hunt sheets as possible. Stress that they should be looking for trends that explain why different groups of animals are endangered.

Session 7: Wrap-up discussion, "Biodiversity! Exploring the Web of Life" video

Using their scavenger hunt sheets and knowledge of the HIPPO dilemma, have the students make generalizations about why different types of animals are endangered. Then, as an introduction to the video, ask, "Are species diversity and biodiversity the same thing?" Show *Biodiversity! Exploring the Web of Life* (30 minutes), then assign "Ask Dr. B!" in *WOW!—A Biodiversity Primer* as homework reading. As part of their homework assignment, have the students write a paragraph describing the various aspects of biodiversity.

Session 8: "Ask Dr. B!" "It's a Wild World"

Your discussion of "Ask Dr. B!" should emphasize that species diversity is only a part of biodiversity. After completing the discussion, students should be able to:

- define biodiversity
- give examples of genetic, species, and ecosystem diversity
- begin to talk about why each level of biodiversity is important

This would be a good time to introduce the action component of this unit (to be carried out starting in session 26) so that your students can begin thinking about possible project topics.

Using the Biodiversity Bulletin Board idea from "It's a Wild World," use the rest of this session to set up a bulletin board in your classroom that focuses on biodiversity issues in your area. Have students bring in news stories from local papers, or gather information from local organizations involved in biodiversity issues. See "Getting Involved!" (pages 358–377) for other suggestions on what students can look for in their bulletin board research. Look specifically at step 1 under "Planning a Project." Establish guidelines and deadlines for student bulletin board contributions and review them as they come in. (You may want to get the class started with a few submissions of your own.)

Session 9: "Super Sleuths"

This activity further illustrates some of the fascinating relationships between living things. Stress the "Levels of Diversity" brainteaser (page 67 in the Student Book) in the discussion. As a review before starting the activities on genetic diversity (sessions 10-14), ask students for examples of how species diversity is important. (They can draw on their experience from "Endangered Species Gallery Walk" and "The Case of the Florida Panther.") Then explain that over the next few days, they will be looking at the importance of genetic diversity and ecosystem diversity.

Sessions 10, 11, and 12: "The Gene Scene"

Use "The Gene Scene" to take a closer look at why genetic diversity is important to a species' survival. In session 10, conduct Part I of the activity. In session 11, play Part II and begin getting ready for the giraffe game (Part III). Have the students read "All About Giraffes" for homework. Then play the giraffe game in session 12. Allow ample time for the discussion questions and the assessment.

Sessions 13 and 14: "Diversity on Your Table"

Introduce this activity as another way to look at the importance of genetic diversity. End at a place in session 13 where it will be easy to pick up again the next day. Assign "Raiders of the Lost Potato" in *WOW!—A Biodiversity Primer* as homework reading.

In session 14, try to bring in a box of amaranth flakes, a bag of carob chips, or other non-typical foods for students to try. During the discussion (steps 7-9), have students make their own lists of key factors in the loss of crop diversity and ways that people are helping to protect the genetic diversity of crop plants. Try to tie together the simulation, real-life scenarios, and the "Raiders" article during the discussion.

To prepare for the upcoming activities focusing on the ecosystem aspect of biodiversity, have students search for pictures of different ecosystems in magazines, family photos, or drawings (in class or for homework). They should bring these in the following day.

Session 15: "Mapping Biodiversity" (Part I)

Display the Global 200 map in the middle of a bulletin board. Have the students create a border for the map using the ecosystem pictures they brought in. You might also suggest that they add the names of their endangered species (from "Endangered Species Gallery Walk") to the appropriate ecoregions.

Session 16: "Secret Services"

Introduce "Secret Services" as an activity that will help students understand the importance of maintaining ecosystem diversity. Have them complete their simulation explanations for homework if they don't do so in class.

Session 17: "Secret Services" presentations

Give the groups time to make final preparations. Set a time limit on presentation length and leave time for a wrap-up discussion.

Session 18: "Mapping Biodiversity" (Part II)

Review Part I of "Mapping Biodiversity." Then introduce Part II by asking, "What are the keys to protecting biodiversity?" During the activity, emphasize the considerations that go into prioritizing conservation efforts.

Session 19: "Mapping Biodiversity" (Part II), Assessment

Have the students use this time to finish drawing map layers and to complete their worksheets. During the discussion of their worksheets, you might want to ask, "What problems did you face coming up with recommendations?" and "Were you all in agreement?"

Session 20: "Thinking About Tomorrow" (Part I)

Make sure to leave plenty of time to discuss the results of the simulations. Have students read "Easter's End" for homework, highlighting dates, changes in biodiversity, and human activities.

Session 21: "Easter's End"

Students can answer the discussion questions provided in the activity as an activity assessment. Assign "Lessons from Jungle Tombs" in WOW!—A Biodiversity Primer as homework reading. Have students keep a list of sustainable and non-sustainable activities mentioned in the article.

Session 22: "Food for Thought"

To introduce this activity ask, "How do you feel the citizens of the United States compare to those of other countries in terms of practicing environmentally sustainable living?" Have them complete the assessment for homework.

Sessions 23 and 24: "Future Worlds"

This activity will help students personalize the information from the unit. Assign "Making It Happen" (step 6) as a homework assignment between sessions 23 and 24.

Session 25: "The Spice of Life"

"The Spice of Life" can be used as a unit wrap-up and as a way to further help students clarify their own values and thinking about protecting biodiversity.

Sessions 26—30: Getting Involved! (action project)

Throughout this unit, students should have gathered information about local biodiversity issues and shared that information on their biodiversity bulletin board. Now it's time to have them look at the different issues they've uncovered and discuss possible ways they might, as a class, get involved in them. Share information from "Getting Involved: Facilitating Action Projects" (page 367) for helpful points students should keep in mind as they move through the process. Also have the students go through steps 2 and 3 under "Getting Involved: Planning a Project" (pages 360–365).

Students should carefully consider the different types of projects they can participate in. (See "Getting Involved: Project Ideas" on pages 367–371.) To help students come up with a realistic project, don't hesitate to impose constraints such as a time line for completion.

After your students have agreed on a topic for their project (step 4 of "Getting Involved: Planning a Project"), lead them through the development of an action plan (step 5). Use the questions in the box titled "Mapping Your Action Project" (page 364) to help mold their plan of action. Keep in mind that the nature of the action project you and your students choose will determine how much class time you'll need to spend on it.

Build "Ideas for Measuring Success" (page 366) into your action project. Decide ahead of time how you will assess student performance on the action project. Make sure to leave time for you and your students to reflect on the action project and the unit as a whole.

Unit at a Glance Introducing Biodiversity–version 1

Session 1	*Session 2*	*Session 3*	*Session 4*	*Session 5*
Endangered Species Gallery Walk	**The Case of the Florida Panther** ────		Work Period (for endangered species posters)	**Ten-Minute Mysteries**
		Discussion: Should we protect the Florida panther?		Endangered species posters due.
Assign poster projects.	*HW: Read "The HIPPO Dilemma" in WOW!—A Biodiversity Primer.*			

Session 6	*Session 7*	*Session 8*	*Session 9*	*Session 10*
Endangered Species Gallery Walk ────	Wrap-up discussion	Discussion: "Ask Dr. B!"	**Super Sleuths**	**The Gene Scene**
	Biodiversity! Exploring the Web of Life (video)	**It's a Wild World** Set up a biodiversity bulletin board.		Part I
	HW: Read "Ask Dr. B!" in *WOW!—A Biodiversity Primer.*	HW: Begin to gather information on local issue. ──		

Session 11	*Session 12*	*Session 13*	*Session 14*	*Session 15*
The Gene Scene ────		**Diversity on Your Table** ────		**Mapping Biodiversity**
Part II, Part III (steps 1–3)	Finish Part III	Introduce and conduct simulation.	Wrap-up discussion	Part I
HW:	Assessment			
• Read "All About Giraffes."		HW: Read "Raiders of the Lost Potato" in *WOW!—A Biodiversity Primer.*	HW: Find pictures of different ecosystems.	
• Research local issue. ──				

Session 16	*Session 17*	*Session 18*	*Session 19*	*Session 20*
Secret Services ────		**Mapping Biodiversity** ────		**Thinking About Tomorrow**
	Presentations	Part II	Part II Assessment	Part I
HW: Research local issue. ──				HW: Read "Easter's End."

Session 21	*Session 22*	*Session 23*	*Session 24*	*Session 25*
Easter's End	**Food for Thought**	**Future Worlds** ────		**The Spice of Life**
HW:	HW: Assessment	HW: "Making It Happen" (step 6)		Use as unit assessment.
• Read "Lessons from Jungle Tombs" in *WOW!— A Biodiversity Primer.*				
• Research local issue. ──				

Session 26	*Session 27*	*Session 28*	*Session 29*	*Session 30*
Getting Involved! ────				
"Planning a Project" (steps 2 and 3)	step 3 and up	step 4, cont'd.	step 5	step 6

*HW = homework

Introducing Biodiversity

version 2 (20 sessions)

Like version 1, this unit is designed to give students an overview of biodiversity issues. It starts with having students think about how they are connected to biodiversity in their day-to-day lives. It's divided into four sections: "What is biodiversity?," "Why is biodiversity important?," "What's the status of biodiversity?," and "How can we protect biodiversity?" By participating in the activities in each section, students will be able to provide increasingly sophisticated answers to these questions.

What Is Biodiversity?

Session 1: "Something for Everyone" (steps 1-3)

This activity will introduce your students to several of the "down-to-earth" ways biodiversity affects people's lives. Many of the students will probably be able to identify with some of the students depicted in this activity. Have the students keep a biodiversity journal for homework (step 4).

Session 2: "Something for Everyone" (step 5), "Biodiversity! Exploring the Web of Life" video

Discuss the students' journals. By the end of the discussion, everyone should realize that biodiversity satisfies many of our needs, including food, medicines, oxygen, shelter, and tools. Reinforce our dependence on biodiversity by showing *Biodiversity! Exploring the Web of Life* (30 minutes). Then assign "Ask Dr. B!" in *WOW! —A Biodiversity Primer* as homework reading. Have students write down at least five interesting facts they learned from the article.

Session 3: "All the World's a Web"

Have the students quiz each other with one or more of the facts they wrote down while reading "Ask Dr. B!" This will "warm up" the students for the web activity. When reviewing the web words with the students, try to use examples from both the video and the article to help bring these terms to life.

Session 4: "Backyard BioBlitz" (Part I)

Now that the students are familiar with biodiversity and some of the issues surrounding it, see how much they know about biodiversity in their own community. Have them begin work on the ecoregional survey in class and finish it for homework. Take the quiz yourself to become familiar with some of the possible answers. Check with local experts to get answers to questions you're unsure of.

Session 5: "Backyard BioBlitz" (Part II, steps 1-3)

Discuss the quiz answers. Then divide the students into groups and have them plan their outdoor study. Look over their plans and help with fine-tuning as necessary. Make a sketch of the study area to pass out to each group in session 6.

Session 6: "Backyard BioBlitz" (Part II, steps 4-7)

Have the groups conduct their BioBlitz and share the results. In your wrap-up discussion, encourage students to analyze their results. Student answers to discussion questions can serve as an informal assessment for the activity.

Why Is Biodiversity Important?

Session 7: "Super Sleuths"

Now that the students have an understanding of what biodiversity is, introduce this activity for a closer look at the ecological importance of species. Pass out the "Super Sleuths" puzzles and give students time to get started on them in class. They can finish the rest for homework.

Session 8: "Super Sleuths," "Biodiversity Performs!"

Hand out "Believe It or Not!" from "Super Sleuths" to show students some more incredible relationships between living things. Then introduce and conduct "Biodiversity Performs!" as a way to illustrate the importance of ecosystem services. (You might also want to refer back to the "Ask Dr. B!" article.) Assign "The Nature of Culture" worksheet (pages 85–90 in the Student Book) for homework.

Sessions 9 and 10: "The Culture/Nature Connection"

Continue your examination of the importance of biodiversity by looking at the cultural importance of species. You might want to assign the family heritage homework (part II) over a weekend to give students plenty of time to interview family members. Students should do the assessment for homework.

What's the Status of Biodiversity?

Sessions 11 and 12: "The Case of the Florida Panther"

Now that students are familiar with biodiversity and its importance, use this case study to investigate the status of biodiversity. At the end of session 11, have students read "The HIPPO Dilemma" in *WOW!—A Biodiversity Primer* (or you can assign it for homework). Save time at the end of session 12 for the assessment.

Session 13: "Putting a Price Tag on Biodiversity"

This activity challenges students to look at other ways that the worth of biodiversity can be evaluated. For homework, have them read "How Much Is a Gray Wolf Worth?" and list three different ways economists try to put a price tag on biodiversity.

Session 14: "Mapping Biodiversity" (Part I)

Introduce this activity by reminding students that an estimated one-half of the species in the world are found in tropical rain forests. As a result, tropical rain forests have gotten a lot of attention from conservationists. Tell the students that, although tropical rain forests are important, many other areas are also biologically rich and in need of protection. This activity will give students a better understanding of some of the world's many diverse ecoregions.

Sessions 15 and 16: "Mapping Biodiversity" (Part II)

Use this activity to help students understand the ecoregional approach to protecting biodiversity. Save time at the end of session 16 for the assessment.

How Can We Protect Biodiversity?

Session 17: "Thinking About Tomorrow" (Part I)

This simulation works well as an introduction to the concept of sustainability. Make sure you save enough time for a wrap-up discussion. Students should read the "Easter's End" article or cartoon (pages 183–192 in the Student Book) for homework, taking note of dates, human activities, and changes in plant and animal populations.

Session 18: "Easter's End" assessment

Discuss the "Easter's End" story. Have the students complete one of the discussion questions as a written homework assignment. Use the assignment as an assessment.

Session 19: "The Spice of Life"

Use this activity to give students a chance to reflect on the material they've learned in this unit and to clarify their thinking about what biodiversity means to them.

Session 20: "The Biodiversity Campaign"

This activity can serve as the unit assessment.

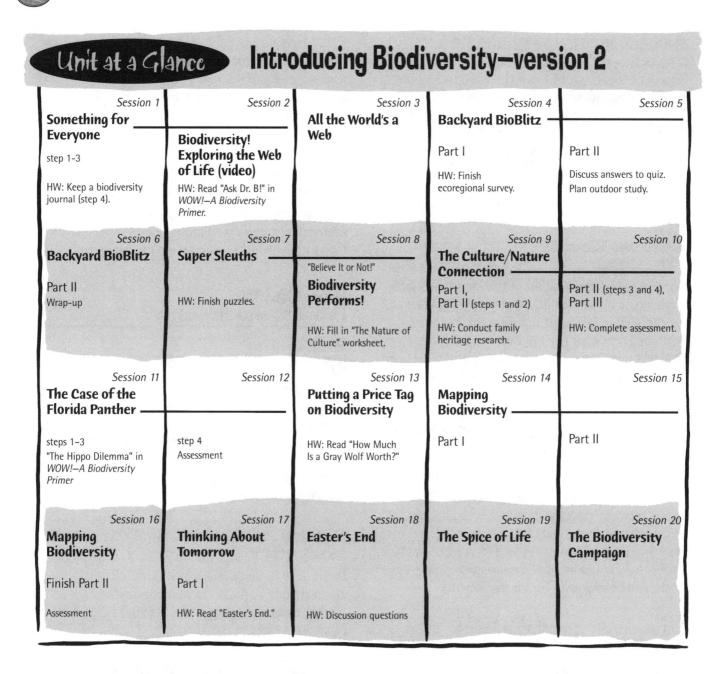

Unit at a Glance — Introducing Biodiversity—version 2

Session 1 **Something for Everyone** step 1–3 HW: Keep a biodiversity journal (step 4).	*Session 2* **Biodiversity! Exploring the Web of Life (video)** HW: Read "Ask Dr. B!" in *WOW!—A Biodiversity Primer.*	*Session 3* **All the World's a Web**	*Session 4* **Backyard BioBlitz** Part I HW: Finish ecoregional survey.	*Session 5* Part II Discuss answers to quiz. Plan outdoor study.
Session 6 **Backyard BioBlitz** Part II Wrap-up	*Session 7* **Super Sleuths** HW: Finish puzzles.	*Session 8* "Believe It or Not!" **Biodiversity Performs!** HW: Fill in "The Nature of Culture" worksheet.	*Session 9* **The Culture/Nature Connection** Part I, Part II (steps 1 and 2) HW: Conduct family heritage research.	*Session 10* Part II (steps 3 and 4), Part III HW: Complete assessment.
Session 11 **The Case of the Florida Panther** steps 1–3 "The Hippo Dilemma" in *WOW!—A Biodiversity Primer*	*Session 12* step 4 Assessment	*Session 13* **Putting a Price Tag on Biodiversity** HW: Read "How Much Is a Gray Wolf Worth?"	*Session 14* **Mapping Biodiversity** Part I	*Session 15* Part II
Session 16 **Mapping Biodiversity** Finish Part II Assessment	*Session 17* **Thinking About Tomorrow** Part I HW: Read "Easter's End."	*Session 18* **Easter's End** HW: Discussion questions	*Session 19* **The Spice of Life**	*Session 20* **The Biodiversity Campaign**

Biodiversity Inquiry
(about 20 sessions)

This unit is designed to give students an opportunity to develop their scientific investigation skills while learning about biodiversity. Using case studies, the unit focuses on environmental health issues and provides several examples of the ways scientists approach their investigations. Students will design and conduct their own scientific investigations, and will take part in an action project focusing either on amphibian populations in your area or on some other topic of interest. Make arrangements early for professionals to be available for interviews for "Career Moves" (see sessions 16 and 17) and gauge student interest throughout the unit to get ideas for the focus of their action project.

Session 1: "The F-Files" (steps 1-3)

Use this activity as an introduction to the unit. Begin by asking, "How do scientists figure things out?" Complete steps 1–3 only. Let students know that you will be returning to the case of the disappearing frogs later in the unit.

Session 2: "Inquiring Minds" (Part I)

This activity points out that making careful observations and asking good questions can lead to scientific discoveries. For homework, have students observe one or more people for about ten minutes. Students should write down their observations.

Session 3: "Inquiring Minds" (Part II, option #3)

To begin your discussion of the differences between objective and subjective observations, have volunteers share their observations of people. Then have the students evaluate each other's observations as either objective or subjective. As a follow-up activity, have students repeat their observations of people—this time trying to make only objective observations.

Sessions 4 and 5: "Inquiring Minds" (Part II, option #1)

Use this activity to engage your students in the process of scientific inquiry. Decide ahead of time whether the students will conduct a single investigation as a class or whether student teams will conduct several different investigations. Encourage students to develop investigations that can be conducted in a short amount of time. (Alternatively, you could have them conduct more time-consuming investigations as an ongoing homework assignment.)

Session 6: "Something for Everyone" (steps 1-3)

Introduce this activity by mentioning that the frog problem discussed in "The F-Files"—and perhaps even the subjects of the students' investigations in "Inquiring Minds"—are part of a much bigger problem: the loss of biodiversity. Have the students keep a biodiversity journal for homework (step 4).

Session 7: "Something for Everyone" (step 5), "Biodiversity! Exploring the Web of Life" video

After discussing the students' biodiversity journals (step 5), show *Biodiversity! Exploring the Web of Life* (30 minutes). For homework, have them read "Ask Dr. B!" in *WOW!—A Biodiversity Primer*.

Sessions 8 and 9: "The Case of the Florida Panther"

You may want to introduce this activity by pointing out that the information in it was drawn from many scientific investigations. Based on investigations such as these, scientists have come to some conclusions about the threats to biodiversity. These conclusions are summed up in "The HIPPO Dilemma" in *WOW!—A Biodiversity Primer*, which you can assign as homework between sessions 8 and 9. The Florida panther is just one example of how the HIPPO Dilemma affects endangered species.

Session 10: "Ten-Minute Mysteries"

Use this activity to help students understand some of the connections that link people to biodiversity. Then have them do the activity extension for homework. Encourage them to draw upon their earlier investigations in coming up with their own mysteries.

Session 11: "The Many Sides of Cotton" (steps 1-4)

Use steps 1–4 to reinforce students' understanding of how pollution can lead to biodiversity loss. Afterward, ask the students if they can think of other examples of the effects of pollution on biodiversity. (If no one mentions the frog problem described in "The F-Files," remind students that this may be one such example.) For homework, assign students different readings from "Cotton Concerns" (pages 148–159 in the Student Book).

Session 12: "The Many Sides of Cotton" (steps 5-7)

Have students get into groups made up of at least one student who read each article. Have them share the information they read and fill out the "Sorting Out the Issues" worksheet and chart (pages 160–161 in the Student Book). Assign step 8 as an assessment, to be done for homework.

Session 13: "The F-Files" (steps 4 and 5)

Now return to "The F-Files." By this time the students should have a better understanding of environmental health.

Sessions 14 and 15: "The F-Files" (research—step 5)

Students will need access to computers and/or library resources so they can continue their research. Since they now have some experience conducting scientific investigations, encourage the students to look critically at investigations done by others. Suggest that they consider the following questions: *Do the investigations seem scientifically sound? If not, how might they be improved?* You might also want to lead a discussion of topics that would be difficult to explore using the scientific method. Assign "Career Profiles" (pages 198–204 in the Student Book) as homework to prepare for session 16.

Sessions 16 and 17: "Career Moves" (steps 1-5)

To wrap up the unit, use "Career Moves" to illustrate the fact that many people are working on biodiversity issues. If necessary, the interviews and presentations could take place at the same time as the action project (session 20)—or even while you're moving into another unit. To save time, arrange for several professionals from a variety of backgrounds to visit the class during session 18, and have different groups of students interview different professionals. If some of the interviewees aren't able to attend in person, try to have them available at a specific time for a telephone interview.

If having a variety of visitors to the class isn't possible, you might want to arrange to have just one or two visitors that the class as a whole can interview. Be sure to set ground rules for all interviews.

Session 18: "Career Moves" (step 6)

If your students will be working in groups, set up places in the school where they can conduct their interviews.

Session 19: "Career Moves" (steps 7 and 8)

Have students share their interviewing experiences, and discuss both the process of conducting interviews and the different ways people are involved with biodiversity. If you have time, have the students work on a short presentation that they can share during the following session.

Session 20: Getting Involved! (action project)

As an extension to the unit, help your students organize an action component that will get them involved in investigating biodiversity in your area. If "The F-Files" caught their attention, they might enjoy participating in one of several ongoing amphibian monitoring projects. The North American Amphibian Monitoring Program Web site <www.im.nbs.gov/amphibs.html> provides a "Teacher's Toolbox" with descriptions of different projects, including resources for each and advice on collecting amphibian data. Your students can report any deformed frogs they find to the North American Reporting Center for Amphibian Malformations <www.npwrc.org/narcam/>.

If a topic other than amphibian deformities caught the students' interest during your biodiversity unit, you might want to suggest that they use that topic as the basis of their action project. ("Getting Involved!" [page 371] lists several groups that are set up to help students collect and share data on a variety of environmental topics.) Once you've decided on a project, set up a time and place for a field trip during which students can collect data. Depending on the project, you may need to do an evening or weekend field trip. Or you could suggest that the students make the project the focus of an after-school science or nature club.

Unit at a Glance — Biodiversity Inquiry

Session 1	Session 2	Session 3	Session 4	Session 5
The F-Files	**Inquiring Minds**			
steps 1–3	Part I	Part II, option #3	Part II, option #1	Part II, option #1
	HW: Observe people.	HW: Observe people.		HW: Conduct investigations.

Session 6	Session 7	Session 8	Session 9	Session 10
Something for Everyone		**The Case of the Florida Panther**		**Ten-Minute Mysteries**
steps 1–3	step 5 **Biodiversity! Exploring the Web of Life (video)**	steps 1–3	step 4, assessment	
HW: Keep a biodiversity journal (step 4).	HW: Read "Ask Dr. B!" in WOW!—A Biodiversity Primer.	HW: Read "The HIPPO Dilemma" in WOW!—A Biodiversity Primer.		HW: Activity extension

Session 11	Session 12	Session 13	Session 14	Session 15
The Many Sides of Cotton		**The F-Files**		
steps 1–4	steps 5–7	steps 4 and 5	Research	steps 6 and 7
HW: Read "Cotton Concerns" (step 5).	HW: Step 8 (assessment)	HW: Continue research.		HW: Read "Career Profiles."

Session 16	Session 17	Session 18	Session 19	Session 20
Career Moves				**Getting Involved!**
steps 1–4	step 5	Interview day (step 6)	Wrap-up discussion (steps 7 and 8)	

Classifying Biodiversity

(11 sessions)

This unit emphasizes the importance of classification in measuring and preserving biodiversity. The activities are designed to give your students opportunities to identify and classify organisms and to understand the processes on which our biological classification system is based. Throughout the unit, students are challenged to collect and analyze their own data from field observations.

Session 1: "Sizing Up Species" (Parts I and II)

Use this activity to introduce your students to classification. The game in Part II encourages students to think about the relative numbers of species in different groups of organisms.

Session 2: "Sizing Up Species" (Part III), "Biodiversity! Exploring the Web of Life" video

Part III of "Sizing Up Species" gives students an opportunity to graphically represent the information they learned in Part II. When showing *Biodiversity! Exploring the Web of Life* (30 minutes), have your students pay particular attention to the Terry Erwin clip.

For homework, have the students read "The Natural Inquirer" in *WOW!—A Biodiversity Primer.* You might also want to have them start thinking about topics for their own biodiversity tabloid.

Session 3: "It's a Wild World" ("You Read It Here First")

Have students complete the "You Read It Here First" option of the activity. Use this assignment to bring some of the organisms from the different classification groups to life. Make sure the students cover a wide range of the groups of organisms. Have resources available to help students generate ideas for possible tabloid-style articles, as well as to help them in their initial research once they've chosen a topic. Possible resources include magazines such as *Ranger Rick,* National Geographic's *World,* and *Discover.* Articles should be completed by session 10.

Session 4: "Biodiversity—It's Evolving" (Part I)

Present this activity as a way of explaining the similarities of organisms belonging to the same classification group. You might start out by asking your students to name as many different kinds of birds (or some other group of animals) as they can. Ask, "What do all of these animals have in common? What are some ways these animals are different from one another?" Tell the students that the simulation they're about to do will help explain these similarities and differences. For homework, have the students read "The Case of the Peppered Moth" (page 60 in the Student Book), to be completed by session 6.

Sessions 5 and 6: "Biodiversity—It's Evolving" (Part II)

Part II challenges your students to apply the concepts of natural selection and evolution introduced in Part I. Understanding the mechanism of natural selection should give your students a context in which classification makes sense. They should come away from this activity knowing that the classification of animals is based on inherited characteristics.

Sessions 7, 8, and 9: "Insect Madness"

This activity will give your students experience classifying organisms. The Sequential Comparison Index will provide a way to quantify classification data. When students are identifying the insects they've collected during session 8, have them put the insects into taxonomic orders using the identification keys and any field guides you find useful. During your wrap-up discussion in session 9, review the characteristics of the different insect orders and the idea that these groups are based on evolutionary relationships.

Sessions 10 and 11: "Space for Species" (Part II), tabloid articles

This activity is another opportunity for your students to practice their classification skills. Emphasize distinguishing differences between species rather than actually identifying particular species. One message that you can leave with your students during the wrap-up discussion is the importance of using classification as a tool in preserving biodiversity.

Collect the tabloid articles that you assigned in session 3. Compile them as a class publication or use them to create a culminating bulletin board display.

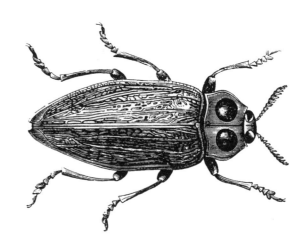

Unit at a Glance — Classifying Biodiversity

Session 1	Session 2	Session 3	Session 4	Session 5
Sizing Up Species Parts I and II	Part III **Biodiversity! Exploring the Web of Life (video)** HW: Read "The Natural Inquirer" in *WOW!—A Biodiversity Primer.*	**It's a Wild World** "You Read It Here First!" HW: Work on articles.	**Biodiversity—It's Evolving** Part I HW: "The Case of the Peppered Moth"	Part II (steps 1–4)
Session 6	Session 7	Session 8	Session 9	Session 10
Biodiversity—It's Evolving Part II (steps 5–6) HW: Continue work on article.	**Insect Madness** steps 1–6	steps 7–9	steps 10–12 Assessment	**Space for Species** Part II, Day 1 Articles due.
Session 11				
Space For Species Part II, Day 2				

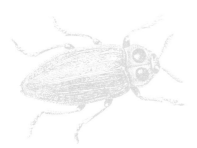

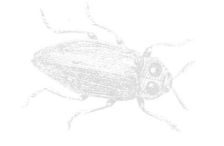

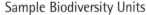

Exploring Geography, Sustainability, and Biodiversity

(17 sessions)

It's easy to think of biodiversity only in scientific terms. But social issues are inextricably tied to both biodiversity loss and biodiversity conservation. This unit addresses some of these social issues through an examination of resource use and the concept of sustainability.

Session 1: "Food for Thought"

This interactive simulation is sure to get your students' attention. In your discussion, you may want to focus on the implications of a small percentage of the population controlling a large percentage of the world's resources. Don't worry if it's too early in the unit for students to see the connection between resource use and biodiversity issues.

Session 2: "All the World's a Web"

This activity will help students make the connection between resource use and biodiversity. You may want to add words from the discussion in session 1 as "web words."

Assign "Ask Dr. B!" in *WOW!—A Biodiversity Primer* as homework. Have the students pay particular attention to how people fit into the biodiversity picture.

Session 3: "Endangered Species Gallery Walk," "Biodiversity! Exploring the Web of Life" video

Briefly discuss "Ask Dr. B!," focusing on Dr. B's description of the HIPPO Dilemma. Assign endangered species posters from "Endangered Species Gallery Walk" (due session 7) and show *Biodiversity! Exploring the Web of Life* (30 minutes) to further emphasize the information in "Ask Dr. B!"

Session 4: Endangered species research

Reserve time in the library and/or computer room, or bring in a collection of resources from your library for the students to use in their research.

Session 5: "Mapping Biodiversity" (Part I)

By doing this activity, your students will learn where some of the world's most biologically diverse regions are. Encourage them to identify ecoregions where "their" endangered species live.

Session 6: "Super Sleuths"

You can link this activity to "Mapping Biodiversity" by emphasizing that the relationships that exist between species are what hold ecoregions together. Focus on seed dispersal by bringing in seeds that illustrate as many different dispersal mechanisms as you can find—or else have the students find seeds as a homework assignment.

Toward the end of the session, give the students time to work on their endangered species posters. (Remind them that the posters will be due during the next session.)

Session 7: "Super Sleuths," "Endangered Species Gallery Walk"

After students finish the three puzzles have them start on the "Super Sleuths" assessment, which they can finish for homework.

Collect the students' endangered species posters and assess them. Design the endangered species scavenger hunt worksheet, then hang the posters in a gallery-type display.

Session 8: "Endangered Species Gallery Walk"

Give the students two-thirds of the period to tour the endangered species gallery and fill out their worksheets. Encourage them to get around to as many of the different posters as they can. The wrap-up discussion should focus on why different kinds of animals are endangered. Have the students read "The HIPPO Dilemma" in *WOW!—A Biodiversity Primer* for homework. As an assessment, have them list at least three examples from the endangered species gallery for each of the five problems stated in "The HIPPO Dilemma."

Sessions 9 and 10: "Mapping Biodiversity" (Part II)

Use this activity to help students understand the ecoregional approach to protecting biodiversity. Students will conduct their own gap analysis of an area they create themselves and experience first-hand the factors involved in developing protection plans for rare species and ecosystems.

Save time at the end of session 10 for the assessment. For homework, have them read "Easter's End." Ask the students to take note of dates and of changes in plant and animal populations that correspond to changes in human activity.

Session 11: "Easter's End"

This activity focuses on what can happen when people engage in unsustainable activities. Any one of the discussion questions could be given to students as a written assessment for homework.

Session 12: "Thinking About Tomorrow" (Part I)

This simulation will give students firsthand experience with sustainability. Afterward, have them read "Lessons from Jungle Tombs" in *WOW!—A Biodiversity Primer*. Also have them make a list of activities that they feel are sustainable, as well as those that aren't.

Sessions 13 and 14: "Thinking About Tomorrow" (Part II)

This activity enables students to apply the concept of sustainability to a real-life situation. They'll get to compare their solutions to a sustainability problem to those that were actually used by a community. In session 14, clearly state the ground rules and expectations for the students' presentations, which will be presented in session 15. If they will be formally assessed, let them know what the assessment criteria are.

Session 15: "Thinking About Tomorrow" (presentations)

Be strict on time limits during the presentations so that you have plenty of time to discuss what Santa Barbara actually did. One of the discussion questions could serve as a written assessment for homework.

Session 16: "Future Worlds" (option #1, steps 1-5)

Introduce this activity by telling students that this will be their chance to personalize the information they learned in this unit and to look to the future. Assign step 6 as homework (this can also be used as an assessment for the entire unit).

Session 17: "Future Worlds" (option #1, step 7)

In the discussion, you may want to ask your students if they think that people in other parts of the world would rank the items in the same way they did. (This will bring them back to the "Food for Thought" activity that started the unit.)

Unit at a Glance — Exploring Geography, Sustainability, and Biodiversity

Session 1	Session 2	Session 3	Session 4	Session 5
Food for Thought	**All the World's a Web**	**Endangered Species Gallery Walk** — Assign species. **Biodiversity! Exploring the Web of Life (video)** HW: Research endangered species.	Continue research on endangered species. HW: Begin endangered species posters.	**Mapping Biodiversity** Part I
	HW: Read "Ask Dr. B" in *WOW!—A Biodiversity Primer.*			

Session 6	Session 7	Session 8	Session 9	Session 10
Super Sleuths —	Endangered species posters due.	**Endangered Species Gallery Walk**	**Mapping Biodiversity** — Part II	Part II and assessment
HW: Finish endangered species posters.	HW: "Super Sleuths" assessment	HW: Read "The HIPPO Dilemma" in *WOW!—A Biodiversity Primer.*		HW: Read "Easter's End."

Session 11	Session 12	Session 13	Session 14	Session 15
Easter's End	**Thinking About Tomorrow** — Part I	Part II	Part II Work period	Presentations and wrap-up discussion
HW: Answer assessment questions.	HW: Read "Lessons From Jungle Tombs" in *WOW!—A Biodiversity Primer.*			

Session 16	Session 17
Future Worlds — option #1 steps 1–5	Wrap-up discussion (option #1, step 7)
HW: "Making It Happen" (step 6)	

Biodiversity and Language Arts
(10 sessions)

Language arts teachers can use this unit either in conjunction with a biodiversity unit being taught by a social studies or science teacher, or as a stand-alone unit. The activities emphasize reading, creative writing, and expression of ideas through oral and dramatic means. They also challenge students to grapple with some of the complex issues surrounding biodiversity conservation.

Session 1: "Something for Everyone" (steps 1 - 4)

Beginning the unit with this activity will introduce your students to several of the "down-to-earth" ways biodiversity affects peoples' lives. Many of the students will probably be able to identify with some of the students depicted in this activity. Have students keep a biodiversity journal for homework (step 4).

Session 2: "Something for Everyone" (step 5), "It's a Wild World" (step 2)

Have students share entries from their biodiversity journals. Then have them decide on a theme for a biodiversity bulletin board to which they can add stories and pictures throughout the unit. Have students set up the bulletin board. If time permits, students can read "Ask Dr. B!" in *WOW!—A Biodiversity Primer*. Have them finish reading the article for homework.

Session 3: "Biodiversity! Exploring the Web of Life" video, "It's a Wild World" (step 2)

After a brief discussion about "Ask Dr. B!," show *Biodiversity! Exploring the Web of Life* (30 minutes). Review the amazing examples of biodiversity highlighted in the video. Then give the students the assignment from "It's a Wild World" titled "You Read It Here First!" (step 2) due in session 10. Give students some of the suggestions in the activity and have issues of *Ranger Rick* and other magazines available for inspiration. Have them read "The Natural Inquirer" in *WOW!—A Biodiversity Primer* for additional inspiration.

Session 4: "Biodiversity Performs!"

Your students will have fun with this activity. Make connections between the ecosystem services described on each card and services covered in the video and the "Ask Dr. B!" article. Have them work on their news stories for homework.

Sessions 5 and 6: "The Nature of Poetry"

By examining poems about biodiversity, your students can explore both a range of feelings about different aspects of nature and the language used to create tone and convey ideas. After reading and discussing poems in class in session 5, have the students find and bring in other examples of poems related to biodiversity.

During session 6, have the students write their own poems (and complete them for homework, if necessary). Encourage them to embellish their poems with artwork. Then have them add their creations to their biodiversity bulletin board.

Sessions 7, 8, and 9: "What's a Zoo to Do?"

This activity will challenge your students to work through biodiversity-related dilemmas that various institutions face. During session 7, divide the class into groups and pass out information on each of the dilemmas. Give them the rest of the session and all of session 8 to discuss their dilemmas and develop their presentations. Have them share their presentations in session 9.

Session 10: "The Spice of Life"

This activity gives students a chance to evaluate why biodiversity is important to them. The assessment for this activity can be used as a homework assignment. Along with the news stories from "It's a Wild World," this assessment can serve as the assessment for the entire unit.

Unit at a Glance — Biodiversity and Language Arts

Session 1	Session 2	Session 3	Session 4	Session 5
Something For Everyone — steps 1–3	step 5 **It's a Wild World** Biodiversity bulletin board	**Biodiversity! Exploring the Web of Life (video)** "You Read It Here First!" "The Natural Inquirer" in _WOW!—A Biodiversity Primer._	**Biodiversity Performs!**	**The Nature of Poetry** steps 1 and 2
HW: Keep a biodiversity journal (step 4).	HW: Read "Ask Dr. B" in _WOW!—A Biodiversity Primer._	HW: Work on news stories.		HW: Bring in a biodiversity poem.

Session 6	Session 7	Session 8	Session 9	Session 10
The Nature of Poetry step 3	**What's a Zoo to Do?** steps 1 and 2	Work period (step 3)	Presentations (step 4)	**The Spice of Life** News stories due.
HW: Work on news stories.				HW: Assessment

Unit at a Glance — Exploring Biodiversity Outside

Session 1	Session 2	Session 3	Session 4	Session 5
All the World's a Web	**"Connect the Creatures" Scavenger Hunt**	**Biodiversity Performs!**	**Backyard BioBlitz** Part II	

Session 6	Session 7	Session 8	Session 9	Session 10
Insect Madness		**Space For Species** Part I	Part II	Part II

Exploring Biodiversity Outside
(10 sessions)

This unit is designed for science educators who work in outdoor settings, as well as for other educators who want their students to take a firsthand look at biodiversity in their own "backyards."

Session 1: "All the World's a Web"

Use this activity as an introduction to biodiversity. Students will learn about some of the many connections that make up the web of life and should be ready to start looking for examples of these connections in the next session.

Session 2: "'Connect the Creatures' Scavenger Hunt"

In the outside version of this activity, your students can look for some of the connections they were introduced to in the previous activity, along with other connections. Encourage and reward careful observations.

Session 3: "Biodiversity Performs!"

Students will have fun acting out ecosystem "services," some of which they learned about in the "'Connect the Creatures' Scavenger Hunt" activity.

Sessions 4 and 5: "Backyard BioBlitz" (Part II)

This activity will encourage your students to start thinking about where organisms live, what their habitat requirements are, and how scientists quickly find out about biodiversity. Continue to stress the importance of careful observation.

Sessions 6 and 7: "Insect Madness"

Capitalize on some of the discoveries your students made in the last two activities by encouraging a closer examination of insects. You might want to focus this study in a different area from the one you used to conduct the "BioBlitz," or you can use the same area if you remind the students that they have just scratched the surface of the area's biodiversity.

Session 8: "Space for Species" (Part I)

This activity can enhance what students have already learned about the resource requirements of different species. It also introduces the concept of habitat fragmentation.

Sessions 9 and 10: "Space for Species" (Part II)

This part will give your students another way to quantify their investigations of biodiversity.

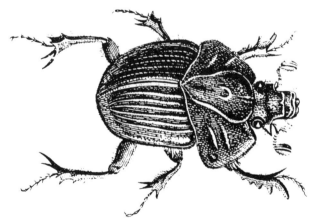

Genes, Species, and Populations
(12 sessions)

This unit is designed to combine the study of biodiversity with a unit on population biology. In it, students will learn about the function of genetics in natural selection and about the difficulty some wild populations have adapting to human-caused environmental changes. At the end of the unit, students will look at possible ways to protect populations through the development of natural reserves.

Session 1: "The Gene Scene" (Part I)

This activity serves as a good introduction to genetics by getting students to look at how human characteristics, particularly those within the class, vary. You might want to introduce the concept of a population during this first activity, using the members of the class as an example of a population. You can carry the concept further in Parts II and III of the activity and in the other activities in this unit.

Session 2: "The Gene Scene" (Part II)

In this part, your students are introduced to the concept that higher genetic variability in a population leads to better adaptability of the population, which leads to a better chance of long-term survival for the population. If time permits, complete step 1 of Part III to get ready for the giraffe simulation game. Have the students read "All About Giraffes" for homework.

Session 3: "The Gene Scene" (Part III)

Review some of the interesting facts about giraffes from the reading and explain how they're connected to the rules of the simulation game. Leave time at the end for a wrap-up discussion and the assessment, even if it means skipping some of the last rounds of the game.

Session 4: "Biodiversity—It's Evolving" (Part I)

By introducing the concept of natural selection, the simulation in Part I of this activity will add another layer to your students' understanding of the role of genetics in population biology.

Sessions 5 and 6: "Biodiversity—It's Evolving" (Part II)

Your students will apply their understanding of the concepts of natural selection and speciation in unraveling the mystery of the "twitomites." At the end of session 6, have students complete "The Case of the Peppered Moth." You can use this worksheet as an assessment for this activity.

Sessions 7 and 8: "The Case of the Florida Panther"

Your students will become aware of the many problems that endangered populations face by looking at this case study of the Florida panther. They should also be able to recognize the fact that, since the population size of the Florida panther is so small, its genetic diversity is also small. Because of this, it's difficult for them to adapt to the many challenges they face.

Session 9: "Space for Species" (Part I, steps 1-10)

This activity deals directly with the effect of habitat fragmentation on species diversity, but the concepts can be applied to populations of organisms as well.

Session 10: "Space for Species" (Part I, steps 11 and 12), "Mapping Biodiversity" (Part II, steps 2-6)

The mapping activity will give students an opportunity to apply the information they learned in "Space for Species" about developing plans for reserves. Use the gap analysis section (Part II) of "Mapping Biodiversity" as a follow-up to "Space for Species."

Session 11 and 12: "Mapping Biodiversity" (Part II, steps 5-8)

Have students finish their map layers if they didn't get a chance to do so in session 10. If time permits, you can show the Global 200 map to your students of as an example of how the mapping skills they learned in the mapping activity can create a picture of biodiversity around the world. If you can, finish the unit by having a local expert in gap analysis come in and speak to your class about his or her work and how it relates to population biology. The assessment in "Mapping Biodiversity" can be used as an assessment for the entire unit.

Unit at a Glance — Genes, Species, and Populations

Session 1	Session 2	Session 3	Session 4	Session 5
The Gene Scene			**Biodiversity– It's Evolving**	
Part I	Part II Part III (step 1) HW: Read "All About Giraffes."	Part III (steps 2–8) Assessment	Part I	Part II (steps 1–4)

Session 6	Session 7	Session 8	Session 9	Session 10
Biodiversity– It's Evolving Part II (steps 5–7) Read "The Case of the Peppered Moth."	**The Case of the Florida Panther** steps 1–3	step 4 Assessment	**Space for Species** Part I (steps 1–10)	Part I (steps 11–12) **Mapping Biodiversity** Part II (steps 2–6)

Session 11	Session 12
Mapping Biodiversity Part II (steps 5–7)	**Guest Speaker** (depending on time and availability) Assessment

Biodiversity Loss: Should We Be Concerned?

(13 sessions)

This social studies unit is designed to help students look at the different ways biodiversity—and its loss—affects their lives. Students will learn about the many connections among living things, as well as the connections between different cultures and the natural environment. They'll also discover some of the economic values of biodiversity, learn about sustainability, and prioritize conditions they'd like to see in the future. The final activity gives students a chance to create their own ad campaigns based on what they feel are the most important actions people need to take to conserve biodiversity.

Session 1: "What's Your Biodiversity IQ?"

To introduce this unit, tell your students that some scientists estimate that we're losing a species every 20 minutes. At this rate, half the species on the planet will become extinct in the next thirty years. Some scientists feel that trying to reverse this trend will be one of the greatest challenges we'll face.

To find out what students already know about biodiversity and its decline, have them take the "What's Your Biodiversity IQ?" quiz. In step 4, don't worry about having the students come up with an exhaustive list of why biodiversity is important. Your goal here should be to get the students thinking.

Session 2: "'Connect the Creatures' Scavenger Hunt"

Complete the indoor version of this activity and let the students know ahead of time that they will be creating a bulletin board with the pictures they find in the scavenger hunt. Have the students create a web of life as described in the Extensions section (page 106) for homework.

Session 3: Create a bulletin board

Use this activity as an assessment for the scavenger hunt. Have each team put together a display for the bulletin board using the pictures they collected from the scavenger hunt. (See the assessment for "'Connect the Creatures' Scavenger Hunt.")

Session 4: "The Culture/Nature Connection" (Part I)

This activity shows students how biodiversity has helped to shape different cultures around the world. During the last part of the session, challenge students to start thinking about connections between biodiversity and their own family heritage by assigning family research for homework (Part II). Give the students at least two nights to complete their research.

Session 5: "The Culture/Nature Connection" (Part III)

Skip ahead to the role-play in Part III to give your students extra time to complete their family heritage homework. (Doing the role-plays first may also help some students feel more comfortable sharing their personal family heritage connections to biodiversity.) You can use the pollination and seed dispersal brainteasers in "Super Sleuths" to reinforce ecological connections.

Session 6: "The Culture/Nature Connection" (Part II, steps 3 and 4)

Give plenty of opportunity for students to explore their families' connections to biodiversity in the discussion. Afterward, if time permits, have students complete the assessment. Otherwise, have them do it for homework.

Session 7: "Ten-Minute Mysteries"

Begin by asking students if they think the family heritage/biodiversity connections they discussed in the previous session are outdated or old-fashioned. If so, the "mysteries" in this activity will show them some modern-day connections between people and biodiversity.

Session 8: "Putting a Price Tag on Biodiversity"

This activity will show students some of the less obvious economic values of biodiversity. Assign "How Much Is a Gray Wolf Worth?" for homework reading.

Session 9: "Putting a Price Tag on Biodiversity," "Easter's End" (Day 1)

Wrap up the discussion of the homework reading by reviewing the values and benefits of biodiversity. Then ask the students what some of the costs of not preserving biodiversity might be. List them on the board. Ask if anyone knows of any instances where biodiversity loss has led to the collapse of a civilization. Use this discussion to introduce "Easter's End." Have the students read the story of Easter Island's collapse for homework. Let them know that they'll be constructing time lines in class during the next session, and suggest that they take notes while reading.

Session 10: "Easter's End" (Day 2)

Keep students moving along with working on the time lines in order to have time at the end of the session for discussion. As an assessment for the activity, have them answer question 4 for homework.

Session 11: "Future Worlds" (option #2, steps 1-5)

Have some of the students share their homework as a review of the concept of sustainability. Then move on to option 2 in "Future Worlds." Have them read "Making It Happen" for homework.

Session 12: "Future Worlds" (wrap up)

Use the discussion of the "Making It Happen" stories as an introduction to the final assignment: designing biodiversity ads. Start by having the students read the ads (for homework, if necessary).

Session 13: "The Biodiversity Campaign"

Have students work on their ads. If necessary, allow an extra day for them to complete their work. The ads can serve as an assessment for the entire unit.

Unit at a Glance — Biodiversity Loss: Should We Be Concerned?

Session 1	Session 2	Session 3	Session 4	Session 5
What's Your Biodiversity IQ?	**"Connect the Creatures" Scavenger Hunt** HW: Have students create a web of life.	Create a biodiversity bulletin board.	**The Culture/Nature Connection** Part I Part II (steps 1 and 2) HW: Conduct family heritage research (Part II).	Part III

Session 6	Session 7	Session 8	Session 9	Session 10
The Culture/Nature Connection Part II (steps 3 and 4)	**Ten-Minute Mysteries**	**Putting a Price Tag on Biodiversity** HW: Read "How Much Is a Gray Wolf Worth?"	Discuss homework. **Easter's End** Day 1 HW: Read "Easter's End."	Day 2 HW: Answer question 4 (step 4).

Session 11	Session 12	Session 13		
Future Worlds option #2, steps 1–5 HW: "Making It Happen" (step 6)	Wrap-up discussion HW: Read "Green Team Ads."	**The Biodiversity Campaign**		

The Value of Biodiversity
(20 sessions)

This interdisciplinary unit, a natural for a science-and-society course, is designed to give your students a thorough look at the many ways biodiversity enriches our lives. Students start out gaining a general sense of why biodiversity is important. Then they take an in-depth look at each of these reasons. The activities should help guide them to develop a deeper understanding of the "services" healthy ecosystems perform, the economics of biodiversity, and the many links between culture and the environment.

Session 1: "What's Your Biodiversity IQ?"

Use the quiz-show format of this activity as a "gee-whiz" introduction to biodiversity. (For more questions, refer to pages 5 and 6 of the *WOW!—A Biodiversity Primer* educator's guide, entitled *Ideas for Using the Biodiversity Primer*.) Have the students read "Ask Dr. B!" in *WOW!—A Biodiversity Primer* for homework. You may also want to have them answer the following questions in writing:

- Who is Dr. B and what does he do?
- What does the term *biodiversity* refer to?
- List some examples of how biodiversity is important to people.
- What are the major threats to biodiversity?
- Who is responsible for preserving biodiversity?

Session 2: Discussion, "Super Sleuths" ("Levels of Diversity"), "Biodiversity! Exploring the Web of Life" video

Use the students' answers to their homework assignment to lead an introductory discussion on biodiversity. The "Levels of Diversity" activity in "Super Sleuths" will reinforce the concept of the three levels of biodiversity (genes, species, and ecosystems). The video *Biodiversity! Exploring the Web of Life* (30 minutes) will round out the introduction to the unit.

Session 3: "Mapping Biodiversity" (Part I)

By participating in this activity, your students will become familiar with some of the important ecoregions around the world. Use this activity to focus attention on ecosystem diversity, which they were introduced to in "Super Sleuths."

Session 4: "It's a Wild World" ("Planet Earth: Your Vacation Destination")

This activity takes the concept of ecosystem diversity one step further by having the students research an ecosystem (or ecoregion) of interest to them. Have research materials available to your students to help them choose a focus for their travel brochures (due in session 13).

Session 5: "Biodiversity Performs!"

This activity provides an interactive way for students to learn about the "services" that species and ecosystems provide for us. In the activity summary, challenge your students to begin to think about different ways to assess the value of biodiversity.

Sessions 6 and 7: "Diversity on Your Table"

Refer back to the "Levels of Diversity" activity as you introduce this activity. In session 6, you should get through the simulation game (step 6) and at least begin the discussion of the importance of genetic diversity (step 7). Jump ahead to step 9 and assign "Raiders of the Lost Potato" in *WOW!—A Biodiversity Primer* as homework. In session 7, have the students start out in small groups to briefly discuss the questions in step 9. Then go through steps 7-8, finishing with a look at alternative foods that are beginning to show up in supermarkets. You may want to bring in some non-typical foods, such as amaranth flakes (cereal) and carob chips, for the students to sample.

Session 8: "Putting a Price Tag on Biodiversity"

This activity introduces efforts being made to assess the economic value of biodiversity. Have students read, "How Much Is a Gray Wolf Worth?" for homework.

Sessions 9, 10, and 11: "Dollars and Sense"

Briefly discuss "How Much Is a Gray Wolf Worth?" before launching into this activity. Part I will introduce your students to the concepts of supply, demand, and price. Parts II and III will allow them to apply these concepts, as well as teach them about hidden costs and the factors of production.

Sessions 12 and 13: "The Many Sides of Cotton"

This activity serves as a good follow-up to "Dollars and Sense." Encourage students to use their newly gained knowledge of economics to help them evaluate the different positions in this activity. Ask, "How does economics factor into the positions of the different players?" Plan on getting through step 4 in session 12. For homework, assign one of the "Cotton Concerns" readings to each student in the class.

In session 13, have the students form groups of at least seven, making sure that at least one person in each group has read one of the seven different readings. Save five minutes at the end of the session for the students to complete step 7. To assess the activity, have them do step 8 for homework. Collect the travel brochures from "It's a Wild World,"which were assigned as homework in session 4.

Session 14: "The Culture/Nature Connection" (Part I)

This activity highlights the cultural values placed on biodiversity. In Part I, students should be encouraged to link their own personal values to biodiversity. Assign the family heritage research in Part II as homework (due in session 16).

Session 15: "The Culture/Nature Connection" (Part III)

This role-play activity will get students thinking about cultural diversity in their area. By doing Part III first, students will have extra time to complete the family heritage research necessary for Part II.

Session 16: "The Culture/Nature Connection" (Part II)

If time permits, have the students do the assessment at the end of class. If not, have them do it for homework.

Sessions 17, 18, and 19: "What's a Zoo to Do?"

This activity will give your students another opportunity to evaluate different positions as they work through some of the biodiversity-related dilemmas various institutions face. During the first session, divide the group into teams of three to five students and pass out information on each of the dilemmas. Give them the rest of the session and the next session to discuss their dilemmas and develop their presentations. Students will give their presentations in session 19.

Session 20: "The Spice of Life"

This activity is a great way for students to evaluate their own values associated with biodiversity, and can serve as an assessment for the entire unit.

Unit at a Glance — The Value of Biodiversity

Session 1	Session 2	Session 3	Session 4	Session 5
What's Your Biodiversity IQ?	Discuss "Ask Dr. B!"	**Mapping Biodiversity**	**It's a Wild World**	**Biodiversity Performs!**
	Super Sleuths	Part I	"Planet Earth: Your Vacation Destination"	
	"Levels of Diversity"			
HW: Read "Ask Dr. B" in *WOW!—A Biodiversity Primer.*	**Biodiversity! Exploring the Web of Life (video)**		HW: Work on travel brochure.	

Session 6	Session 7	Session 8	Session 9	Session 10
Diversity on Your Table		**Putting a Price Tag on Biodiversity**	**Dollars and Sense**	
steps 1–7	steps 7–19		Part I	Part II
HW:				
• Read "Raiders of the Lost Potato" in *WOW!— A Biodiversity Primer.*		HW: Read "How Much Is a Gray Wolf Worth?"		
• Work on travel brochure.				

Session 11	Session 12	Session 13	Session 14	Session 15
Dollars and Sense	**The Many Sides of Cotton**		**The Culture/Nature Connection**	
Part III	steps 1–4	steps 5–7	Part I, Part II (step 1)	Part III
HW: Work on travel brochure.		Travel brochures due. HW: step 8	HW: Conduct family heritage research.	
	HW: Read "Cotton Concerns."			

Session 16	Session 17	Session 18	Session 19	Session 20
The Culture/Nature Connection	**What's a Zoo to Do?**			**The Spice of Life**
Part II (steps 3 and 4) Assessment	steps 1 and 2	step 3	Presentations (step 4)	

©Richard I'Anson/Lonely Planet Images

"We will be known by
the tracks we leave behind."

–Dakota Proverb

Appendices

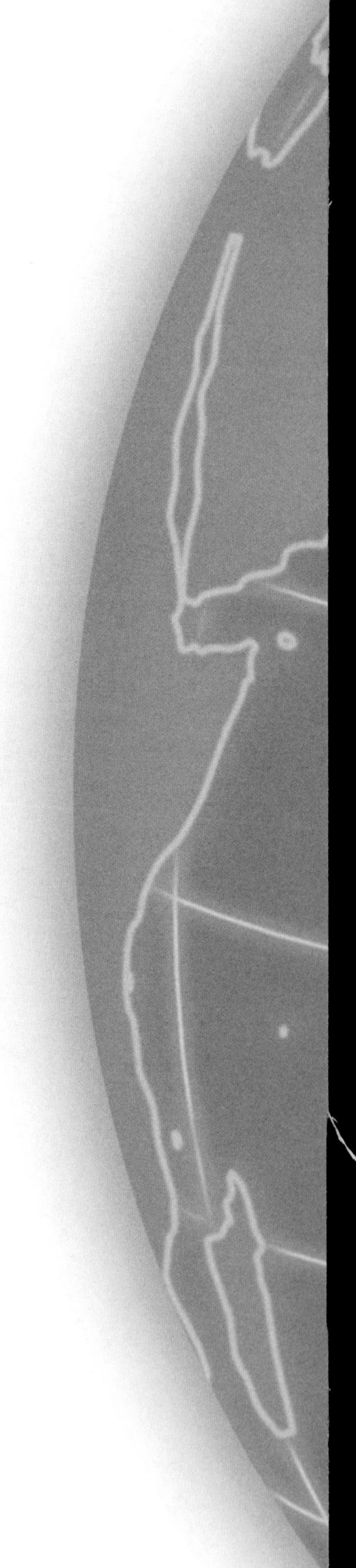

"Conservation is a

way of living and

an attitude that

humanity must

adopt if it wants

to live decently

and permanently

on Earth."

**–Paul Bigelow Sears,
paleoecologist**

Appendix A
Glossary

abiotic: refers to the nonliving components of the environment (such as rock types, slope, and geographic setting) that affect ecological functions.

adaptive radiation: the evolution of a single ancestor species into several new species within a relatively short period of time and in a certain geographic area. The plants and animals of the Galápagos Islands are a result of adaptive radiation, where one plant or one animal species diversified into many species that fill a variety of ecological roles. For example, more than a dozen species of finches evolved from a single founding species that colonized the islands from the mainland of South America.

allele: one of the possible forms of a given gene responsible for different genetic traits. For example, an allele might result in blue or brown eye color.

bacteria: a large and diverse group of microorganisms that perform many important functions, such as enriching soil fertility. Bacteria are some of the most primitive life forms on Earth.

biodiversity: the variety of life on Earth, reflected in the variety of ecosystems and species, their processes and interactions, and the genetic variation within and among species.

biogeography: the study of living systems and their distribution. Biogeography is important to the study of the Earth's biodiversity because it helps with understanding where animals and plants live, where they don't, and why.

biotic: refers to the living components of the environment (such as plants, animals, and fungi) that affect ecological functions.

decomposers: organisms such as bacteria, fungi, earthworms, and vultures that feed on dead animals and plants, as well as other organic wastes and cause them to break down physically and chemically.

ecological or ecosystem services: valuable services provided by natural systems. Examples of ecological services include flood control, air purification, and climate control.

ecoregion: a relatively large unit of land that is characterized by a distinctive climate, ecological features, and plant and animal communities. North America has more than 100 different ecoregions, such as the Everglades Flooded Grasslands and the Great Basin Lakes and Streams.

ecosystem: a community of plants, animals, and microorganisms that are linked by energy and nutrient flows and that interact with each other and with the physical environment. Rain forests, deserts, coral reefs, grasslands, and a rotting log are all examples of ecosystems.

endangered species: a species threatened with extinction. The Florida panther and the California condor are endangered species.

endemic: describes an animal or plant species that naturally occurs in only one area or region. For example, the redfin darter is a fish endemic to the rivers of the Ozark forests, and the Joshua tree is a plant endemic to the Mojave desert.

evolution: the process of change in the traits of organisms or populations over time. Evolution, through the process of natural selection, can lead to the formation of new species.

externalities: in economics, benefits or costs that are not included in the market price of goods or services. For example, the cost of natural resource depletion, pollution, and other environmental and social factors are externalities that often are not factored into the market price of the product.

extinct: refers to a species that no longer exists. *Local extinction* occurs when every member of a particular population has died. *Global extinction* occurs when every member of a species has died. The passenger pigeon and the dodo are examples of globally extinct birds.

fauna: the animals that live in a particular area.

flora: the plants that live in a particular area.

food chain: a lineup of organisms from producers (plants) to consumers (other plants, animals, and fungi), with each organism feeding on or getting nutrients from the previous organism.

fragmentation: the breaking up of large habitats into smaller, isolated chunks. Fragmentation is one of the main forms of habitat destruction, which is the primary reason biodiversity is in decline.

fungi: organisms that use living or dead organisms as food by breaking them down and then absorbing the substances into their cells. Fungi make up one of the five kingdoms of living things on Earth. Mushrooms, yeast, and molds are types of fungi.

gene: a segment of DNA that includes the coded information in an organism's cells that makes each species and individual unique. Genes contain the hereditary characteristics that are transmitted from one generation to the next and determine how organisms look and behave. Genes are responsible for features such as hair color and texture, and resistance to disease.

gene bank: a facility that stores genetic material. For example, the U.S. Department of Agriculture's gene banks store seeds and other plant parts for future use.

genetic diversity: the genetic variation present in a population or species. For example, the genetic diversity in the hundreds of varieties of potatoes can be seen by their differences in size, shape, color, taste, and rate of growth.

global warming: the hypothesis that the Earth's atmosphere is warming because of the release of "greenhouse gases," such as carbon dioxide. These gases are released into the air from burning gas, oil, coal, wood, and other resources and trap heat in an action similar to that of the walls of a greenhouse.

greenhouse effect: the trapping of heat in the Earth's atmosphere by certain gases such as CO_2, CH_4, and N_2O. Some scientists predict that the temperature and sea level rise associated with global warming could adversely affect biodiversity.

habitat: the area where an animal, plant, or microorganism lives and finds the nutrients, water, sunlight, shelter, living space, and other essentials it needs to survive. Habitat loss, which includes the destruction, degradation, and fragmentation of habitats, is the primary cause of biodiversity loss.

hypothesis: a statement consisting of an action that can be tested and a predicted result. Making a hypothesis is part of scientific inquiry.

introduced species: an organism that has been brought into an area where it doesn't naturally occur. Introduced species can compete with and cause problems for native species. Introduced species are also called *exotic, non-native,* and *alien* species.

invertebrate: an organism that does not have a backbone.

microorganism: a living organism too small to be seen with the naked eye. Bacteria, protozoans, viruses, microscopic algae, and some types of fungi are all microorganisms.

migration: the movement of animals in response to seasonal changes or changes in the food supply. Examples of animals that migrate include ruby-throated hummingbirds, salmon, monarch butterflies, buffalo, and elephants.

mutation: a spontaneous change in the genetic information of a cell. Mutations can provide genetic diversity within populations and allow evolutionary change to occur, although many mutations are either neutral or harmful to an individual.

native species: a species that occurs naturally in an area or a habitat. Also called *indigenous* species.

natural resource: any aspect of the environment that species depend on for their survival. People depend on natural resources such as land, soil, energy, and fresh water.

natural selection: the process by which genetic traits are passed on to each successive generation. Over time, natural selection helps species become better adapted to their environment. Also known as "survival of the fittest," natural selection is the driving force behind the process of evolution.

organic: (1) describes matter that is living or was once living; (2) describes agricultural products grown or raised without pesticides or other synthetic chemicals.

over-consumption: the use of resources at a rate that exceeds the ability of natural processes to replace them.

pesticides: chemicals that kill or inhibit the growth of organisms that people consider undesirable. Fungicides (which kill fungi), herbicides (which kill plants), and insecticides (which kill insects) are types of pesticides.

photosynthesis: the process by which green plants, algae, and other organisms that contain chlorophyll use sunlight to produce carbohydrates (food). Oxygen is released as a byproduct of photosynthesis.

phylogeny: the evolutionary history of a group of organisms.

poaching: hunting, trapping, or fishing illegally.

pollination: the process by which pollen is transferred from the male part of a flower to the female part of the same or another plant. Insects, birds, mammals, and other creatures, as well as wind or water, can all pollinate plants.

range: the area in which an organism may travel in its lifetime. Range also refers to the geographic distribution of a particular species.

rapid assessment: a quick scientific survey or count that helps measure local biodiversity.

reintroduce: to return members of a species to their historical range. This strategy is sometimes used when a species has become locally extinct or if its population is threatened.

scientific inquiry: the process of designing and conducting scientific investigations.

slash and burn agriculture: an agricultural system in which farmers periodically clear land for farming by cutting and burning patches of forest. Traditionally, patches used for agriculture were allowed to revert to forests for a number of years before being replanted, thereby causing minimal impact. Today, however, intensive slash and burn agriculture damages many tropical forest ecosystems.

speciation: the process by which one or more populations of a species become genetically different enough to form a new species. This process often requires populations to be isolated for a long period of time.

species: (1) a group of organisms that have a unique set of characteristics (like body shape and behavior) that distinguishes them from other organisms. If they reproduce, individuals within the same species can produce fertile offspring; (2) the basic unit of biological classification. Scientists refer to species using both their genus and species name. The house cat, for example, is called *Felis catus.*

sustainable: meeting the needs of the present without diminishing the ability of people, other species, or future generations to survive.

symbiotic: refers to an ecological relationship between two organisms. The relationship may be beneficial or detrimental to one or both organisms. For example, termites have a symbiotic relationship with protozoans that live in their gut and digest cellulose. The protozoans benefit from a safe and nutrient-rich environment.

systematics: the study and classification of the relationship between life forms and evolution.

taxonomy: the process and study of classifying organisms.

temperate rain forest: a type of forest found in only a few places around the world, such as the Pacific temperate rain forest on the west coast of North America. These forests are often dominated by conifers adapted to wet climates and cool temperatures.

trait: a genetic feature or characteristic, such as hair color or blood type, that may be passed on from one generation to the next.

tropical dry forest: a type of forest found near the equator that has distinct rainy and dry seasons. Many tropical dry forest plants are adapted to withstand high temperatures and seasonal droughts.

tropical rain forest: a type of wet forest found near the equator that harbors the richest diversity of terrestrial plant and animal species.

watershed: a geographic area that drains into a single river system and its tributaries.

wetlands: areas that, at least periodically, have waterlogged soils or are covered with a relatively shallow layer of water. Bogs, freshwater and saltwater marshes, and freshwater and saltwater swamps are examples of wetlands.

Appendix B
Legislation Protecting Biodiversity

For your reference, we have compiled a short list of laws, conventions, and treaties that support the protection of biodiversity in the United States and around the world. Federal laws apply throughout the United States, and international agreements apply to countries that have accepted the terms of the agreement. For more detailed information about any of these laws, or about laws in your state, see the section of this Appendix entitled "For More Information" (page 417). The U.S. Fish and Wildlife Service is the agency primarily responsible for implementing and enforcing most of the following laws.

Federal Laws

Lacey Act (1900)

Originally passed in 1900 and amended in 1981, this is one of the strongest laws in the world designed to curb illegal wildlife trade. It prohibits the importation of animals that were illegally killed, or products that were illegally collected or exported from another country. The Lacey Act also allows the government to seize illegally imported goods.

Migratory Bird Treaty Act (1918)

This act mandates that all migratory birds and their parts, including eggs, nests, and feathers, are fully protected. In the 19th century, trade in migratory birds severely reduced the populations of many native species. The Migratory Bird Treaty Act was created to end the commercial trade of these species as well as that of all common wild birds found in the United States, except the house sparrow, starling, feral pigeon, and resident game birds. Since the adoption of this act, four international agreements between the United States and Canada, Russia, Mexico, and Japan have been reached to protect migratory birds.

Migratory Bird Hunting and Conservation Stamp Act (1934)

Created to supplement the Migratory Bird Treaty Act, this law provides funds for the acquisition of sanctuaries and breeding grounds for the protection of migratory birds. Migratory bird hunters are required to purchase a Federal Migratory Waterfowl Hunting Stamp (called a Duck Stamp) and a state hunting license. Funds from the sale of Duck Stamps are used to purchase land for migratory bird habitat protection.

Federal Aid in Wildlife Restoration Act (1937)

Better known as the Pittman-Robertson Act, this act provides states with federal aid generated from an excise tax on ammunition and sporting arms. Funds are used for wildlife conservation work, including wildlife surveys, research, land acquisition, and technical assistance. This act was followed by the Federal Aid in Sport Fish Restoration Act (the Dingell-Johnson Act) in 1950, which provides federal aid (from a tax on sport fishing equipment) for the restoration of fish species.

Federal Insecticide, Fungicide, and Rodenticide Act (FIFRA) (1947)

Originally passed in 1947 and significantly amended during the 1970s and 1980s, this act prohibits the use of pesticides that have an unreasonably adverse effect on the environment. Pesticides must be evaluated by the Environmental Protection Agency (EPA) for their effects on nonpest species, especially threatened or endangered wildlife. All phases of pesticide use (including sale, use, handling, and disposal) are regulated by the EPA.

National Environmental Policy Act (NEPA) (1969)

Called "the basic national charter for protection of the environment" by the EPA, this law requires that an environmental impact statement be conducted for federal government projects that could have a significant impact on the environment. Federal agencies must carefully consider the environmental effects of their proposed actions and must restore and enhance environmental quality as much as possible.

Clean Water Act (Federal Water Pollution Control Act) (1972)

The main objective of this act is to restore and maintain the chemical, physical, and biological quality of the nation's waters, including its seas, lakes, rivers, streams, and wetlands. With strict permit requirements, the act limits the amount of pollutants released into the nation's waters. The ultimate goal of this law is to maintain water quality that will support both healthy wildlife populations and recreational use by people.

Marine Mammal Protection Act (1972)

This act makes it illegal for any person under the jurisdiction of the United States to kill, hunt, injure, or harass any species of marine mammal, regardless of its population status. The act also makes it illegal to import marine mammals, or products made from them, into the United States. Certain numbers of marine mammals can be collected for scientific and display purposes, as accidental by-catch in commercial fishing operations, and in subsistence hunting by natives of the North Pacific and Arctic coasts. The National Marine Fisheries Service is primarily responsible for enforcement of this act.

Endangered Species Act (1973)

This act aims to ensure the survival of endangered species, defined as those "in danger of extinction throughout all or a significant portion of their range," and threatened species, defined as those "likely to become endangered within the foreseeable future." According to this act, the term "species" includes species and subspecies of fish, wildlife, and plants. This includes any member of the animal kingdom (any mammal, fish, bird, amphibian, reptile, mollusk, crustacean, arthropod, or other invertebrate) and any member of the plant kingdom. Habitats that are critical to the conservation of a threatened or endangered species may be covered under this act.

Magnuson-Stevens Fishery Conservation and Management Act (1976)

This act was passed to establish a zone from 3 nautical miles to 200 nautical miles off the coast of the United States where overfishing by foreign fishing fleets could be controlled. As a result of this act, a conservation and management structure of eight regional fishery management councils was established. The councils' primary responsibility is to develop fishery management plans that prevent overfishing, promote conservation, and protect the long-term stability of fisheries.

Marine Plastic Pollution Research and Control Act (MPPRCA) (1987)

This act prohibits the disposal of plastic, or garbage mixed with plastic, into any U.S. waters. MPPRCA works in conjunction with the Marine Pollution Treaty (MARPOL) (1973/1978), which prohibits the discharge of oil, sewage, plastics, and garbage into coastal and ocean waters.

Wild Bird Conservation Act (1992)

This act is designed to protect wild populations of exotic birds from the international pet trade. More than 1,000 species of birds living throughout the world are threatened with extinction from excessive trade and habitat destruction. Restrictions are placed on the importation of bird species listed on the Appendices to the Convention on International Trade in Endangered Species (CITES) (page 417). Bans on the trade of 10 species of birds and limiting quotas have proven successful in the protection of endangered or threatened birds.

International Treaties and Conventions

Pan American Convention (Convention on Nature Protection and Wildlife Preservation in the Western Hemisphere) (1940)

The objective of this treaty is to preserve all species of flora and fauna that are native to the Americas as well as all areas of extraordinary beauty, striking geological formations, and historic or scientific value. Parties to the treaty, which include the United States and 18 other nations of the Western Hemisphere, agree to establish national parks and wilderness reserves, to conserve species listed for special protection, and to cooperate in the field of research.

Antarctic Treaty (Agreed Measures for the Conservation of Antarctic Fauna and Flora) (1959)

While it guarantees freedom of scientific research and exchange of data in the Antarctic region (the region south of 60° south latitude), the Antarctic Treaty prohibits all military activity, nuclear explosions, and disposal of radioactive waste in Antarctica. The Agreed Measures for the Conservation of Antarctic Fauna and Flora is an annex to the treaty. It prohibits the killing, wounding, capturing, or molesting of native birds or mammals without a permit, and it provides for the establishment of protected areas of scientific interest or ecological uniqueness.

Ramsar Convention (Convention on Wetlands of International Importance Especially as Waterfowl Habitats) (1971)

Adopted in Ramsar, Iran, this convention seeks to protect wetlands to maintain the ecological functions and recreational opportunities they provide. The convention maintains a list of wetlands of international importance and encourages the wise use of all wetlands. Member nations must protect listed wetlands within their territories and are expected to promote the protection of all wetlands.

World Heritage Convention (Convention Concerning the Protection of the World Cultural and Natural Heritage) (1972)

This convention maintains a list of global sites that are of particular cultural or ecological significance. By signing the convention, countries pledge to conserve listed sites within their territories. Sites may be of particular architectural or religious significance, they may represent the traditional way of life of a certain culture, they may represent ongoing ecological processes or geological phenomena, or they may contain important and significant natural habitats needed for the conservation of biological diversity.

CITES (Convention on International Trade in Endangered Species of Wild Fauna and Flora) (1973)

This convention was formed to help prevent the depletion of wild plant and animal populations that are frequently traded on the international market. Products that caused a need for protection of traded species include ivory, rhinoceros horn, tortoise-shell jewelry, and animal pelts. More than 130 nations have become parties to the treaty, which means they have agreed to develop wildlife protection laws within their territories for species listed for protection.

Convention on Biological Diversity (1992)

Signed by 156 countries and the European Union during the 1992 Earth Summit in Rio de Janeiro, Brazil, this convention recognizes how valuable—and endangered—biodiversity is around the world. By ratifying the treaty, governments agree to accept responsibility to safeguard the species, genetic material, habitats, and ecosystems of the natural world. They commit themselves to development that sustainably uses biological resources, and they agree to recognize every nation's sovereignty over its biodiversity.

For More Information

Federal Wildlife and Related Laws Handbook

This handbook contains an overview of U.S. wildlife law and wildlife issues, statutes, cooperative agreements, and treaties. It also contains a glossary, a bibliography, and selected acronyms. It was produced by the Center for Wildlife Law at the University of New Mexico and is available from Government Institutes, 4 Research Place, Suite 200, Rockville, MD 20850.

State Wildlife Laws Handbook

This handbook was produced by the Center for Wildlife Law at the University of New Mexico. It contains overviews of wildlife law and wildlife poaching in the United States, summaries of state wildlife laws, and comparisons of state laws. It is available from Government Institutes, 4 Research Place, Suite 200, Rockville, MD 20850.

The Evolution of National Wildlife Law

This book by Michael J. Bean and Melanie J. Rowland is a project of the Environmental Defense Fund and the World Wildlife Fund-U.S. (Praeger Publishers, 1997). It examines the historical foundation of wildlife conservation, laws governing species conservation, wildlife habitat laws and regulations, and the laws and treaties that regulate worldwide wildlife conservation.

Web Sites

Much of the information contained in this appendix can be found in greater detail on the Internet. Using any search engine, enter the name of the law you would like more information about as your search term. An excellent resource for more information on wildlife laws is the U.S. Fish and Wildlife Service (USFWS). It is responsible for enforcing most of the laws listed in this Appendix. Visit the USFWS Web site <www.fws.gov>. For information about the laws it administers, go to <www.fws.gov/laws>.

Appendix C

A Biodiversity Education Framework
Key Concepts and Skills

Biodiversity is a challenging and engaging topic. It offers an exciting opportunity for creative teaching and a framework for exploring the world around us and our place in it. Encompassing everything from endangered species to the social implications of global warming, biodiversity offers a variety of "windows on the wild"—opportunities for people to see the world from different perspectives.

This biodiversity education framework is designed to break down the topic of biodiversity into teachable concepts and important skills that can help educators build an effective education program. Part I outlines about 70 key concepts for understanding biodiversity—focusing on what biodiversity is, why it's important, why we're losing it, and how we can conserve and maintain it for the future. Part II highlights the skills, grouped into eight major categories, that we feel are key to learning about biodiversity and making responsible decisions about how best to conserve it. These include the skills students need to gather, process, and act upon information, as well as those they need to become effective citizens.

Although this framework is not a curriculum in itself, the concepts and skills it outlines can provide the basis for developing a comprehensive program that uses biodiversity as the organizing theme. This framework can also help educators develop an effective unit, course, or public education program that focuses on one aspect of biodiversity, such as the science of biodiversity or the ways biodiversity issues play out in a community.

We drew from the work of many organizations and individuals to help develop this framework, including the following: Dr. Bora Simmons (Northern Illinois University) and the North American Association for Environmental Education (*Environmental Education Guidelines for Excellence: What School-Age Learners Should Know and Be Able to Do*, © NAAEE, 1997), Dr. Ben Peyton (*Environmental Education Module on Biological Diversity for Secondary Educators—A Teacher Training Workshop Text*, UNESCO, 1992), Dr. Randy Champeau and his team's energy education framework (*K–12 Energy Education Program—A Conceptual Guide in Wisconsin*, © The Energy Center of Wisconsin, 1996), Bill Andrews and the California Department of Education (*California Guide to Environmental Literacy*, California Department of Education, draft version, 1995), the Association for Supervision and Curriculum Development, the American Association for the Advancement of Science, many environmental science textbooks and science and social studies frameworks, and the advice of individual scientists and educators. We also added concepts to reflect current thinking in the field of biodiversity.

This framework has evolved with the development of *Windows on the Wild* and will continue to evolve as the program grows. Along the way, many people have provided feedback and review—including scientists, social scientists, evaluation specialists, teachers, environmental educators, business leaders, and curriculum developers.

Part I: The Conceptual Framework

Because the issues surrounding biodiversity can be complex, the topic can be challenging to understand—and to teach. That's why we've broken the topic down into the following teachable concepts and organized them in a way that will help you see, and communicate, the relationships among the different levels of biodiversity, the ecological principles behind it, and how we relate to it. We've also linked the concepts to the *Biodiversity Basics* activities that are designed to teach them. While no single activity can teach these concepts completely, each can contribute to your students' growing understanding. At the beginning of each activity you'll find the numbers of the framework concepts that are touched upon in the activity. And listed here, you'll find the numbers of the activities that explore each concept.

The concepts are organized under four themes posed as questions: What is biodiversity? Why is biodiversity important? What is the status of biodiversity? How can we protect biodiversity? Each theme is followed by concepts that address the question, and the concepts are further organized into subthemes. The themes are arranged so that they build on each other—starting with ecological foundations and moving to broader societal issues. Although there is some overlap in this framework, we have tried to identify key concepts within each theme that will help learners better understand biodiversity issues.

What Is Biodiversity?

- **Definition of Biodiversity**
- **Basic Ecological Principles**
- **Key Ecological Definitions That Help Us Understand Biodiversity**

Why Is Biodiversity Important?

- **Quality of the Environment**
- **Quality of Life**

What's the Status of Biodiversity?

- **General Factors Affecting Biodiversity**

How Can We Protect Biodiversity?

- **Studying Biodiversity**
- **Conserving Biodiversity**
- **Future Outlooks for Maintaining and Restoring Biodiversity**

What Is Biodiversity?

The concepts within this theme provide students with a fundamental knowledge and appreciation of biodiversity. These concepts also help students understand the characteristics of living systems and the fact that the environment is made up of systems nested within larger systems.

Definition of Biodiversity

Understanding these concepts helps students identify biodiversity as a way of organizing our thinking about nature and ecology and helps them describe our relationship with nature.

1. **Biological diversity,** also called *biodiversity,* encompasses the variety of all life on Earth, including life on land, in the oceans, and in freshwater ecosystems such as rivers and lakes. People often analyze biodiversity at many levels, ranging from large to small. The three most common levels of analysis focus on ecological diversity, species diversity, and genetic diversity.
1, 2, 3, 4, 6, 13, 22

2. **Ecological diversity** refers to the variety of Earth's ecosystems, habitats, and biological communities. It also refers to the variety of ecological processes that sustain life and contribute to its evolution. Ecosystems vary from place to place because of physical differences (type of soil, amount of water, climate) as well as human and natural disturbances. Examples of ecosystems include rain forests, pine forests, coral reefs, deserts, and savannas.
2, 4, 13, 15, 19, 23

3. **Species diversity** describes the number and variety of species that live on Earth. Species diversity can refer to the diversity within specific groups of organisms, such as sparrows, elephants, whales, snakes, mushrooms, and ticks, as well as the total diversity of organisms on Earth and the relationships among them.
2, 3, 4, 6, 7, 9, 10, 12, 13, 21, 23, 27

4. **Genetic diversity** refers to the variety of genetic information contained in the genes of individuals, species, populations within a species, or evolutionary lineages.
11, 12, 13, 20

5. **Cultural diversity** describes the richness of knowledge, beliefs, and traditions within our own species. Cultural diversity influences biodiversity and is influenced by biodiversity. 18

6. The study of biodiversity is inherently **interdisciplinary** because it cannot be sufficiently described or understood through any single discipline. The study of biodiversity uses knowledge from the natural sciences, the physical sciences, the geological sciences, and the social sciences. The three levels of biodiversity are measured and monitored by geneticists, botanists, zoologists, conservation biologists, and other experts. 32, 33

7. The biological importance of an area is based on broad measures of species richness (the number of different kinds of species in an area), endemism (the number of unusual, rare, and geographically unique species), unusual ecological and evolutionary phenomena (monarch butterfly and caribou migrations), and the global rarity of habitat types found there (the number of examples of a particular type of habitat that exist worldwide). 23

Basic Ecological Principles

Students need to understand basic ecological principles to understand how natural systems function and why biodiversity is important. They also need to understand that people are part of nature, not separate from it.

8. The source of the radiant energy that sustains life on Earth is the sun. 15

9. The cycling of elements between living and nonliving components of ecosystems is driven by energy. (Examples of natural cycles include the carbon cycle, the water cycle, and the nitrogen cycle.) Human societies, like natural ecosystems, need energy to organize and maintain themselves.

10. Plants use radiant energy to produce food energy—a process called *photosynthesis*. Energy passes from plants to other living things through food chains and food webs. 15

11. Nutrients that sustain life on Earth are continuously recycled and converted from one useful form to another by a combination of biological, geological, and chemical processes. 15

12. All of the components of an ecosystem, both living and nonliving, are connected. The loss of one species can threaten the survival of other species, and thus can undermine the integrity of an entire ecosystem.
1, 6, 9, 10, 12, 13, 14, 15, 19, 21, 23, 27, 30

13. Terrestrial, freshwater, and marine ecosystems interact with each other in complex ways. For example, runoff from terrestrial ecosystems such as agricultural fields can affect the water quality of nearby rivers and streams, as well as marine ecosystems such as coastal estuaries. Also, some species spend different parts of their lives in different ecosystems and can play important roles in each of the ecosystems they inhabit. 1, 26

14. The structure, and therefore the biodiversity, of an ecosystem changes over time. Succession is a progressive, often predictable, change in ecosystem structure over time.

15. Natural systems are dynamic, and natural disturbances are an ongoing part of ecosystem functioning. For example, small-scale natural disturbances, such as a single tree falling in a forest, can maintain or increase biodiversity by providing nutrients and shelter for fungi, insects, and other organisms, and by allowing sunlight to reach the forest floor. Large-scale natural disturbances, such as hurricanes or fire, create a mosaic of different habitat types across the larger landscape.

16. Diversity of species in a system increases the likelihood that at least some species will survive changes in environmental conditions.

17. Biodiversity is not distributed evenly on the planet. Some areas are naturally very diverse in species and habitat types (tropical forests, coral reefs) and others are not as naturally diverse (tundra, temperate forests). In general, biodiversity decreases from the equator to the poles. Diversity also tends to decrease in areas with less rainfall and at increasing elevations. 23

18. A high number of species in an ecosystem doesn't necessarily mean that the ecosystem is healthier than those with fewer species. Some ecosystems naturally have fewer species than others. 10, 18

19. Environmental changes over time help shape the evolution of species on Earth. Evolution occurs when the gene pool of a population is exposed to new environmental conditions and results in differential reproduction among individuals (certain traits get passed on to future generations; other traits eventually die out). 11, 12, 20

20. The species and individuals that are alive today provide the basis for ongoing evolution. The more diversity there is today, the more there can be in the future. 11, 12, 20

Key Ecological Definitions That Help Us Understand Biodiversity

By understanding the following definitions and underlying concepts, students will be able to communicate more effectively about biodiversity issues and to understand background information and research related to biodiversity issues.

21. A group of organisms that is evolving separately from other groups is called a **species**. For organisms that reproduce sexually, a species can also be defined as organisms that interbreed only among themselves.
7, 9, 10, 12, 21, 22, 23, 27

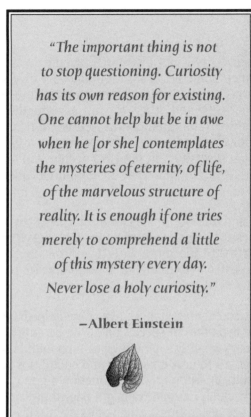

"The important thing is not to stop questioning. Curiosity has its own reason for existing. One cannot help but be in awe when he [or she] contemplates the mysteries of eternity, of life, of the marvelous structure of reality. It is enough if one tries merely to comprehend a little of this mystery every day. Never lose a holy curiosity."

–Albert Einstein

22. Each species exists in **populations** which are groups of individual organisms of the same species living in the same area. For example, all the bluegills in a lake, the gray squirrels in a forest, the maple trees in a forest, or the people in a country are examples of populations.
11, 12, 22

23. The word **habitat** refers to a place that provides whatever the individuals of a species need to survive, such as nutrients, water, sunlight, and living space. "Black bear habitat" and "sagebrush habitat" mean places that support those specific organisms.
12, 21, 22, 23, 27

24. Every species has an **ecological niche**, which is the species' unique function or role in its ecosystem. A niche describes all the physical, biological, and chemical factors that a species needs to survive, stay healthy, and reproduce. To help distinguish between a species' habitat and its niche, scientists often use this analogy: a species' habitat is its address and its niche is its profession.

25. When talking about an area and all the species that live and interact in it, we use the word **community**. When talking about the sum of interactions between living things in a community and nonliving components of that community—such as bodies of water, soil, and energy sources—we use the word **ecosystem**. Ecosystems can be large, such as a desert or a rain forest ecosystem, or they can be small, such as a vernal pond or a rotting log.
15, 19, 23

26. Scientists use the terms **endangered**, **threatened**, and **extinct** to describe the status of species. *Endangered* species are those species that are in immediate danger of becoming extinct. (Bengal tigers, giant pandas, and manatees are endangered species.) *Threatened* species are those whose numbers are low or declining and whose gene pool is becoming too small to ensure variation in offspring. A threatened species is not in immediate danger of extinction but is likely to become endangered if it isn't protected. (Northern spotted owls and grizzly bears are threatened.) *Extinct* species are no longer living. (Passenger pigeons are extinct.)

Why Is Biodiversity Important?

Concepts in this section can help students investigate how biodiversity affects their lives and supports life on Earth. Recognizing the importance of biodiversity increases students' awareness of why and how people's actions affect biodiversity and why it's important to maintain and restore biodiversity.

Quality of the Environment

Biodiversity can be a unifying theme to help students understand that ecological systems and processes maintain life on Earth. By understanding the following concepts, students can see how biodiversity is linked to the quality of the environment and the health of all living things.

27. Biodiversity contributes to the restoration of the landscape after large-scale natural events such as floods, droughts, earthquakes, hurricanes, and volcanic eruptions.

28. Each level of biodiversity is essential to fundamental life processes (life support systems):

    • Genetic diversity within species allows species to adapt to changes in the environment over time.

    • Species diversity provides a variety of interactions that contribute to energy flow and nutrient cycling in ecosystems.

    • Ecological diversity provides habitat for different species, as well as essential "services" that maintain the biosphere, including water and air purification, micro-climate control, and soil formation and stability.
      11, 12, 15, 19, 20

29. Extracting and using natural resources (through development, timber harvesting, mining, and so on) can alter environmental conditions, often leading to reduced air and water quality, loss of habitat, and disrupted nutrient cycles. 22, 27, 30, 34

Quality of Life

People value biodiversity for many reasons, including aesthetic, moral, spiritual, educational, economic, and recreational reasons. By understanding the following concepts, students will see how biodiversity directly affects their lives and the lives of people throughout the world. They will also begin to understand how individual and societal actions affect biodiversity.

Economics

30. Human economies depend on biodiversity. For example, biodiversity benefits agriculture, provides opportunities for medical discoveries, and stimulates industrial innovations. 20, 28

31. Our economic system is based on resources—both natural and human. Economic health depends on maintaining ecosystems that function well and that deliver the goods and services that society depends on. 28

32. Many economists are thinking about how to factor the services biodiversity provides (water purification, climate control, and pollination) into the price of goods. They're also factoring in the costs of depleting resources or causing pollution. This approach will help people understand the economic value of healthy ecosystems and encourage consumers, suppliers, and policy makers to consider environmental impacts when setting prices. 24, 28

> *"Buried within the human species lies a deep and enduring urge to connect with living diversity. . . . We evolved in the company of other creatures and in a matrix of conditions making this varied existence possible. . . . Our identity remains rooted in our connections with the natural world."*
>
> **—Stephen R. Kellert, writer, professor**

Health and Safety

33. The health and well-being of individuals or groups of people can be adversely affected by loss of biodiversity at local or global levels. 3, 5, 15, 19

34. Some biologists, psychologists, sociologists, and others believe that people have a deeply ingrained need to connect with nature for our mental well-being. According to this view, it's important to sustain a relationship with nature, especially as we become more urbanized.

35. Reduced air and water quality, deforestation, and changes in land use because of road building and other development can pose risks to the health and well-being of people and other living things. 14, 21, 22, 27, 30

36. Genetic diversity in plants and animals is essential for the development of new varieties of crop plants and animals that are bred for human use. These new varieties and breeds may be resistant to disease or able to tolerate extreme environmental conditions and can be important sources of food. 20

Lifestyles

37. Human values can be affected by a variety of factors, including wealth, health, religion, ecology, and culture. These factors influence the development of lifestyles that may or may not be supportive of maintaining biodiversity. 17, 29

Sociopolitics

38. Increasing human populations have produced many modifications to the natural world and its components and processes. Often these modifications or changes have resulted in problems, such as decreasing gene pools in wild species, loss of ecosystems, and the decline and change of local human societies whose existence is closely related to the natural world. 22, 27, 30, 34

39. Environmental problems associated with biodiversity loss don't necessarily affect all people in a region or country equally. Often, they adversely affect populations living in low-income communities or less industrialized countries. 25

40. The decision to protect biodiversity (or not to protect it) is the result of choices people make as families, community members, voters, consumers, employees, politicians, and neighbors. These choices can reflect values and beliefs (making a choice based on religious beliefs), knowledge of the issues and the consequence of a choice (making a choice based on inadequate information), a need to satisfy basic human needs (making a choice to cut firewood for heat or to poach wildlife for food), or other factors. An understanding of biodiversity issues can help us predict future trends and determine the positive and negative effects of our choices and the values they reflect. 17, 18, 26, 29, 30, 33

Culture

41. Culture can be closely linked to biodiversity. Human cultures are shaped in part by the characteristics of local ecosystems. Human cultures interact with, understand, and value nature in many different ways, and efforts to conserve biodiversity are often developed within the constructs of local cultures. Conversely, conserving local or regional biodiversity often helps to protect cultural integrity and values. For example, protecting marine species in the Bering Sea can help protect the cultural traditions of the native people who depend on marine species for their survival. 18, 30

42. The ways different cultures around the world feel about and use the natural world are expressed through art, architecture, urban planning, music, language, literature, theater, dance, sports, religion, and other aspects of their lives. 3, 16, 18

43. Because a society's understanding of and relationship with nature changes over time, cultural expressions of nature also change over time. 16, 18

What's the Status of Biodiversity?

Concepts in this theme can help students understand the status of biodiversity and why biodiversity is declining around the world. By learning about the causes and consequences of biodiversity loss, students will be able to participate in maintaining biodiversity in the future.

General Factors Affecting Biodiversity

Biodiversity is reduced by changes in the environment that exceed the ability of populations of plants, animals, and other living things to adapt. This inability to adapt to changing environmental conditions leads to the extinction of species—either locally or globally.

44. The five major causes of biodiversity decline are human population growth; loss, degradation, and fragmentation of habitat; introduced species; over-consumption of natural resources; and pollution. 21, 22, 27, 30

Human Population Growth

45. Human populations continue to grow at a high rate. As human populations grow, they use resources and alter the environment at ever-faster rates. Continued human population growth is likely to lead to continued loss of biodiversity, even though experts predict that human numbers could level off by the middle of the 21st century. 25

Loss, Degradation, and Fragmentation of Habitat

46. The loss, degradation, and fragmentation of habitats, such as forests, wetlands, and coastal waters, is the single most important factor behind species extinction. This large-scale loss is the result of human population growth, pollution, and unsustainable consumption patterns. 22, 27

47. The loss of biodiversity has sometimes been the unintended consequence of certain technologies (for example, building dams and roads; development of intensive agriculture made possible by pest control, hybridization, and, more recently, biotechnology). 20, 22, 27

Introduced Species

48. When non-native species are introduced into new areas, they can reduce the biodiversity of those areas. In some cases, these *introduced* or *alien* species outcompete native species for nutrients, water, sunlight, and living space. Introduced species usually arrive without the predators, diseases, or parasites that keep their populations in check in their natural habitats. In addition, introduced species can bring new diseases to an area that can harm native species, and they can sometimes survive better in areas that have been disturbed than the native species can. 22

49. In healthy ecosystems, most introduced species die out or don't cause major problems in their new habitats. However, introduced species are threatening biodiversity in many areas, especially in fragile ecosystems and on islands. And in light of increasing international trade and travel, many experts are concerned that the threats will increase.

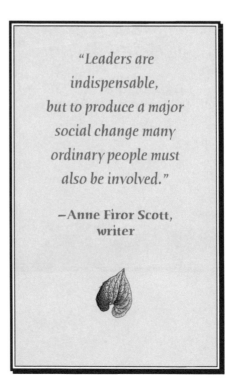

> *"Leaders are indispensable, but to produce a major social change many ordinary people must also be involved."*
>
> **—Anne Firor Scott, writer**

Over-Consumption of Natural Resources

50. Over-consumption results when human use of natural resources exceeds the capacity of natural systems to replenish the resources through natural processes. Increased demand for some products and more efficient ways to take and harvest resources have led people to consume fossil fuels, forests, fish, minerals, water, and other resources at a much higher rate than those resources can be replaced. **30, 34**

51. The majority of the world's resources are being used by people in industrialized areas such as the United States, Europe, and Japan. Many experts believe that this high rate of resource use in industrialized countries is not sustainable. In addition, many people believe it is unethical for people in some countries to use so many resources while others use so few. **25**

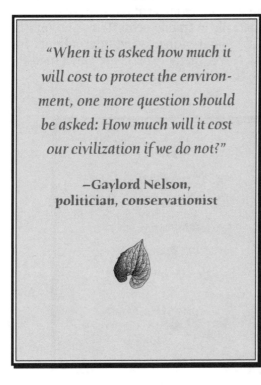

"When it is asked how much it will cost to protect the environment, one more question should be asked: How much will it cost our civilization if we do not?"

—Gaylord Nelson, politician, conservationist

Pollution

52. Pollution is the release of harmful substances, such as pesticides and sewage, into the environment. Biodiversity is threatened by many types of pollution, including the build-up of carbon dioxide and other "greenhouse" gases in our atmosphere (which many scientists think is causing global warming), acid rain (caused by sulfur and nitrogen compounds that are released into the air when fossil fuels are burned), and toxic chemicals (released into the air, soil, or water during manufacturing, farming, building, mining, transportation, and many other activities). **14, 22, 26**

How Can We Protect Biodiversity?

Concepts in this section help students identify ways to ensure that adequate biodiversity will be maintained for future generations. For students to willingly and effectively take action to protect biodiversity, they must have a thorough understanding and appreciation of what biodiversity is, why it's important, why we're losing it, and what people can do to help maintain and conserve it. Students should also begin to understand that ecological integrity, social equity, and economic prosperity are connected and are important components of a sustainable society.

Studying Biodiversity

These concepts will help students better understand how we study biodiversity and how much we still need to learn in order to protect biodiversity while meeting human needs. Students will also see that there are many opportunities for new discoveries and understandings in science, technology, and the social sciences.

53. Science and technology contribute to our understanding of biodiversity and the reasons for its loss. **8, 23**

54. Scientists can evaluate trends in biodiversity in a particular ecosystem by measuring changes in the number and kinds of species in that ecosystem over time. **9, 10, 27**

55. We are still learning about ecological systems and the consequences of human actions on these systems. As a result, many people differ in their interpretations of scientific evidence and other data. This often happens in fields that are actively engaged in research. **14, 22, 26**

56. Technology (satellite imagery, advanced computer modeling programs, genetic mapping techniques, and so on) has helped us better understand what generates and maintains biodiversity. Future technological advances could change the way we think about, study, maintain, and restore biodiversity. **23**

57. Because issues related to biodiversity are complex and require the synthesis of information gathered by investigators in different fields, biodiversity research involves professionals with backgrounds in science, sociology, demographics, technology, planning, history, anthropology, mathematics, geography, and other disciplines. **32**

Conserving Biodiversity

An understanding of these concepts can help students learn that although issues surrounding biodiversity loss are complex, they must be resolved by individuals, organizations, and governments with conflicting beliefs and values. Students will recognize their ability to make decisions regarding resource use and the ways those decisions influence biodiversity. They also can identify actions they can take based on their decisions. Finally, students should learn that protecting biodiversity on a large scale takes creativity, innovation, and collaborative thinking.

Role of Values

58. The natural world influences people's core values and how they view their place in the world. Biodiversity supports core values such as health, cultural integrity, economic prosperity, and spirituality. **16, 18**

59. Individuals, organizations, and governments base decisions and actions on underlying values and beliefs. The current and future relationship between the quality of human life and the quality of the natural world will be determined by these decisions. **17, 25, 26**

60. Conflicts in personal values and beliefs often create obstacles to solving problems related to biodiversity use, loss, and conservation. **17, 26, 33**

Roles of Citizens

61. Citizens in a democracy have both a right and a responsibility to participate in the development of policies that influence biodiversity.

62. The issues surrounding biodiversity conservation, use, and loss are often complex and involve many factors. Conserving biodiversity requires input from individuals who have a strong understanding of ecology, genetics, economics, geography, and other subjects, as well as an ability to communicate with and understand others. **31, 32**

63. A citizen, acting individually or as part of a group or organization, can make lifestyle choices (riding a bicycle instead of driving a car) and take actions (writing a letter to an elected official or planting a butterfly garden) that help protect biodiversity. Citizens can also affect the actions of other individuals, families, groups, or organizations to determine how we manage and protect biodiversity. Strategies for protecting biodiversity include ecomanagement (physical action such as cleaning up rivers), education, persuasion, consumer action, political action, and legal action.

Role of Civil Society, Government, and Industry

64. Democratic governments, institutions, and organizations provide opportunities for citizens to participate in collective decision making regarding biodiversity issues.

65. Loss or use of biodiversity can have political, economic, and social consequences. **24, 25, 28**

66. The collective decisions we make (in communities and as a society) reflect a public consensus about the course and conduct of future interactions between humans and the natural world. **29**

67. Educational, legal, economic (business and industry), political, religious, and cultural institutions can help educate people about biodiversity and solve problems related to biodiversity conservation. **26, 33**

68. Conservation of biodiversity takes creativity, innovation, and collaborative thinking on the part of individuals, organizations, governments, and industry.

Future Outlooks for Maintaining and Restoring Biodiversity

By understanding these concepts, students can evaluate how their actions affect the quality of life and the environment in their community, their country, and the world. Students can also predict how scientific, technological, and social changes can influence biodiversity and affect ecosystem health in the future.

69. Choices made today about biodiversity will affect the future quality of life and the environment. **29**

70. It takes less energy and fewer resources to protect biodiversity than it does to restore ecological systems after they have been altered.

71. New ways of managing and protecting biodiversity over the long term, including plans to manage biodiversity at the ecoregional level and new technologies to study and restore biodiversity, are now being developed. **23**

72. All sectors of society influence biodiversity to some extent and can work to protect biodiversity through policy initiatives, media campaigns, corporate mission statements, and other public activities. **31, 32**

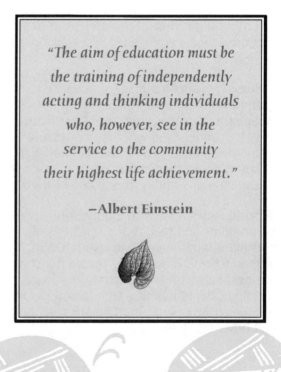

"The aim of education must be the training of independently acting and thinking individuals who, however, see in the service to the community their highest life achievement."

—Albert Einstein

Part II: The Skills Framework

The activities in this *Biodiversity Basics* promote a variety of life-long learning skills—from how to identify bias in information to how to plan and implement a community action project. As you develop teaching units, we encourage you to think about how to combine the concepts and skills in ways that make sense for your group and your objectives.

To help you integrate skill development into the activities, we have emphasized the following eight general skill categories, and the specific skills they contain, throughout this module. The skill categories reflect the current thinking on how students gather, process, and act upon information and what students need to be able to do to become responsible citizens. We believe that all of these skills are necessary for students to be able to think critically, independently, and creatively about biodiversity.

The skills were adapted from those advocated by the Association for Supervision and Curriculum Development (ASCD) and the American Association for the Advancement of Science (AAAS). We also referred to the list of skills in Project WET, a national water curriculum, which used the ASCD and AAAS skills in its materials. In addition, we adapted the civics standards of the Center for Civic Education to come up with the citizenship development skill category, and we used NAAEE's *Environmental Education Guidelines for Excellence: What School-Age Learners Should Know and Be Able to Do* to provide insights into those skills that can help lead to a more environmentally literate and responsible citizenry. If you'd like to focus on certain skills or skill categories, use the matrix beginning on page 432 to identify the major skill categories used in each of the activities. Each activity also lists the key skills in the introductory information. Also, a complete list of skills used in each activity is on pages 446–450.

Gathering Information
reading comprehension, observing, listening, simulating, collecting, researching, interviewing, measuring, recording, identifying main ideas, brainstorming

Organizing Information
matching, plotting data, graphing, sorting, arranging, sequencing, prioritizing, listing, classifying, categorizing, mapping, drawing, charting, manipulating materials

Analyzing Information
identifying components and relationships among components, identifying patterns, comparing and contrasting, questioning, calculating, discussing

Interpreting Information
generalizing, summarizing, translating, relating, inferring, making models, drawing conclusions, defining problems, identifying cause and effect, confirming, reasoning, elaborating

Applying Information
planning, designing, building, constructing, composing, experimenting, restructuring, inventing, estimating, predicting, hypothesizing, proposing solutions, problem solving, decision making, developing and implementing investigations and action plans, synthesizing, creating

Evaluating Information
establishing criteria, verifying, testing, assessing, critiquing, identifying bias

Presenting Information
demonstrating, writing, illustrating, describing, public speaking, acting, reporting, persuading, debating, articulating, explaining, clarifying, making analogies and metaphors

Developing Citizenship Skills
working in a group, debating, compromising, seeking consensus, evaluating a position, taking a position, defending a position, evaluating the need for citizen action, planning and taking action, evaluating the results of action, becoming involved in community decision making

Appendix D

Language Learning Tips

If you have been working with students who need special language assistance, the following tips are probably very familiar to you. However, if you are a new teacher or community educator working for the first time with language learners, you might want to read these tips before starting a biodiversity unit.

1. Developing Strong Listening and Speaking Skills

Students' literacy begins with developing strong listening and speaking skills. To encourage speaking fluency, linguists recommend that you first ignore the students' grammatical mistakes if they do not interfere with understanding. Instead, encourage expression. The following are specific ideas to help you develop your students' ability to express themselves:

- Have students express concepts in more than one way by using both scientific and ordinary language.

- Help students "unpack" the meaning of sentences that have abstract terms by restating them in concrete, everyday language.

- Use structured organizers such as outlines, charts, graphs, symbols, and diagrams; demonstrate what you are discussing whenever possible; and show visually the relationship among terms, processes, or steps by using concrete objects.

- Ask questions often and rephrase the students' questions and answers to summarize what they've learned.

- Use cooperative group activities to help students process difficult information before they tackle individual assignments. Students can ask one another for summaries and explanations of the material, or they can practice posing and responding to specific questions.

2. Building Vocabulary

As students acquire new vocabulary related to biodiversity, it's essential that they know more than a simple definition of words. Terms such as "introduced species" or "conservation" carry with them a set of ideas and values that students need to know before they can truly understand the material. Here are some ways to help students develop vocabulary skills:

- Preview topics and vocabulary when presenting new material. Pictures and other visuals are important aids for understanding.

- Provide vocabulary games or worksheets so students can practice working with key terms.

- Focus attention on technical terms, defining them in language that is familiar. Go beyond defining, making sure students understand the range of issues relevant to the term.

- Pay attention to "cultural" vocabulary that might be unfamiliar to students from some backgrounds. For example, names of foods, current trends, and familiar expressions ("sweet tooth," "green thumb") are often unfamiliar.

- Help students see connections between verbs and nouns (consume/consumption) or adjectives and nouns (extinct/extinction).

- Have students use new terms often. Encourage them to use technical terms when they ask questions, contribute to a discussion, or write about biodiversity topics.

- Have students develop vocabulary journals in which they list any new words they encounter.

3. Creating Readers

Many of the activities in this module contain text that could be a challenge for some of your students. Several ways to help your students get through material that's dense or that contains a lot of technical terms are described below:

- Read aloud to students, and allow them to read aloud to one another (not necessarily in front of the whole class). Hearing a text read aloud while they follow along gives students a chance to hear how words are pronounced and how the bits of information are parceled out. This method can help students make associations between spoken and written forms of language.

- Provide reading strategies for tackling difficult reading passages. For example, you might focus students' attention on headings, provide preview questions, or highlight key points.

- Provide outlines or other reading aids for students to complete as they read.

- Pay attention to terms that occur more often in written language than in spoken language and therefore may be unfamiliar to students. Words used to compare or make causal connections (such as *produces, leads to, causes,* and *results in*) and abstract vocabulary (such as *principle, example, property, explain,* and *generalize*) are important for academic literacy.

4. Developing Writing and Editing Skills

By middle school, students are ready to go beyond the narrative and imaginative writing they did in elementary school to more advanced writing. Many of the activities in this module help develop students' skills in classifying, analyzing cause and effect, expressing their opinions, and justifying their hypotheses. You'll want to focus on different aspects of their writing, depending on the assignment. For example, you might focus on content alone for journal writing, or you might help students with organization, grammar, and vocabulary on more formal reports. In addition, consider the following writing tips:

- Brainstorm with students before they write, helping them develop the vocabulary that they will need for the writing task.

- Assign writing tasks after listening activities and oral discussions. Give students practice in writing descriptions, comparisons, captions, definitions, reports, and other types of writing. Provide examples to help them compare their writing with that of others.

- Present models and outlines of the kind of writing you expect from students. Point out some of the key features of the grammar in the models (for example, that the verbs are in present tense or that there are numerous causal conjunctions such as *if, because,* or *so*).

- Use a process approach to writing that allows students to draft, review, revise, and rewrite the most important assignments. This approach gives them a chance to develop their writing skills by focusing on the same piece of text in different ways.

Appendix E

Activities Overview

What Is Biodiversity?

Activities	At a Glance	Objectives	Subjects
1. All the World's a Web (pages 74-79)	Create a "word web" that illustrates the interconnections in nature.	Define biodiversity and create a word web that illustrates some of the complex connections in the web of life. Discuss at least one way biodiversity affects people's lives.	science, social studies, language arts
2. What's Your Biodiversity IQ? (pages 80-89)	Take a "gee-whiz quiz" to find out how much you know about biodiversity.	Define biodiversity, discuss facts and issues related to biodiversity, and list reasons why biodiversity is important.	science, language arts, social studies
3. Something for Everyone (pages 90-93)	Read a mystery story that introduces biodiversity, match mysterious notes from the story with facts about biodiversity, and list ways biodiversity affects people's everyday lives.	Define biodiversity. Identify and list ways biodiversity affects people.	language arts, science, social studies
4. It's a Wild World (pages 94-97)	Take part in one or more creative activities that introduce the incredible variety of life on Earth.	Define biodiversity, research and discuss amazing biodiversity facts, and describe some of the ways biodiversity affects people's lives.	science, social studies, language arts, art
5. Ten-Minute Mysteries (pages 98-101)	Solve "mysteries" as you discover some amazing and little-known connections among people, species, and ecosystems.	Explain several ways biodiversity is connected to human lives.	language arts, science, social studies
6. "Connect the Creatures" Scavenger Hunt (pages 102-107)	Search for examples of connections among living things and their habitats by taking part in a scavenger hunt.	Describe some of the ways organisms are connected both to their habitats and to other organisms, and give examples of these connections. Discuss specific ways that people are connected to the natural world.	science
7. Sizing Up Species (pages 108-121)	Classify organisms using a classification flow chart, play a team game to find out how many species may exist within different groups of organisms, and make a graph to illustrate the relative abundance of living things.	Use a classification flow chart to classify organisms. Name the major groups of organisms and the relative number of species identified in each group. Construct bar graphs that compare the number of species in different groups of organisms.	science, mathematics
8. Inquiring Minds (pages 122-133)	Focus on scientific inquiry by asking questions, reading about scientific discoveries, making observations, and designing an investigation.	Describe the process of scientific inquiry. Explain how good questions and accurate observations can lead to new discoveries in science. Discuss the difference between objective and subjective observations. Develop a hypothesis and an investigation to test it.	science, social studies

Time	Framework Links	Key Skills	Connections
one session	1, 12, 13	organizing (arranging), analyzing (identifying components and relationships among components), interpreting (relating), applying (creating)	This introductory activity works well either before or after "'Connect the Creatures' Scavenger Hunt" and "Something For Everyone."
one session	1, 2, 3	analyzing (discussing), interpreting (inferring, reasoning), applying (synthesizing)	This introductory activity can be followed by any of the other activities in the module.
two sessions	1, 3, 33, 42	gathering (reading comprehension), organizing (matching), applying (synthesizing)	This introductory activity works well before or after "All the World's a Web." "Ten-Minute Mysteries," "Biodiversity Performs!," and "The Spice of Life" are good follow-ups to this activity.
one session to two weeks	1, 3	gathering (researching), applying (creating), presenting (illustrating, writing)	This introductory activity can be followed by any of the other activities in the module.
one session	33	analyzing (identifying components and relationships among components, identifying patterns)	For a fun way to introduce your students to biodiversity, try "What's Your Biodiversity IQ?" or "Something for Everyone." To learn more about amazing biodiversity connections, try "Super Sleuths," "'Connect the Creatures' Scavenger Hunt," or "The Culture/Nature Connection."
one session	1, 3, 12	gathering (collecting), organizing (manipulating materials), analyzing (identifying components and relationships among components), presenting (describing, illustrating)	Before this activity, you can introduce the concept of biodiversity with "Something for Everyone" or "All the World's a Web." Try "Super Sleuths" to explore more connections among living things. For more on habitats and the species that use them, try "Backyard BioBlitz," "Mapping Biodiversity," or "Space for Species."
two sessions	3, 21	organizing (classifying, estimating, graphing), analyzing (calculating), interpreting (relating)	Use "Biodiversity—It's Evolving" to show how closely related species develop. To explore arthropod diversity in your area, try "Insect Madness." For more on graphing, try "Space for Species."
three sessions	53	gathering (reading comprehension), organizing (arranging, sequencing), analyzing (comparing and contrasting, questioning), citizenship (working in a group)	To reinforce the role of observation in scientific inquiry, try "Backyard BioBlitz" after this activity. You might also want to do this activity before you try "'Connect the Creatures' Scavenger Hunt," "Space for Species," or any of the other outdoor activities in this module because this activity can help your students focus on how to make good observations.

Activities Overview — What Is Biodiversity?

Activities	At a Glance	Objectives	Subjects
9. Backyard BioBlitz (pages 134-143)	Answer an ecoregional survey, then take a firsthand look at biodiversity in your community.	Name several native plants and animals and describe your local environment. Design and carry out a biological inventory of a natural area.	science
10. Insect Madness (pages 144-157)	Create a diversity index for insects in your schoolyard.	Use a simple chart to identify insects from different orders. Measure insect diversity of a site using the Sequential Comparison Index. Discuss how diversity indices are used by scientists.	science, mathematics
11. The Gene Scene (pages 158-167)	Play several different games that introduce genetic diversity and highlight why it's important within populations.	Identify and classify genetic traits using a genetic wheel. Explain why genetic diversity may by necessary for the long-term survival of a population of animals or plants. Explain that lack of genetic diversity is one of the reasons why small and fragmented populations are vulnerable to extinction.	science
12. Biodiversity— It's Evolving (pages 168-179)	Take part in a short simulation about natural selection and analyze a set of fictional journals to see how natural selection can lead to new species.	Define natural selection, evolution, and speciation. Describe how natural selection can lead to changes in a population over time and eventually to the evolution of new species. Describe the relationships between genetic diversity and natural selection, and between evolution and biodiversity.	science, language arts, social studies

Time	Framework Links	Key Skills	Connections
Part I—long term project Part II—two sessions	3, 12, 21, 54	gathering (collecting, observing, researching), citizenship (working in a group)	Follow up with "Insect Madness" to learn more in-depth techniques to measure diversity, or "Inquiring Minds" to explore the scientific inquiry process. To learn more about global classification and surveys, try "Sizing Up Species" or "Mapping Biodiversity."
variable—at least three sessions	3, 12, 21, 54	gathering (collecting), organizing (classifying, manipulating materials), analyzing (calculating), interpreting (drawing conclusions)	For a good introductory outdoor activity, try "Backyard BioBlitz." Use "Sizing Up Species" before this activity to give students background on the numbers and types of species that live on the planet as well as an introduction to the taxonomic system and how insects fit into it.
three sessions	4, 19, 20, 22, 28	gathering (simulating), analyzing (identifying patterns), interpreting (inferring, identifying cause and effect)	For more on the importance of genetic diversity, follow up with "Diversity on Your Table" or "Biodiversity— It's Evolving." To investigate how the loss of genetic diversity affects species, try "The Case of the Florida Panther."
Part I—one to two sessions Part II—one session	3, 4, 12, 19, 20, 21, 22, 23, 28	gathering (simulating), organizing (charting), analyzing (identifying components and relationships among components), interpreting (identifying cause and effect), presenting (explaining)	For an introduction to genes and the importance of genetic diversity, try "The Gene Scene." For more on genetic diversity in food crops, try "Diversity on Your Table." Follow up with "Sizing Up Species" to give your group a sense of how many species inhabit the planet today.

Activities Overview **Why Is Biodiversity Important?**

Activities	At a Glance	Objectives	Subjects
13. Super Sleuths (pages 182-187)	Solve a series of logic problems, word games, and other brain teasers to learn about some of the fascinating relationships that exist among living things.	Give several examples of how species are connected to one another and their habitats. Give an example of a plant that has evolved in a way that ensures pollination and seed dispersal. Draw a food web. Describe the three levels of biodiversity and several adaptations of desert plants and animals.	science, social studies, language arts
14. The F-Files (pages 188-193)	Explore the connection between toxic chemicals and living things as you read a comic strip about an investigation of frogs in Minnesota.	Explain some of the ways biodiversity can be affected by toxic chemicals and other environmental threats. Research environmental health topics using books, journals, interviews, and the Internet. Produce a summary and position statement on an environmental health-related research topic.	science, social studies
15. Biodiversity Performs! (pages 194-197)	Play a charades-like game to learn about some of the "free" services that ecosystems provide.	Work in teams to act out different ecosystem services. Describe several "free" services that biodiversity provides to humans and explain how these services make life on Earth possible.	science, art (drama), language arts
16. The Nature of Poetry (pages 198-205)	Read and discuss several poems related to biodiversity, then write original biodiversity poetry.	Compare several poems that relate to the theme of biodiversity. Recognize the tone, meaning, rhythm, and use of specific language and imagery in poems. Write poetry to convey personal reflections on biodiversity.	language arts
17. The Spice of Life (pages 206-213)	Explore feelings about why biodiversity is important and why it should be protected.	Explain personal beliefs about protecting biodiversity.	language arts, science, social studies (ethics)
18. The Culture/Nature Connection (pages 214-225)	Through a role-play, explore some of the intriguing connections between culture and biodiversity.	Describe connections between nature and human culture. Explore personal heritage and traditions that are connected to the natural world. Examine culture/nature connections by taking part in a role-playing activity.	language arts, social studies
19. Secret Services (pages 226-229)	Perform simulations that demonstrate some of the important ecosystem services that biodiversity provides.	Perform a series of simulations that demonstrate ecosystem services. Identify and discuss the services illustrated in the simulations.	science
20. Diversity on Your Table (pages 230-237)	Play a game to understand the importance of protecting the genetic diversity of food plants, and learn about the ways scientists are working to protect plant diversity.	Role-play a farmer and make decisions about what crops to plant. Identify several reasons that protecting genetic diversity in plants is important, including the consequences to food production when genetic diversity is lost. List examples of how people are protecting genetic diversity in plants today.	science, social studies (economics)

Time	Framework Links	Key Skills	Connections
one session	1, 2, 3, 4, 12	gathering (reading comprehension), interpreting (drawing conclusions, reasoning), presenting (designing)	"Something for Everyone," "All the World's a Web," or "'Connect the Creatures' Scavenger Hunt" can provide good introductions to this activity. For more on the relationship between species and their habitats, try "Space for Species," "Mapping Biodiversity," or "Backyard BioBlitz."
two sessions	12, 35, 52, 55	gathering (reading comprehension, researching), analyzing (questioning), interpreting (identifying cause and effect, relating), applying (problem solving)	The video "Biodiversity! Exploring the Web of Life" introduces the topic of environmental health. Try using "Easter's End" and "The Many Sides of Cotton" to further illustrate connections between environmental health and human health.
one session	2, 8, 10, 11, 12, 25, 28, 33	gathering (reading comprehension), analyzing (identifying patterns), applying (creating), presenting (acting)	After "acting out" ecosystem services in this activity, let your students perform a series of simulations to find out more in "Secret Services." To learn about other ways that humans benefit from biodiversity, try "Something for Everyone," "Ten-Minute Mysteries," "The Culture/Nature Connection," or "Diversity on Your Table."
two sessions	42, 43, 58	gathering (reading comprehension), analyzing (discussing), presenting (writing)	This activity can be enhanced by introducing your students to any of the activities in the "What Is Biodiversity?" or "Why Is Biodiversity Important?" chapters.
one session	37, 40, 59, 60	organizing (prioritizing), analyzing (discussing), presenting (articulating), citizenship (debating, evaluating a position, taking a position, defending a position)	This activity works best once the students have become fairly familiar with biodiversity issues. It is a good culminating activity for a biodiversity unit.
three sessions	5, 18, 40, 41, 42, 43, 58	gathering (researching), analyzing (comparing and contrasting, discussing), presenting (acting)	For more on the ways humans depend on biodiversity, try "Something for Everyone," "Ten-Minute Mysteries," "Biodiversity Performs!," "The Nature of Poetry," or "Diversity on Your Table."
two sessions	2, 12, 25, 28, 33	gathering (simulating), organizing (manipulating materials), interpreting (identifying cause and effect, inferring, making models), presenting (demonstrating, explaining), citizenship (working in a group)	"Biodiversity Performs!" works well before doing this activity. Try "Super Sleuths" to help your students explore some other important services—either before or after this activity. To learn about other ways that humans benefit from biodiversity, try "Something for Everyone," "Ten-Minute Mysteries," "The Culture/Nature Connection," or "Diversity on Your Table."
two sessions	4, 19, 20, 28, 36, 47	gathering (reading comprehension, simulating), analyzing (calculating, identifying patterns), interpreting (inferring), citizenship (working in a group, evaluating the need for citizen action, planning and taking action)	To extend learning about the importance of genetic diversity, use "The Gene Scene" or "Biodiversity—It's Evolving."

Activities Overview   # What's the Status of Biodiversity?

Activities	At a Glance	Objectives	Subjects
21. Endangered Species Gallery Walk (pages 240-245)	Conduct research to create a poster about an endangered species, then take a walk through a poster "gallery" to find out more about endangered species around the world.	Identify groups of animals that are endangered and name some species within each group. Research one species, describe why it's endangered, and compare its problems to those of other endangered species.	science, language arts, social studies
22. The Case of the Florida Panther (pages 246-251)	Work in small groups to discover how the Florida panther's decline is tied to the major causes of biodiversity loss around the world, and discuss what people are doing to help protect the panther.	Describe how habitat loss, introduced species, pollution, population growth, and over-consumption are threatening Florida panthers and biodiversity in general. Discuss ways people are trying to protect the Florida panther.	science, social studies, language arts
23. Mapping Biodiversity (pages 252-265)	Take part in mapping activities to explore the world's ecoregions, and learn how experts make decisions about which natural areas to protect.	Give several examples of how maps help scientists protect biodiversity. Compare the following terms: biome, ecoregion, and habitat. Give several examples of how plants and animals are adapted to specific ecoregions.	science, social studies (geography), art (graphic)
24. Putting a Price Tag on Biodiversity (pages 266-269)	Read an article about re-introducing wolves to Yellowstone National Park as you learn about new thinking in natural resource economics.	Discuss economic and non-economic values of natural resources, and how personal economic values about natural resources influence our decisions.	social studies (economics, ethics)
25. Food for Thought (pages 270-277)	Play an interactive game to compare and contrast how population density, distribution, and resource use affect biodiversity.	Explain how the distribution of population and resources varies among different regions of the world, and how population size and growth combine with resource distribution and use to affect biodiversity.	social studies (economics, ethics, geography), science, mathematics
26. The Many Sides of Cotton (pages 278-285)	Investigate the pros and cons of conventional and organic cotton farming, analyze conflicting opinions, and develop a personal position.	Compare methods for growing organic and conventional cotton. Analyze various essays on the pros and cons of organic cotton. Identify the problem, issues, parties involved, and positions, beliefs, and values of each party. Defend a personal position in support of either traditional or organic cotton or describe what additional information you would need to make a decision.	social studies (economics, ethics), science, language arts,
27. Space for Species (pages 286-301)	Play an outdoor game, conduct a survey of plant diversity, and analyze current research to explore the relationship between habitat size and biodiversity.	Describe factors that affect the relationship between habitat fragmentation and biodiversity loss. Create a graph that demonstrates the relationship between biodiversity and the size of a habitat. Describe different strategies for designing reserves that could help lessen the effects of fragmentation.	science, mathematics
28. Dollars and Sense (pages 302-321)	Using chewing gum as an example, explore some of the connections between economics and biodiversity.	Explain the role of consumers and producers in a market economy, and how supply, demand, and price are interrelated. Describe the connection between what we buy and biodiversity, and give examples of how culture, biodiversity, and economics are linked.	mathematics, social studies (economics), language arts

Time	Framework Links	Key Skills	Connections
two to three sessions	3, 12, 21, 23, 26, 35, 44	gathering (collecting, researching), analyzing (comparing and contrasting), interpreting (defining problems, generalizing), applying (creating, designing), presenting (illustrating, writing)	For more on endangered species, try "The Case of the Florida Panther." To look more closely at the link between species and their habitats, try "Space for Species" or "Mapping Biodiversity." Use "The Spice of Life" to help your students examine their feelings about endangered species protection.
two sessions	1, 21, 22, 23, 29, 35, 38, 44, 46, 47, 48, 52, 55	gathering (reading comprehension), analyzing (comparing and contrasting, discussing), applying (proposing solutions)	Use "Endangered Species Gallery Walk" to learn about other endangered species around the world. To investigate the importance of habitat to species, try "Space for Species." "The Gene Scene" can help your students understand the importance of genetic diversity in small populations. And "The Spice of Life" can help your students think about why it is important to protect species like the Florida panther.
Part I—one to two sessions Part II—one to two sessions	2, 3, 7, 12, 17, 21, 23, 25, 53, 56, 71	gathering (reading comprehension), organizing (manipulating materials, mapping, matching), interpreting (relating), applying (proposing solutions, synthesizing), citizenship (working in a group)	For a look at more species from around the world, try "Endangered Species Gallery Walk," and explore the major causes of biodiversity loss with "The Case of the Florida Panther." To look at some of the other factors important in designing reserves, use "Space for Species."
one session	32, 65	gathering (reading comprehension), applying (decision making), citizenship (evaluating a position, evaluating the results of action, taking and defending a position)	Take your study of economic principles one step further by completing "Dollars and Sense." Use "The Case of the Florida Panther" for a look at a species whose value is being debated.
one to two sessions	39, 45, 51, 59, 65	gathering (simulating), analyzing (comparing and contrasting, identifying patterns), interpreting (identifying cause and effect, relating), applying (decision making, proposing solutions, restructuring), citizenship (working in a group)	Before doing this activity, you might want to try "The Spice of Life" to get your students thinking about why biodiversity is important. After exploring some of the social and economic issues affecting biodiversity, use "Future Worlds" to help your students think about their future and what's important to them.
two sessions	13, 40, 52, 55, 59, 60	gathering (reading comprehension), organizing (charting), analyzing (comparing and contrasting, identifying components and relationships among components), evaluating (identifying a bias), citizenship (evaluating a position, taking and defending a position)	To explore economics and biodiversity, try "Putting a Price Tag on Biodiversity" or "Dollars and Sense." For more on ethical dilemmas and issues analysis, try "What's a Zoo to Do?" And for more on connections between health and the environment, see "The F-Files."
Part I—one to two sessions Part II—two sessions	3, 12, 21, 23, 29, 35, 38, 44, 46, 47, 54	gathering (collecting, simulating), organizing (charting, graphing), analyzing (calculating, identifying patterns), interpreting (inferring), applying (proposing solutions), citizenship (evaluating the need for citizen action, planning and taking action)	For an introductory outdoor activity, try "Backyard BioBlitz." Use "Insect Madness" for more on measuring biodiversity. Follow up with "The Case of the Florida Panther" to investigate the effects of habitat loss and other factors that can cause species to become endangered. For more on designing reserves, try the gap analysis activity in "Mapping Biodiversity."
Part I—one session Part II—one session Part III—one session	30, 31, 32, 65	gathering (reading comprehension), organizing (plotting data, graphing, sequencing), analyzing (calculating, identifying components and relationships among components), interpreting (identifying cause and effect)	For students without an economic background, this could be a challenging activity. Both "Putting a Price Tag on Biodiversity" and "The Many Sides of Cotton" directly relate to different aspects of economics and biodiversity. However, if you want to build a unit that looks at economics and the environment, check out some of the resources listed at the end of this activity that go into more depth on the relationships between economics and environmental issues.

Activities Overview

How Can We Protect Biodiversity?

Activities	At a Glance	Objectives	Subjects
29. Future Worlds (pages 324–329)	Build a pyramid to reflect personal priorities for the future, and investigate ways people are working to protect the environment and improve the quality of life on Earth.	Create a personal vision for the future, especially as it relates to biodiversity. Reach group consensus using negotiation and conflict resolution skills. Brainstorm about ways of arriving at the envisioned future. Analyze various approaches people are taking to arrive at those futures.	social studies (ethics) language arts
30. Easter's End (pages 330–333)	Read an article or a comic strip about Easter Island to get a historical perspective on connections between biodiversity and human survival.	Create a time line that shows changes in Easter Island's biodiversity and human society. Describe the conditions that some people think led to the decline of the human civilization and biodiversity on Easter Island, and discuss the relevance of the story to people today.	language arts, social studies
31. The Biodiversity Campaign (pages 334–337)	Review educational ads focusing on biodiversity, then create your own biodiversity ads to educate others.	Discuss and analyze a variety of public service announcements about biodiversity. Design and create ads to educate others about biodiversity. Apply concepts of good advertising to messages about biodiversity.	art, language arts
32. Career Moves (pages 338–343)	Read profiles of people who work in biodiversity-related professions, then interview people in your community with similar occupations.	Describe various careers related to biodiversity. Identify and discuss biodiversity-related careers held by members of the community. Conduct a successful interview.	science, social studies, language arts
33. What's a Zoo to Do? (pages 344–347)	After exploring various perspectives on controversial issues, come up with ideas to deal with some of the difficult dilemmas that zoos, aquariums, botanical gardens, and other community institutions face.	Explore the ethical issues and challenges surrounding biodiversity dilemmas faced by institutions such as zoos, aquariums, and botanical gardens. Discuss differing viewpoints and articulate personal views regarding how to address these dilemmas. Make recommendations to resolve a particular dilemma.	science, social studies (ethics), language arts
34. Thinking About Tomorrow (pages 348–357)	Take part in a demonstration of unsustainable use of a resource, and use a case study that focuses on drought as you develop a plan to manage a scarce resource.	Define sustainable use of natural resources. Describe several consequences of unsustainable use of natural resources for people and for other species. Recognize the difficulty in identifying ways to use resources when demand and supplies fluctuate. Develop ideas for using water in more sustainable ways.	science, social studies (geography)

Time	Framework Links	Key Skills	Connections
two sessions	37, 40, 66, 69	gathering (brainstorming), organizing (prioritizing), analyzing (comparing and contrasting, discussing), interpreting (relating), presenting (articulating), citizenship (evaluating the result of citizen action, debating, planning citizen action, working in a group)	Introduce your students to the concept of sustainability with "Thinking About Tomorrow." Try "Career Moves" or "The Biodiversity Campaign" to help your students think about the roles they can play in making their futures reality.
two sessions	12, 29, 38, 40, 41, 44, 50	gathering (reading comprehension), organizing (sequencing), analyzing (identifying components and relationships among components), interpreting (relating), presenting (illustrating)	This activity works well after "Thinking About Tomorrow." "Future Worlds" and "Food for Thought" are good follow-up activities that will help you explore the challenges of sustainable resource use.
one to two sessions	62, 72	gathering (reading comprehension), analyzing (identifying components and relationships among components), applying (creating, designing), evaluating (critiquing), presenting (illustrating, persuading, writing), citizenship (evaluating the need for action, planning and taking action)	For more on what people are doing to help protect biodiversity in their personal and professional lives, try "Career Moves" or "Future Worlds" before or after this activity.
variable (much of the activity done outside class time)	6, 57, 62, 72	gathering (interviewing, reading comprehension), analyzing (questioning), interpreting (summarizing), presenting (public speaking)	To help your students think about the future and to introduce them to some of the actions people are taking to create a more sustainable society, try "Future Worlds" before or after this activity.
two to three sessions	6, 40, 60, 67	gathering (reading comprehension), analyzing (identifying components and relationships among components), interpreting (defining problems, summarizing), applying (problem solving, proposing solutions), evaluating (identifying bias), presenting (articulating), citizenship (debating, evaluating, seeking consensus, taking and defending a position, working in a group)	To introduce your students to a variety of viewpoints about biodiversity protection, try "The Spice of Life" before this activity. For other activities dealing with challenging issues and personal opinions, see "The Many Sides of Cotton," "Food for Thought," and "Future Worlds."
Part I—one session Part II—two sessions	29, 38, 50	gathering (simulating, reading comprehension), organizing (sequencing), interpreting (reasoning), applying (planning, problem solving, proposing solutions), presenting (explaining), citizenship (working in a group, evaluating the need for citizen action, planning action, compromising)	This activity provides an introduction to resource use. To further explore the problems and challenges of resource use, try "Food for Thought," "Future Worlds," or "Easter's End."

Appendix E: Subject Matrix — What Is Biodiversity?

Activity	Science	Social Studies	Mathematics	Language Arts	Art
1. All the World's a Web (pages 74-79)	X	X		X	
2. What's Your Biodiversity IQ? (pages 80-89)	X	X		X	
3. Something for Everyone (pages 90-93)	X	X		X	
4. It's a Wild World (pages 94-97)	X	X		X	X
5. Ten-Minute Mysteries (pages 98-101)	X	X		X	
6. "Connect the Creatures" Scavenger Hunt (pages 102-107)	X				
7. Sizing Up Species (pages 108-121)	X		X		
8. Inquiring Minds (pages 122-133)	X	X			
9. Backyard BioBlitz (pages 134-143)	X				
10. Insect Madness (pages 144-157)	X		X		
11. The Gene Scene (pages 158-167)	X				
12. Biodiversity— It's Evolving (pages 168-179)	X	X		X	

Appendix E: Subject Matrix — Why Is Biodiversity Important?

Activity	Science	Social Studies	Mathematics	Language Arts	Art
13. Super Sleuths (pages 182-187)	✕	✕		✕	
14. The F-Files (pages 188-193)	✕	✕			
15. Biodiversity Performs! (pages 194-197)	✕			✕	✕
16. The Nature of Poetry (pages 198-205)				✕	
17. The Spice of Life (pages 206-213)	✕	✕		✕	
18. The Culture/Nature Connection (pages 214-225)		✕		✕	
19. Secret Services (pages 226-229)	✕				
20. Diversity on Your Table (pages 230-237)	✕	✕			

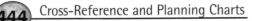

Appendix E: Subject Matrix

What's the Status of Biodiversity?

Activity	Science	Social Studies	Mathematics	Language Arts	Art
21. Endangered Species Gallery Walk (pages 240-245)	✕	✕		✕	
22. The Case of the Florida Panther (pages 246-251)	✕	✕		✕	
23. Mapping Biodiversity (pages 252-265)	✕	✕			✕
24. Putting a Price Tag on Biodiversity (pages 266-269)		✕			
25. Food for Thought (pages 270-277)	✕	✕	✕		
26. The Many Sides of Cotton (pages 278-285)	✕	✕		✕	
27. Space for Species (pages 286-301)	✕		✕		
28. Dollars and Sense (pages 302-321)		✕	✕	✕	

Appendix E: Subject Matrix — How Can We Protect Biodiversity?

Activity	Science	Social Studies	Mathematics	Language Arts	Art
29. Future Worlds (pages 324-329)		X		X	
30. Easter's End (pages 330-333)		X		X	
31. The Biodiversity Campaign (pages 334-337)				X	X
32. Career Moves (pages 338-343)	X	X		X	
33. What's a Zoo to Do? (pages 344-347)	X	X		X	
34. Thinking About Tomorrow (pages 348-357)	X	X			

Appendix E: Skills Matrix — What Is Biodiversity?

Activity	Gathering Information	Organizing Information	Analyzing Information	Interpreting Information	Applying Information	Evaluating Information	Presenting Information	Developing Citizenship Skills
1. All the World's a Web (pages 74-79)		**arranging,*** classifying, drawing	**identifying components and relationships among components,** discussing, identifying patterns	relating, drawing conclusions, inferring	creating, restructuring			
2. What's Your Biodiversity IQ? (pages 80-89)	identifying main ideas, reading comprehension	categorizing, listing	**discussing**	**inferring, reasoning**	**synthesizing**			
3. Something for Everyone (pages 90-93)	**reading comprehension,** recording	**matching**	comparing and contrasting, discussing		**synthesizing**		writing	
4. It's a Wild World (pages 94-97)	**researching,** collecting		identifying components and relationships among components		**creating,** designing, planning		**illustrating, writing,** reporting	
5. Ten-Minute Mysteries (pages 98-101)	listening, reading comprehension		identifying components and relationships among components, **identifying patterns,** discussing	summarizing				
6. "Connect the Creatures" Scavenger Hunt (pages 102-107)	collecting	manipulating materials	identifying components and relationships among components				describing, illustrating	
7. Sizing Up Species (pages 108-121)	reading comprehension	**classifying, estimating, graphing**	calculating, comparing and contrasting, identifying patterns	**relating, reasoning**				working in a group
8. Inquiring Minds (pages 122-133)	**reading comprehension,** observing, researching	**arranging, sequencing**	**comparing and contrasting, questioning,** discussing, identifying components and relationships among components	generalizing	experimenting, hypothesizing, developing and implementing investigations, predicting, synthesizing	testing	describing, illustrating	**working in a group,** seeking consensus
9. Backyard BioBlitz (pages 134-143)	**collecting, observing, researching**	classifying	discussing, questioning		decision making		reporting	**working in a group**

*Key skills are in **boldfaced** type.

Appendix E: Skills Matrix — What Is Biodiversity?

Activity	Gathering Information	Organizing Information	Analyzing Information	Interpreting Information	Applying Information	Evaluating Information	Presenting Information	Developing Citizenship Skills
10. Insect Madness (pages 144-157)	**collecting,** measuring, observing, recording	**classifying, manipulating materials**	**calculating,** comparing and contrasting, discussing	**drawing conclusions**				
11. The Gene Scene (pages 158-167)	**simulating,** observing, reading comprehension		**identifying patterns,** comparing and contrasting, discussing	**identifying cause and effect, inferring,** drawing conclusions				
12. Biodiversity— It's Evolving (pages 168-179)	**simulating,** observing, reading comprehension	**charting,** categorizing	**identifying components and relationships among components,** comparing and contrasting, discussing	**identifying cause and effect,** elaborating, reasoning, relating			**explaining,** describing	

Appendix E: Skills Matrix — Why Is Biodiversity Important?

Activity	Gathering Information	Organizing Information	Analyzing Information	Interpreting Information	Applying Information	Evaluating Information	Presenting Information	Developing Citizenship Skills
13. Super Sleuths (pages 182-187)	**reading comprehension**, brainstorming	charting, drawing, matching, sequencing, sorting		**drawing conclusions, reasoning**			**designing**, illustrating	
14. The F-Files (pages 188-193)	**reading comprehension**, researching		**questioning**, comparing and contrasting	**identifying cause and effect, relating**, inferring	**problem solving**	identifying bias	explaining	evaluating a position, taking a position, working in a group
15. Biodiversity Performs! (pages 194-197)	**reading comprehension**, identifying main ideas		**identifying patterns**	drawing conclusions	**creating**		**acting**	working in a group
16. The Nature of Poetry (pages 198-205)	**reading comprehension**		**discussing, comparing and contrasting, identifying components and relationships among components**				writing	
17. The Spice of Life (pages 206-213)		**prioritizing**, classifying	discussing	reasoning			**articulating**, explaining	debating, defending a position, evaluating a position, taking a position, seeking a consensus, working in a group
18. The Culture/ Nature Connection (pages 214-225)	researching, brainstorming, reading comprehension		**comparing and contrasting**, discussing	relating			**acting**	
19. Secret Services (pages 226-229)	**simulating**, measuring, observing, recording	**manipulating materials**, matching	comparing and contrasting	**identifying cause and effect, inferring, making models**, drawing conclusions, relating, summarizing		testing	**demonstrating**, explaining	**working in a group**
20. Diversity on Your Table (pages 230-237)	reading comprehension, simulating		**calculating, identifying patterns**, comparing and contrasting, discussing	**inferring, defining problems**, drawing conclusions				evaluating the need for citizen action, planning and taking action, working in a group

*Key skills are in **boldfaced** type.

Appendix E: Skills Matrix **What's the Status of Biodiversity?**

Activity	Gathering Information	Organizing Information	Analyzing Information	Interpreting Information	Applying Information	Evaluating Information	Presenting Information	Developing Citizenship Skills
21. Endangered Species Gallery Walk (pages 240-245)	collecting, researching, reading comprehension		comparing and contrasting	defining problems, generalizing, drawing conclusions, identifying cause and effect	relating	creating, designing, constructing		illustrating, writing, explaining
22. The Case of the Florida Panther (pages 246-251)		reading comprehension, brainstorming	categorizing	comparing and contrasting, discussing		proposing solutions		evaluating the need for citizen action, planning action, working in a group
23. Mapping Biodiversity (pages 252-265)	reading comprehension	manipulating materials, mapping, matching, drawing	comparing and contrasting, identifying patterns	relating, inferring	proposing solutions, synthesizing, decision making		describing, public speaking	**working in a group,** evaluating the need for citizen action, evaluating the results of action, planning action
24. Putting a Price Tag on Biodiversity (pages 266-269)	reading comprehension, brainstorming		discussing	comparing and contrasting, defining problems, inferring, reasoning	decision making			evaluating a position, evaluating the results of action, taking and defending a position, seeking consensus
25. Food for Thought (pages 270-277)	simulating		comparing and contrasting, identifying patterns, calculating, discussing, identifying cause and effect	identifying cause and effect, relating, inferring	decision making, proposing solutions, restructuring, predicting			working in a group, compromising, debating
26. The Many Sides of Cotton (pages 278-285)	reading comprehension	charting, categorizing	comparing and contrasting, identifying components and relationships among components			identifying bias		evaluating a position, taking and defending a position
27. Space for Species (pages 286-301)	collecting, simulating, listening	charting, graphing, classifying	calculating, identifying **patterns,** comparing and contrasting	inferring, relating	proposing solutions, decision making	assessing		evaluating the need for citizen action, planning and taking action
28. Dollars and Sense (pages 302-321)	reading comprehension	graphing, plotting data, sequencing, manipulating materials, matching	calculating, identifying components and relationships among components, comparing and contrasting	identifying cause and effect				

Appendix E: Skills Matrix — How Can We Protect Biodiversity?

Activity	Gathering Information	Organizing Information	Analyzing Information	Interpreting Information	Applying Information	Evaluating Information	Presenting Information	Developing Citizenship skills
29. Future Worlds (pages 324–329)	brainstorming, identifying main ideas, reading comprehension	prioritizing, manipulating materials	comparing and contrasting, discussing	relating		establishing criteria	articulating	debating, evaluating the results of citizen action, planning citizen action, seeking consensus, working in a group, compromising
30. Easter's End (pages 330–333)	reading comprehension	sequencing, drawing	identifying components and relationships among components, discussing	relating	hypothesizing		illustrating	
31. The Biodiversity Campaign (pages 334–337)	reading comprehension, brainstorming		identifying components and relationships among components, discussing		creating, designing	critiquing, establishing criteria	illustrating, persuading, writing	evaluating the need for action, planning and taking action
32. Career Moves (pages 338–343)	interviewing, reading comprehension, brainstorming		questioning, discussing	summarizing			public speaking	
33. What's a Zoo to Do? (pages 344–347)	reading comprehension		identifying components and relationships among components, discussing, questioning	defining problems, summarizing, reasoning	problem solving, proposing solutions	identifying bias	articulating, describing	debating, evaluating, seeking consensus, taking and defending a position, working in a group
34. Thinking About Tomorrow (pages 348–357)	reading comprehension, simulating, brainstorming, measuring	sequencing, charting, sorting	calculating, comparing and contrasting, discussing, identifying components and relationships among components	reasoning, inferring, relating	planning, problem solving, proposing solutions		explaining	compromising, evaluating the need for citizen action, planning action, working in a group, debating

Cristy Holloway

*"As far as we know, life is the one feature that makes
our planet unique, and life's vast diversity is perhaps its most
impressive trait. Most of the world's political, economic,
and health problems, present and future, are intimately linked to the
way we manage the world's immense variety of wildlife and natural resources.
It is not too late to save a large amount of the Earth's remaining
biodiversity, but time is running very short."*

—**Andrew P. Dobson, professor, writer**

Appendix F

Resources

The following resources can help you design and enhance a biodiversity program or unit. We've included organizations, books, curriculum guides, audiotapes, posters, Web sites, CD-ROMs, and videodiscs. Keep in mind that this resource list includes the best materials we have found or used; however, there are many other resources available on biodiversity.
Please note that we've included prices when available; but all costs are subject to change and don't include shipping or handling fees. Books that are out of print may be found by checking the Internet, libraries, and book stores. (We accessed the Web sites in 1998, but Web addresses change often.)

Organizations

The following organizations are involved in activities that protect biodiversity and educate people about biodiversity issues. They can help you and your students find up-to-date information, curriculum guides, research projects, and other important environmental education resources.

Nonprofit Conservation Organizations

American Museum of Natural History
illuminates the natural history of our planet and promotes universal scientific literacy. Central Park West at 79th St., New York, NY 10024. (212) 769-5000. Web site: <www.amnh.org>.

Bat Conservation International
promotes bat conservation projects worldwide. P.O. Box 162603, Austin, TX 78716. (512) 327-9721. Web site: <www.batcon.org>.

Center for Marine Conservation, Inc.
focuses on protecting marine wildlife and habitats and on conserving coastal and ocean resources. It publishes many types of educational materials for classrooms and young people. 1725 DeSales St., NW, Ste. 600, Washington, DC 20036. (202) 429-5609.
Web site: <www.cmc-ocean.org>.

Center for Plant Conservation
is committed to saving rare and endangered plants native to the United States. P.O. Box 299, St. Louis, MO 63166. (314) 577-9450. Web site: <www.mobot.org/CPC/>.

Chesapeake Bay Foundation
focuses on the management of the Chesapeake Bay. It operates an environmental education program for students, teachers, and the general public. 162 Prince George St., Annapolis, MD 21401. (410) 268-8816. Web site: <www.savethebay.cbf.org>.

The Field Museum
is concerned with the diversity and relationships in nature and among cultures. It provides collection-based research and learning for greater public understanding and appreciation of the world. Roosevelt Rd. at Lake Shore Dr., Chicago, IL 60605. (312) 922-9410.
Web site: <www.fmnh.org>.

Conservation International
is dedicated to the preservation of tropical and temperate ecosystems. 2501 M St., NW, Ste. 200, Washington, DC 20037. (800) 429-5660. Web site: <www.conservation.org.htm>.

The Cousteau Society, Inc.
is dedicated to the protection of marine biodiversity and to the improvement of the quality of life for present and future generations.

It publishes *Dolphin Log* for young readers. 870 Greenbrier Cir., Ste. 402, Chesapeake, VA 23320. (757) 523-9335. Web site: <www.scubaworld.com/cousteau>.

Defenders of Wildlife
focuses on endangered species, habitat conservation, and predator protection. 1101 14th St., NW, Ste. 1400, Washington, DC 20005. (202) 682-9400. Web site: <www.defenders.org>.

Ducks Unlimited, Inc.
focuses on the conservation of waterfowl and wetlands. Children's memberships include *Puddler* magazine. One Waterfowl Way, Memphis, TN 38120. (800) 453-8257. Web site: <www.ducks.org>.

Earth Foundation
strives to empower educators and students to work toward a sustainable economy, a just society, and a healthy environment. It publishes the *Rainforest Rescue Campaign Teacher Update*. 5151 Mitchelldale, Ste. B-11, Houston, TX 77092. (800) 566-6539. Web site: <www.earthfound.com>.

Environmental Defense Fund, Inc.
focuses on global issues such as ocean pollution, rain forest destruction, and global warming. 257 Park Ave. South, New York, NY 10010. (212) 505-2100. Web site: <www.edf.org>.

Friends of the Earth

is dedicated to protecting the planet from environmental disaster and to preserving biological, cultural, and ethnic diversity. The Global Building, 1025 Vermont Ave., NW, 3rd Floor, Washington, DC 20005-6303. (202) 783-7400. Web site: <www.foe.org>.

Greenpeace, Inc.

is dedicated to preserving the Earth and the life it supports, and to creating a green and peaceful world. 1436 U St., NW, Washington, DC 20009. (800) 326-0959. Web site: <www.greenpeaceusa.org>.

International Crane Foundation

works for the preservation of cranes through research, conservation, captive propagation, and public education. It publishes several curriculum packets and other educational materials. P.O. Box 447, Baraboo, WI 53913-0447. (608) 356-9462. Web site: <www. baraboo.com/bus/icf/whowhat.htm>.

International Wolf Center

supports worldwide environmental education about wolves and produces various educational pamphlets. 1396 Hwy. 169, Ely, MN 55731. (218) 365-4695 or (800) ELY-WOLF. Web site: <www.wolf.org>.

Izaak Walton League of America

promotes public education aimed at conserving and responsibly enjoying natural resources. 707 Conservation La., Gaithersburg, MD 20878. (800) 453-5463. Web site: <www.iwla.org>.

Monterey Bay Aquarium

seeks to stimulate interest, increase knowledge, and promote stewardship of Monterey Bay and the world's ocean environment. 886 Cannery Row, Monterey, CA 93940. (831) 648-4888. Web site: <www.mbayaq.org>.

National Audubon Society

is a national organization that uses its state offices to promote wildlife conservation and the protection of endangered ecosystems throughout the world. 700 Broadway, New York, NY 10003. (212) 979-3000. Web site: <www.audubon.org>.

National Environmental Trust

strives to educate the public and Congress about the status of the environment. 1200 18th St., NW, Ste. 500, Washington, DC 20036. (202) 887-8800. Web site: <www.envirotrust.,com>.

National Geographic Society

promotes geography education. It publishes *National Geographic World* for children and provides classroom materials. 1145 17th St. NW, Washington, DC 20036. (202) 857-7000. Web site: <www.nationalgeographic.com>.

National Parks and Conservation Association

a citizen organization with regional offices, is dedicated to preserving, protecting, and enhancing the U.S. National Park Service. 1776 Massachusetts Ave., NW, Ste. 200, Washington, DC 20036. (800) NAT-PARK. Web site: <www.npca.org>.

National Wildlife Federation (NWF)

is a national organization with regional offices and resource centers dedicated to the conservation of wildlife and other natural resources. It has many programs for children and educators, including *Ranger Rick, NatureScope* (environmental education activity guide series), and outdoor programs for families, students, and educators. 8925 Leesburg Pike, Vienna, VA 22184. (703) 790-4000. For a catalog of educational materials, call (800) 822-9919. Web site: <www.nwf.org>.

The Nature Conservancy (TNC)

is a national organization with regional offices committed to preserving biological diversity. It works with states through its natural heritage programs to identify and protect important natural areas. 4245 N. Fairfax Dr., Ste. 100, Arlington, VA 22203–1606. (703) 841-5300. Web site: <www.tnc.org>.

Rainforest Action Network

works internationally with other environmental organizations to protect rain forests and the human rights of people living in them. It publishes educational materials. 221 Pine St., Ste. 500, San Francisco, CA 94104. (415) 398-4404. Web site: <www.ran.org>.

Rainforest Alliance

is an international organization dedicated to the conservation of the world's endangered tropical forests. 65 Bleecker St., New York, NY 10012. (888) MY-EARTH. Web site: <www.rainforest-alliance.org>.

Rocky Mountain Institute

uses research and education to foster the efficient and sustainable use of resources. 1739 Snowmass Creek Rd., Snowmass, CO 81654. (970) 927-3851. Web site: <www.rmi.org>.

Sierra Club

uses its many chapters throughout North America to advocate the preservation of wilderness and open spaces. 85 Second St., 2nd Fl., San Francisco, CA 94105-3441. (415) 977-5500. Web site: <www.sierraclub.org>.

The Wilderness Society

is an organization devoted to preserving wilderness and wildlife. 900 17th St., NW, Washington, DC 20006. (202) 833-2300. Web site: <www.wilderness.org>.

Wildlife Preservation Trust International, Inc.

works to preserve endangered species through captive breeding, research, education, and fieldwork. It provides classroom materials and a children's membership, which includes *Dodo Dispatch* magazine. 1520 Locust St., Ste. 704, Philadelphia, PA 19102. (215) 731-9770. Web site: <www.cc.columbia.edu/cu/cerc /wpti.html>.

World Resources Institute (WRI)

is a policy research center that aids governments, international organizations, and the private sector in addressing vital issues of environmental integrity, natural resource management, and economic growth. 1709 New York Ave., NW, Washington, DC 20006. (202) 638-6300. Web site: <www.wri.org>.

World Wildlife Fund (WWF)

works worldwide in more than 100 countries to protect the diversity of life on Earth. WWF also supports a variety of education and community outreach programs, including *Windows on the Wild*. 1250 24th St., NW, Washington, DC 20037. (202) 293-4800. Web site: <www.worldwildlife.org>.

Zero Population Growth, Inc.

works to educate people about global population growth. It publishes a variety of educational materials. 1400 16th St., NW, Ste. 320, Washington, DC 20036. (202) 332-2200. Web site: <www.zpg.org>.

Note: Many zoos, aquariums, and other organizations offer a variety of environmental education programs, materials, and information about biodiversity.

Professional Associations

American Association of Museums (AAM),

in order to better serve the public, is dedicated to promoting excellence within the museum community through advocacy, professional education, information exchange, accreditation, and guidance on current professional standards of performance. 1575 Eye St., NW, Ste. 400, Washington, DC 20005. (202) 289-1818. Web site: <www.aam-us.org>.

American Zoo and Aquarium Association

promotes the advancement of zoos and aquariums. It focuses on the conservation of wildlife and the preservation and propagation of endangered and rare species. 8403 Colesville Rd., Ste. 710, Silver Spring, MD 20910. (301) 562-0777. Web site: <www.aza.org>.

National Association of Biology Teachers

focuses on improving the teaching of biology in schools across the country. 11250 Roger Bacon Dr. #19, Reston, VA 20190-5202. (703) 471-1134. Web site: <www.nabt.org>.

National Association for Interpretation (NAI)

is dedicated to the advancement of the profession of interpretation (on-site informal education programs at parks, zoos, nature centers, museums, and aquaria). P.O. Box 2246, Fort Collins, CO 80522-9937. (888) 900-8283. Web site: <www.interpnet.com>.

National Council for Geographic Education

promotes geography education in the schools and colleges of the United States and Canada. 16A Leonard Hall, Indiana University of Pennsylvania, Indiana, PA 15705. (724) 357-6290. Web site: <www.ncge.org>.

National Science Teachers Association

is an education organization that works to improve the teaching of science in the United States. 1840 Wilson Blvd., Arlington, VA 22201. (703) 243-7100. Web site: <www.nsta.org>.

North American Association for Environmental Education

supports the work of professional environmental educators and students in North America and more than 55 countries around the world. It publishes a bimonthly newsletter and many other environmental education resources, and it sponsors an annual conference. 1825 Connecticut Ave., NW, Ste. 800, Washington, DC 20009. (202) 884-8912. Web site: <www.naaee.org>.

Wilderness Education Association

promotes wilderness education and preservation through wilderness leadership training. 1101 Otter Creek Rd., Nashville, TN 37220. (615) 331-5739. Web site: <www.ebl.org/wea>.

Government Agencies and Organizations

Bureau of Land Management

is an agency within the U.S. Department of the Interior that manages a portion of the public domain land. It promotes biological diversity (nature's variety of life forms and processes) and sustainable development (the long-term use of natural resources without damaging the environment). 1849 C St., NW, Rm. 406-LS, Washington, DC 20240. (202) 452-5125. Web site: <www.blm.gov>.

National Park Service (NPS)

is an agency within the U.S. Department of the Interior that administers parks, monuments, the Wild and Scenic Rivers System, the National Trail System, and other areas of national significance. The NPS Web site contains information for educators on locating curriculum materials, videos, accredited teacher training seminars, traveling trunks and kits, and teacher and student resource packets. Interior Bldg., P.O. Box 37127, Washington, DC 20013. (202) 208-4747. Web site: <www.nps.gov>.

Smithsonian Institution

is an independent trust established to develop national collections in natural history and anthropology, to conduct scientific research, and to produce publications and programs in conservation, education, and training. Washington, DC 20560. (202) 357-2700. Web site: <www.si.edu>.

U.S. Environmental Protection Agency (EPA)

sets and enforces environmental standards and conducts research on the causes, effects, and control of environmental problems. It supports environmental education programs and projects through the Office of Environmental Education. 401 M St., SW, Washington, DC 20460. (202) 260-2090. Web site: <www.epa.gov>.

U.S. Fish and Wildlife Service

is part of the Department of the Interior and was created to conserve fish and wildlife resources in the United States. The Web site provides information for educators on wildlife laws, environmental education, and the national wildlife refuge system. It also includes extensive database access for searches on wildlife information, including a current list of endangered and threatened species. 1849 C St., NW, Washington, DC 20240. (202) 208-4131. Web site: <www.fws.gov>.

U.S. Forest Service

is an agency of the Department of Agriculture that manages public lands in national forests and grass-lands. The Web site includes information on forests and forestry for educators and students, as well as publications, maps, research information, and links to other federal Web sites. Auditors Bldg. 201 14th St. at Independence Ave., SW, Washington, DC 20250. (202) 205-1760. Web site: <www.fs.fed.us>.

Environmental Education Organizations and Programs

Center for Environmental Information, Inc.

provides an on-call reference and referral service, current information, and educational materials to educators, scientists, and other interested groups. It maintains a special library collection that can be reached at (716) 271-3550. 55 St. Paul St., Rochester, NY 14604. (716) 262-2870. Web site: <www.awa.com/nature/cei>.

Council for Environmental Education (CEE),

formerly the Western Regional Environmental Education Council (WREEC), is an educational organization that produces educational materials and sponsors workshops (including Project WILD). It comprises representatives of state-level departments of education and natural resource-related agencies in the United States. 5555 Morningside Dr., Ste. 212, Houston, TX 77005. (713) 520-1936.

Earth Force

is an educational organization that focuses on environmental action projects, service learning, and national education campaigns. 1908 Mt. Vernon Ave., Alexandria, VA 22301. (703) 299-9400. Web site: <www.earthforce.org>.

Environmental Resource Center

provides resources and educational programs to the public about local, regional, and global environmental issues. P.O. Box 819, Ketchum, ID 83340. (208) 726-4333. Web site: <www.basemountain. com/erc_kids.htm>.

Global Rivers Environmental Education Network (GREEN),

an international network of students, teachers, and institutions, conducts comprehensive watershed education programs designed to improve water quality around the world. 206 S. Fifth Ave., Ste. 150 Ann Arbor, MI 48104. (734) 761-8142. Web site: <www.igc.org/green>.

The Institute for Earth Education

develops and disseminates educational programs to promote understanding of the Earth's natural systems and communities. It also publishes books and program materials. Cedar Cove, Greenville, WV 24945. (304) 832-6404. Web site: <www.slnet.com/cip/iee>.

Project Learning Tree (PLT)

is an award-winning, interdisciplinary, environmental education program designed to educate young people about the environment. 1111 19th St., NW, Ste. 780, Washington, DC 20036. (202) 463-2462. Web site: <www.plt.org>.

Project WET

facilitates and promotes awareness, appreciation, knowledge, and stewardship of water resources through the development and dissemination of curriculum materials and through the establishment of state and internationally sponsored Project WET programs. 201 Culbertson Hall, Montana State University, Bozeman, MT 59717. (406) 994-5392. Web site: <www.montana.edu/wwwwet>.

Project WILD

is a national environmental education program for educators that emphasizes wildlife conservation, ecology, and interdisciplinary teaching. 707 Conservation La., Ste. 305, Gaithersburg, MD 20878. (301) 527-8900. Web site: <www.projectwild.org>.

Windows on the Wild (WOW)

is an international environmental education program from WWF designed to educate people of all ages about biodiversity issues and to stimulate critical thinking, discussion, and informed decision making on behalf of the environment. Through educational materials, training programs, and special events and programs, *Windows on the Wild* uses biodiversity as the organizing theme to help create an environmentally literate citizenry that has the knowledge, skills, and commitment to make responsible decisions about the environment. 1250 24th St., NW, Washington, DC 20037. (202) 293-4800. Web site: <www.wwf.org/windows/frame_windows.htm>.

World Resources Institute (WRI)

produces a variety of educational materials for high school educators on topics such as biodiversity, sustainablility, communities, and population. 1709 New York Ave., NW, Washington, DC 20006. (202) 638-6300. Web site: <www.wri.org>.

Biodiversity Resources

The following are general books about different aspects of biodiversity to provide background information for educators. Some might be appropriate for older students, but most are adult resources.

Atlas of Endangered Resources

by Steve Pollock discusses the status of resources such as energy, water, air, soil, minerals, and biodiversity. (Facts on File, 1995). $16.95

Atlas of the Environment

by Roger Coote looks at global environmental problems. (Raintree Steck-Vaughn, 1992). $28.54

Conservation and Biodiversity

by Andrew P. Dobson provides an introduction to the issues surrounding biodiversity. (Scientific American Library, 1996). $32.95

Conserving the World's Biodiversity

by Jeffrey McNeely, Kenton Miller, and Walter Reid explains what biodiversity is, why it's important, and how to conserve it. (Island Press, 1990). $14.95

The Diversity of Life

by Edward O. Wilson provides an excellent introduction to biodiversity, looking at species diversity, the causes and consequences of biodiversity loss, and the ways people can address the crisis. (Harvard University Press, 1992). $22.25

The Endangered Animals Series

by Dave Taylor includes books about animals from endangered grasslands, forests, wetlands, mountains, islands, oceans, deserts, and savannas. (Crabtree Publishing, 1992–93). $6.36/species

Endangered Species: Opposing Viewpoints

edited by David Bender, Bruno Leone, and Brenda Stalcup explores viewpoints concerning endangered species in a series of papers by various authors. (Greenhaven Press, 1996). $16.20

Environment

by Peter H. Raven, Linda R. Berg, and George B. Johnson is a college-level text that provides an interdisciplinary overview of environmental issues. (Saunders College Publishing, 1998).

Ghost Bears: Exploring the Biodiversity Crisis

by R. Edward Grumbine explores the causes and effects of habitat destruction and of species loss by examining the plight of the grizzly bear in the Greater North Cascade Mountains. (Island Press, 1992). $12.57

Green Planet Rescue: Saving the Earth's Endangered Plants

by Robert R. Halpern presents plants as a renewable energy resource, a source for medicine, and a solution to world hunger. It also provides an explanation of pollination and of the relationships among plants, as well as the relationships among plants, insects, birds, and bats. (Watts, 1993). $23.60

The Idea of Biodiversity: Philosophies of Paradise

by David Takacs analyzes what biodiversity represents to the biologists who work in the field of conservation. (Johns Hopkins University Press, 1996). $35.95

The Last Harvest: The Genetic Gamble That Threatens to Destroy American Agriculture

by Paul Raeburn explains how our food supply is threatened by genetic manipulation, and how losing biodiversity in the world's food supply affects us. (Simon & Schuster, 1995). $9.60

Living in the Environment: Principles, Connections, and Solutions

by G. Tyler Miller Jr. examines environmental science as an interdisciplinary study and presents a general idea of how nature works and how things are interconnected. (Wadsworth Publishing Company, 1996). $83.95

Lost Wild America: The Story of Our Extinct and Vanishing Wildlife

by Robert M. McClung chronicles the past, present, and future of American species in the context of history, moral attitudes, and politics. (Linnet Books, 1993). $27.50

National Geographic: Biodiversity—The Fragile Web

highlights various aspects of bio-diversity, including an article on the basics of what biodiversity is and why it's being lost, as well as articles focusing on the genetic, species, and ecosystem levels of biodiversity. Also includes full-color map insert. (*National Geographic* 195, No. 2 February 1999).

Nature's Services: Societal Dependence on Natural Ecosystems

by Gretchen C. Daily compiles writings on the relationship between humans and the biosphere. It also describes the services provided by ecosystems, ranging from flood control and soil fertility to new pharmaceuticals and food resources. (Island Press, 1997). $49.95

Our Ecological Footprint: Reducing Human Impact on the Earth

by Mathis Wackermagel and William Rees converts seemingly complex eco-logical concepts into an easy-to-understand format. (New Society, 1996). $11.96

Rescue Mission— Planet Earth: A Children's Edition of Agenda 21

in association with the United Nations, is a follow-up of the 1992 Earth Summit. It also displays the artwork, writings, and visions of children about the environment and sustainable development. This resource addresses the natural world, the human world, and the world of science and policy, and offers ideas for young people to get involved. (Kingfisher Books, Griswood & Dempsey, 1994). $9.95 (paperback), $14.95 (hardback)

Saving Nature's Legacy: Protecting and Restoring Biodiversity

by Reed F. Noss and Allen Y. Cooperrider provides a thorough and readable introduction to issues of land management and conservation biology by two leading conservation biologists. (Island Press, 1994). $29.95

The Value of Life: Biological Diversity and Human Society

by Stephen Kellert draws on 20 years of research to consider how factors such as gender, age, ethnicity, occupation, and geographic location affect our values regarding nature and biodiversity. (Island Press, 1996). $13.56

General Education Resources

The following resources highlight special topics in education that promote active citizenship, teamwork, and life-long learning.

Cooperative Learning

Cooperative Learning and the Collaborative School

by Ronald Brandt, ed., compiled from articles previously published in *Educational Leadership*, examines cooperative learning as it has affected teachers, students, and administrators. (Association for Supervision and Curriculum Development, 1991). $21.95

Cooperative Learning in the Classroom

by David W. Johnson, Roger T. Johnson, and Edythe J. Holubec. Based on years of research and experience, provides a wealth of strategies for what to do before, during, and after cooperative learning lessons to maximize learning. (Association for Supervision and Curriculum Development, 1994). $13.95

Constructivism

A Room with a Different View: First Through Third Graders Build Community and Create Curriculum

by Jill Ostrow describes a case study of a successful constructivist classroom and demonstrates constructivist learning theories in practice. It also illustrates how cooperative learning and problem solving take place in a multi-age classroom. (Stenhouse Publishers, 1995). $20.00

In Search of Understanding: The Case for Constructivist Classrooms

by Jacqueline Grennon Brooks and Martin G. Brooks provides a thorough introduction to constructivist education along with suggestions for ways to create constructivist settings in a classroom. (Association for Supervision and Curriculum Development, 1993). $16.95

Education for Sustainability

Learning for a Sustainable Environment: A Professional Development Guide for Teacher Educators

by John Fien, Debbie Heck, and Jo-Anne Ferreira, eds., provides action research strategies for educating for sustainability. Topics include experiential learning, storytelling for the environment, indigenous knowledge, values education, learning outside the classroom, community problem solving, and more. (UNESCO Asia-Pacific Centre of Educational Innovation for Development, 1997).

Service Learning and Action Research

Enriching the Curriculum Through Service Learning

by Carol W. Kinsley and Kate McPherson, eds., provides examples of effective service learning programs from elementary, middle, and secondary schools across the United States. The book focuses on how service learning can create opportunities for students to gain new skills and knowledge while connecting with the community. (Association for Supervision and Curriculum Development, 1995). $15.95

Environmental Education for Empowerment: Action Research and Community Problem Solving

by William B. Stapp, Arjen E. J. Wals, and Sheri L. Stankorb enables students, teachers, administrators, and others to effectively participate in the planning, implementation, and evaluation of educational activities aimed at resolving an environmental issue that they have identified. (Kendall/Hunt Publishing, 1996). $19.95

Character and Values Education

The Character Education Partnership

has a publications list and brochure titled, "Character Education Questions and Answers." 809 Franklin Street, Alexandria, VA 22314-4105. (703) 739-9515.

Environmental Values Education: An Exploration of its Role in the School Curriculum

by William Scott and Chris Oulton discusses the role and contribution of environmental values education in the moral development of individuals and society. The paper comments on and challenges conventional thinking about environmental education's place within a social framework. (Journal of Moral Education, 1998).

A Guide on Environmental Values

by Michael Caduto explains why environmental values education

is important and provides a basic understanding of human values and behavior. It also contains practical tips for educators in designing, implementing, and evaluating an environmental values education program. (UNESCO, 1985).

In Accord with Nature: Helping Students Form an Environmental Ethic Using Outdoor Experience and Reflection

by Clifford E. Knapp is a book designed to help educators assist their students in forming an environmental ethic. It presents a combination of information, theory, and practice dealing with environmental values and ethics, and includes strategies to illustrate how outdoor experiences and reflection can contribute to the process. (WVA: ERIC/CRESS, 1999). $19.00

Their Best Selves—Building Character Education and Service Learning Together in the Lives of Young People

by Bruce O. Boston is a report from the Council of Chief State School Officers, the Association for Supervision and Curriculum Development, the Character Education Partnership, the Close-Up Foundation, Earth Force, and the National Society for Experimental Education. (Character Education Partnership, 1997). First copy is free, additional copies are $3.50; to order call (202) 296-7748.

Futures Education

Educating for the Future: A Practical Classroom Guide

by David Hicks provides theoretical and practical information on futures education for K–12 teachers. It includes activities that help students develop a more future-oriented perspective, imagine alternative sustainable futures, exercise critical thinking and decision-making skills, and engage in active and responsible citizenship. (WWF-UK, 1994). $23.00

Multiple Intelligences

Frames of Mind: The Theory of Multiple Intelligences

by Howard Gardner describes the theory of multiple intelligences, explains the basic intelligences originally studied by Gardner, and offers ideas for incorporating these in classroom applications. (HarperCollins Publishers, 1983). $14.00

Multiple Intelligences in the Classroom

by Thomas Armstrong helps educators identify, nurture, and support the unique capabilities of every student. It provides clear explanations and practical advice on introducing students to Howard Gardner's theory of multiple intelligences. (Association for Supervision and Curriculum Development, 1994). $17.95

Learning Styles

4MAT and Science: Toward Wholeness in Science Education

by Bob Samples, Bill Hammond, and Bernice McCarthy explains McCarthy's 4MAT learning styles system and how it can be used to teach science holistically. It provides sample activities and units along with tips on how to implement the 4MAT system with various science topics. (EXCEL, Inc., 1985).

How to Implement and Supervise a Learning Style Program

by Rita Dunn is a practical guide for educators interested in understanding and implementing a learning style program. It explains learning styles and describes strategies for teaching students with diverse styles. It also discusses strategies for staff development and introducing staff, community, and students to learning styles. (Association for Supervision and Curriculum Development, 1996). $10.95

Critical Thinking and Problem Solving

Environmental Problem Solving: Theory, Practice, and Possibilities in Environmental Education

by L. Bardwell, M. Monroe, and M. Tudor, eds., draws from top leaders in the field. This book unites problem solving theory with its practice in EE, provides a useful framework to explain how learners approach problems, and details how to teach the skills needed to become problem solvers. (North American Association for Environmental Education, 1994). $19.00

Teaching Thinking: Readings from Educational Leadership

by Ronald Brandt, ed., includes a variety of articles from leading educators in the U.S. focusing on innovative strategies for teaching students how to improve thinking skills. It provides a good overview of the field, with tips for measuring success, developing lesson plans, and integrating thinking skills in all subject areas. (Association for Supervision and Curriculum Development, 1989). $21.95

Language Learning

The CALLA Handbook: Implementing the Cognitive Academic Language Learning Approach

by Anna Uhl Chamot and J. Michael O'Malley provides practical advice on how to help language learners acquire the language skills they need for academic learning. It includes lots of tips and aids to help teachers design lessons in a variety of content areas. (Addison-Wesley, 1994). $38.00

The Multicultural Classroom: Readings for Content-Area Teachers

by Patricia A. Richard-Amato and Marguerite Ann Snow, eds., showcases articles by leaders in the field of ESL, covers topics that include classroom interaction with language learners, cultural issues, and the teaching of listening, speaking, reading, and writing. It has chapters on making language comprehensible in various subject areas, including science and social studies. (Longman, 1992). $20.63

Teaching Science to English Learners, Grades 4–8 (Program Information Guide Series, No. 11)

by Ann K. Fathman, Mary Ellen Quinn, and Carolyn Kessler provides tips on teaching science and includes model lessons. Drawing on principles and guidelines from the AAAS on scientific literacy, it reviews the constructivist approach to science teaching. (National Clearinghouse on Bilingual Education, 1992). $4.50

Curriculum Resources

The following curriculum resources cover a wide range of biodiversity-related topics and are appropriate for elementary through secondary students. For a more complete review of many of them, see *The Biodiversity Collection— A Review of Biodiversity Resources for Educators.* (Available for $11.95 from Acorn Naturalists. (800) 422-8886.)

Activities for the Changing Earth System (Middle School–Secondary)

comprises a set of activities on global-change issues, including biodiversity, greenhouse effect and global warming, ozone depletion, freshwater resources, deforestation, climate modeling, volcanic eruptions, and global climate change. (1993). Available from Earth Systems Education Program, The Ohio State University, 2021 Cotley Rd., Columbus, OH 43210-1078. (614) 292-1078. $10.00

Adaptations (Middle School)

is one unit of the Regional Environmental Education Program developed by 200 teachers and 6 nature/environmental centers in southeastern Pennsylvania. It contains 10 lessons and examines adaptations in the context of threatened, endangered, and extinct species in tropical rain forests. (1995). Available from The Schuylkill Center for Environmental Education, 8480 Hagy's Mill Rd., Philadelphia, PA 19128-1998. (215) 482-7300. $30.00

Alaska's Ecology (All levels)

is a multidisciplinary, activity-based curriculum that focuses on general ecological concepts. Although specific to Alaska, it can be adapted to other areas. (1994). Available from Wizard Works, P.O. Box 1125, Homer, AK 99603. (907) 235-8757. $12.95

Alberta's Threatened Wildlife (Elementary)

consists of a series of 10 fact sheets and teacher's guides that contain specific details on individual species such as peregrine falcons, swift foxes, and whooping cranes. Information provided includes status, description, habits, reproduction, food, limiting factors, management, and outlook. (1990–96). Available from Alberta Environmental Protection, Information Center, Main Fl., 9920 108th St., Edmonton, AB T5K 2M4, Canada. (403) 422-2813. $15.00

Animal Tracks (Middle School)

comprises a collection of articles, games, and classroom activities that focus on conservation issues such as backyard wildlife, water conservation, air quality, composting, and endangered species. It contains 30 student books, an educator's guide, and a habitat action pack. (1995–96). Available from National Wildlife Federation, 8925 Leesburg Pike, Vienna, VA 22184. (703) 790-4100. Also from the National Education Association, Professional Library, P.O. Box 509, West Haven, CT 06516-9904. $24.95

Backyard Biodiversity and Beyond (Middle School)

by Deanna Binder, Stewart Guy, and Briony Penn is a resource kit that focuses on the nature and value of biodiversity in British Columbia. Five modules present activities on a variety of subject areas. (1995). Available from Ministry of Environment, Lands, and Parks, Public Affairs and Communications Branch, First Fl., 810 Blanshard St., Victoria, BC V8V 1X4, Canada. (250) 387-1161. $20.00

The Biodiversity Debate: Exploring the Issue (Secondary–Adult)

the fourth book in the Environmental Issues Forums series, produced by the North American Association for Environmental Education in collaboration with World Wildlife Fund, delves into the complex issue of the loss of biological diversity and presents three different perspectives on what the problem is and how it might be resolved. The book helps people discuss controversial viewpoints and learn more about underlying values and beliefs that influence our thinking. (1997). Available from Kendall/Hunt Publishing Company, 4050 Westmark Dr., Dubuque, IA 52002-1840. (800) 228-0810. $7.00

Biodiversity: The Florida Story (Middle School)

comprises a set of five teacher's guides in newsletter format that are linked to the Florida Sunshine State Standards. Topics include urbanization and economics;

agriculture and land management; water quality and its relationship to land use, management, and public/private partnerships; and non-native species, indicator species, and human uses. Teachers may obtain a free set of the materials with a colorful poster and instructions for their use by attending a workshop sponsored by the Regional Service Project. (1998). Available from the Office of Environmental Education/Florida Gulf Coast University, 1311 Paul Russell Rd., Ste. 201A, Tallahassee, FL 32301-4880. (850) 487-7900. Web site: <www.polaris.net/~oee>.

Biodiversity: Understanding the Variety of Life (Middle School)

is one unit in the Scholastic Science Place Program that focuses on the choices people make that affect the survival of our own and other species. It includes a student book and a teacher's guide. (1997). Available from Scholastic Inc., 2931 E. McCarty St., Jefferson City, MO 65101. (800) 724-6527. Student book $7.95; teacher's guide $27.00

Biodiversity Works for Wildlife. You Can Too! (Middle School)

is a book of background information, ideas, and simple instructions to help students carry out wildlife improvement projects at their schools or in their communities under the supervision of a teacher or a group leader. (1994). Available from Canadian Wildlife Federation, 4027 Queensview Dr., Ottawa, ON K2B 1A2, Canada. $3.00

Biological Diversity Makes a World of Difference: A Curriculum for Teachers and Interpreters (Middle School)

a publication by the National Park Service, National Parks and Conservation Association, and

Minnesota Environmental Education Board, contains activities and worksheets to help students understand what biodiversity is and what role parks play in supporting biodiversity. (1995). Available from National Parks and Conservation Association, 1776 Massachusetts Ave., NW, Washington, DC 20036. (800) 628-7275 ext. 216. $19.95

Bottle Biology (All levels)

is a set of activities that describes ways to use recyclable containers to learn about science and the environment. Bottles are used to create an ecosystem, explore the concept of a niche, model a lakeshore, create a simulated tornado, breed fruit flies and spiders, observe the adventures of slime molds, make a microscope, and much more. (1993). Available from Kendall/Hunt Publishing Company, 4050 Westmark Dr., Dubuque, IA 52002-1840. (800) 228-0810. $17.95

A Child's Place in the Environment (Elementary)

is a series designed to provide elementary school teachers with examples of an interdisciplinary, thematic environmental education program. The grade-level specific units are conceptually correlated to the "Science Framework for California Public Schools." Each grade-level unit is activity based and is designed around four major themes: Valuing the Environment, Systems and Interactions, Patterns of Change, and Conservation. (1994). Available from Department of Education, Bureau of Publications, P.O. Box 271, Sacramento, CA 95812-0271. (800) 995-4099. $55.00

Connections: The Living Planet (Middle School)

consists of 12 student books and an educator's guide organized around three environmental themes: ecology and life on Earth, ways human activities influence changes in the Earth's ecosystems, and requirements of human life. (1993–95). Available from Ginn Publishing Canada, Inc., 3771 Victoria Park Ave., Scarborough, ON M1W 2P9, Canada. (416) 497-4600 or (800) 361-6128. $9.35

Conservation Biology
(Secondary)

is a module containing activities that reflect the problems and questions facing conservation biologists as well as politicians, economists, and citizens. Major themes include the value of biodiversity, the principles of island biogeography, habitat fragmentation, and causes of species extinction. (1996). Available from Kendall/Hunt Publishing Company, 4050 Westmark Dr., Dubuque, IA 52002-1840. (800) 228-0810. $49.90

Eco-Inquiry: A Guide to Ecological Learning Experiences (Middle School)

comprises a set of three modules that focus on food webs, decomposition, and nutrient recycling. It includes hands-on science within thematic, multi-dimensional learning experiences. (1994). Available from Kendall/Hunt Publishing Company, 4050 Westmark Dr., Dubuque, IA 52002-1840. (800) 228-0810. $31.45

Ecological Citizenship
(Elementary)

is an urban environmental education program designed to engage students, parents, teachers, and the community. Its multidisciplinary, action-oriented ecology curriculum involves hands-on explorations of environmental issues such as nature's cycles, ecosystems and the effects of human activities, solid waste and recycling, and air pollution. (1996). Available from The Chicago Academy of Sciences, 2060 N. Clark St., Chicago, IL 60614. (773) 549-0606 ext. 3067. $34.95

Ecology for All Ages: Discovering Nature Through Activities for Children and Adults (All levels)

through a variety of activities that focus on habitats, natural systems, and species, emphasizes how all three are linked. It presents examples of some of the environmental problems that directly affect different habitats, along with suggestions for what people can do to help solve the problems. (1994). Available from The Globe Pequot Press, 6 Business Park Rd., Old Saybrook, CT 06475. (800) 395-0440 or (800) 243-0495. $16.95

Economics and the Environment (Secondary)

consists of a set of activities that highlight the relationship between the environment and the economy and that introduce students to an economic approach to analyzing environmental issues. (1996). Available from National Council on Economic Education, Attn: Order Department, 1140 Ave. of the Americas, New York, NY 10036. (800) 338-1192. $49.95

Eco Sense—An Economic Environmental Learning Kit (All levels)

contains interactive lessons that teach students, through the integration of environmental and economic principles, about the profound impact that they, their families, and their communities have on the environment. Elementary and secondary guides are available. (1993). Available from Business Economics Education Foundation, 123 N. Third St., Ste. 504, Minneapolis, MN 55401. (612) 337-5252. $25.00

Ecosystem Matters: Activity and Resource Guide for Environmental Educators (All Levels)

is a collaborative effort of the U.S. Department of Agriculture and the Rocky Mountain Region branch of the U.S. Forest Service. It focuses on ecosystem management and is designed to supplement existing courses and programs for classroom teachers, scout leaders, nature camp instructors, forest rangers, and naturalists. Contact the nearest Government Printing Office for more information, or call (202) 512-1800.

Endangered Species: On Common Ground— Critical Issues/Critical Thinking Experiences for Youth Series (Middle School)

comprises a set of activities to help students enhance their critical thinking skills while learning about endangered species. Topics include an introduction to endangered species, the ways extinction affects the world, balancing human needs with those of plants and animals, and ideas for what can be done to help endangered species. (1996). Available from National 4-H Council Supply Service, c/o Crestar Bank, P.O. Box 79126, Baltimore, MD 21279-0126. (301) 961-2934. $3.00

Environomics—Exploring the Links Between the Economy and the Environment (Secondary)

is a teaching kit designed to help educators teach about the relationship between the environment and the economy. It endorses and supports the goal of sustainable development and holds out the hope that we can be innovative, creative, and concerned enough to achieve a sustainable environment without forcing dramatic economic sacrifices. (1996). Available from Canadian Foundation for Economic Education, 2 St. Clair Ave. West, Ste. 501, Toronto, ON M4V 1L5, Canada. (416) 968-2236. $20.00

Global Environmental Change Series (Secondary)

comprises a series of case studies designed to investigate global environmental changes from an interdisciplinary perspective, with an emphasis on science. The *Biodiversity* section uses Costa Rica as a case study in balancing economic growth and resource conservation. *Deforestation* focuses on Washington State's Olympic Peninsula as a case study to provide a model for addressing deforestation's ecological and economic impacts. (1997). Available from National Science Teachers Association, 1840 Wilson Blvd., Arlington, VA 22201-3000. (800) 722-6782. $12.95 each

Global Environmental Education Resource Guide (Middle School)

is a selection of activities that help frame and clarify key issues associated with the global environment. Topics covered include acid rain, biodiversity, deforestation and desertification, greenhouse gases, marine and estuarine pollution, overpopulation, ozone depletion, and sea level rise. (1996).

Available from University of Southern Mississippi, Institute of Marine Sciences, and the J. L. Scott Marine Education Center and Aquarium, P.O. Box 7000, Ocean Springs, MS 39566-7000. (601) 374-5550. $10.00

Global Issues in the Middle School (Middle School)

is a set of activities focusing on the local and global environment, and is designed to help students see the world as an integrated, interdependent system. (1994). Available from Social Science Education Consortium, P.O. Box 21270, Boulder, CO 80308-4270. (303) 492-8154. $21.95

Global Systems Science Series (Secondary)

comprises nine student guides on topics such as changing climate, ecosystem change, energy flow, human population impact, and losing biodiversity. The series focuses on how people interact with the natural environment and on what we can do to achieve a more sustainable world. Published by Lawrence Hall of Science, University of California, Global Systems Science, Berkeley, CA 94720-5200. Available from NASA.

Global Warming & the Greenhouse Effect (Middle School–Secondary)

contains a variety of laboratory activities, simulations, and discussions designed to help students focus on global warming issues and to explore the related social and ecological consequences. It shows how scientific knowledge influences public debate and policy. (1990). Available from Great Explorations in Math and Science Program (GEMS), Lawrence Hall of Science, University of California, Berkeley, CA 94720-5200. (510) 642-7771. $16.00

Green Inheritance Teaching Pack (Middle School)

produced by WWF-UK, includes teacher background materials, student sheets, resource lists, and wall charts that cover a range of plant conservation issues. To order or for a complete catalog of WWF-UK teaching materials, write to WWF-UK, Publishing Unit, Panda House, Weyside Park, Godalming, Surrey GU7 1XR, England. Also available from Green Brick Road, 429 Danforth Ave., Ste. 408, Toronto, Ontario, M4K1P1, Canada. (800) 477-2665. $37.50

GrowLab: Activities for Growing Minds (Elementary–Middle School)

is a hands-on curriculum guide to promote discovery learning through indoor and outdoor gardens, and includes one section on biodiversity. (1990). Contact the National Gardening Association, 180 Flynn Ave., Burlington, VT 05401. (800) 538-7476. $24.95

Habitat and Biodiversity (Middle School–Secondary)

uses the school as a research laboratory where students study the importance of biological diversity, landscape management, xeriscaping, composting, and integrated pest management. This is one of six modules in the Environmental Action Program. The program's mission is to empower students with the knowledge and skills necessary to make meaningful changes that can be carried into the future. (1998). Available from Dale Seymour Publications, P.O. Box 5026, White Plains, NY 10602-5026. (800) 872-1100. Student edition $5.95; teacher's edition $13.95

Heath Environmental Literacy Program (Secondary)

is a series of books designed to help students acquire the understanding and skills needed to become environmentally literate citizens by recognizing bias and learning to make more environmentally sound choices. *Protecting the Ozone Layer* and *Understanding Global Warming* are two titles in this series. (1994–95). Available from ITP Nelson Canada, 1120 Birchmount Rd., Scarborough, ON M1K 5G4, Canada. (800) 268-2222 or (416) 752-9100 ext. 380. $17.65 each

Mud, Muck, and Other Wonderful Things (Elementary)

is designed to help children understand basic ecological concepts, the effect their actions have on a habitat, and the responsibilities that compose good stewardship. (1995). Available from National 4-H Council, 7100 Connecticut Ave., Chevy Chase, MD 20815. (301) 961-2934. $6.00

The New Explorer's Series: Test Tube Zoo
(Middle School–Secondary)

is a classroom extension of the PBS program "The New Explorer," which profiles scientists and their research. These activities are geared for use in conjunction with videotapes of the program. (1992). For more information, contact Lincoln Park Zoological Gardens, Department of Education, 2200 N. Cannon Dr., Chicago, IL 60614. (312) 742-2000. Web site: <www.lpzoo.org>.

Ocean News (Middle School–Secondary)

is a series of newsletters, teacher's guides, and computer disks that focus on five marine science issues: Exploring the Fluid Frontier, Marine Mammals, Seabirds, Marine Pollution, and Marine Biodiversity. (1996). Available from Bamfield Marine Station, Bamfield, BC V0R 1B0, Canada. (604) 728-3301. $73.00

Our Oceans, Ourselves: Marine Biodiversity for Educators (All levels)

is a marine biodiversity framework for educators, and provides a template to help educators include marine biodiversity education in their programs, background information on marine biodiversity, and a sample of activities for use with students. (1995). Available from Environment Canada, 351 St. Joseph Blvd., 5th Fl., Hull, QC K1A 0H3, Canada. (819) 953-4374. Free

Our Only Earth: A Curriculum for Global Problem Solving
(Elementary–Secondary)

offers a collection of teacher and student background materials and activities that focus on seven global problems, including tropical rain forests, poverty and population, and endangered species. (1991). Contact Zephyr Press, P.O. Box 66006, Tucson, AZ 85738. (602) 322-5090. $19.95/book or $125.65/series

Population Reference Bureau, Inc. (All levels)

produces curriculum guides filled with materials and strategies to teach about population and environmental issues. They are designed to help students understand the complexity of these issues, the relationship between population growth and environmental degradation, and the interdependence of the world's nations and people. Elementary, middle, and secondary guides are available. (1991–97). Available from Population Reference Bureau, Inc., 1875 Connecticut Ave. NW, Ste. 520, Washington, DC 20009-5728. (202) 483-1100 or (800) 877-9881. $10.00–$15.00

Rainforest Lifeline (Middle School)

is a resource guide to tropical forests with a focus on action. It includes a

70-page activity book, a teacher's guide, a full-color book for kids about a rain forest in Costa Rica, and a six-minute video. The guide was developed and published by WWF-Canada. Available from Scholastic Canada. (800) 268-3848. Order number 96C-225036. $56.25

Ranger Rick's NatureScope Series (Elementary–Middle School)

consists of a series of curriculum guides designed to help educators incorporate science and environmental education into their teaching. Each guide focuses on a particular topic and provides educators with a comprehensive, flexible teaching kit. Titles include *Amazing Mammals; Birds, Birds, Birds!; Endangered Species: Wild & Rare; Let's Hear It for Herps!;* and *Rain Forests: Tropical Treasures.* (1987–92). Available from McGraw-Hill Companies, Inc., 11 W. 19th St., New York, NY 10011-4285. $12.95 each

Relationships of Living Things (Middle School)

is one of five grade 3 units in the Macmillan/McGraw-Hill Science series. This activity-based series uses a four-step lesson cycle (engage, explore, develop, extend/apply) built from a constructivist point of view. The focus of the unit is the interaction of living things in an ecosystem. (1995). Available from MacMillan/McGraw-Hill School Division, 220 E. Danieldale Rd., DeSoto, TX 75115. (800) 442-9685. $36.75/teacher's guide or $5.87/student book. Web site: <www.mmhschool.com>.

Schools for Wildlife (Middle School)

is an educational packet produced by WWF-Canada three times a year. Each packet focuses on a specific conservation issue: wilderness, species, or rain forests. Contact WWF-Canada at (416) 489-8800. Free

Science is Elementary
(Elementary–Middle School)

is a resource magazine that provides activities, background information, extensions, and suggestions for authentic assessment congruent with national benchmarks. Volume seven includes four issues: Biodiversity, Habitats and Ecosystems, Our Changing World, and Populations. (1995-1996). Available from MITS, 79 Milk St., Ste. 210, Boston, MA 01929-3903. (617) 695-9771. $22.00/year

Taking Action: An Educator's Guide to Involving Students in Environmental Action Projects (All Levels)

is a collaborative project of World Wildlife Fund and Project WILD that helps educators plan, implement, and evaluate environmental education projects. It includes examples of successful action projects, as well as a step-by-step planning process and a comprehensive bibliography. (1995). Available from Acorn Naturalists, 17300 E. 17th St., #J236, Tustin, CA 92780. (800) 422-8886. $7.00

Teacher's Guide to World Resources: Biodiversity (Secondary)

comprises an array of teaching suggestions and presentation materials based on the most recent and authoritative information on the global environment. Information and analyses are drawn from World Resources Institute publications and databases. (1994). Available from World Resources Institute Publications, P.O. Box 4852 Hampden Station, Baltimore, MD 21211. (800) 822-0504 or (202) 662-2596. $6.95

Threatened and Endangered Animals: An Extended Case Study for the Investigation and Evaluation of Issues Surrounding Threatened and Endangered Animals of the United States
(Middle School–Secondary)

focuses on issues surrounding threatened and endangered species in the United States and promotes the development of environmental action skills. (1993). Available from Stipes Publishing L.L.C., 10 Chester St., Champaign, IL 61820. (217) 356-8391. Student book $9.80; teacher's guide $20.80

What Is It? A Guide to Biological Identification
(All levels)

is a reproducible sourcebook of activities that give students and teachers a hands-on introduction to classification and field work. (1992). Available from J. Weston Walch, 321 Valley St., P.O. Box 658, Portland, ME 04104-0658. (800) 341-6094. $19.95

Wildlife (Middle School)

is a unit of the Adopt-a-Watershed Program, an integrated K–12 science curriculum that uses a local watershed as a focal point for bringing theory into application. Themes covered in the wildlife unit include evolution, patterns of change, and stability. (1997). Available from Adopt-a-Watershed Program, P.O. Box 70, Hayfork, CA 96041. (916) 628-5294 or (916) 628-4608. $67.00

Wildlife Diversity (Middle School)

includes *The Links of Life* (1994) and *Forests Forever* (1992)—two of many activity guides related to the Minnesota region's biodiversity. Created by and available from The University of Minnesota, Minnesota Extension Service, Room 20, Coffey Hall, 1420 Eckles Ave., St. Paul, MN 55108-6069. (800) 876-8636. $6.00/guide

Wildlife for the Future
(Elementary–Middle School)

is an interdisciplinary curriculum that investigates biodiversity connections and focuses on both terrestrial and aquatic biology. It also focuses on the economic,

ecological, and spiritual value of wildlife. Although specific to Alaska, it can be adapted. (1995). Available from Circumpolar Press, P.O. Box 221995, Anchorage, AK 99522. $13.95

W.I.Z.E. Wildlife Inquiry Through Zoo Education
(Middle School–Secondary)

features a two-part curriculum developed by the Bronx Zoo Education Department. *Diversity of Lifestyles* (Module 1) is designed for middle school. *Survival Strategies* (Module 2) is for secondary students. For more information, contact: Manager of National Programs, Education Department, Bronx Zoo/Wildlife Conservation Park, Bronx, NY 10460. Workshop required. (800) 937-5131. Web site: <www.wcs.org/education>.

WOW!: The Wonders of Wetlands (All levels)

contains a curriculum guide that provides background material for teachers, and activities that focus on general wetland concepts and definitions, wetland communities of plants and animals, the role of water in wetlands, the role of soils in a wetland environment, and the interactions between humans and wetlands. (1995). Available from The Watercourse, 201 Culbertson Hall, Montana State University, Bozeman, MT 59717-0057. (406) 994-5392. $15.95

Zero Population Growth
(All levels)

produces readings and activities that show the connections among people, living things, and the environment and focuses on population dynamics and environmental impacts. They are designed to broaden students' knowledge of trends and connections among population change, natural resource use, global economics, gender equity, and community health. Elementary, middle, and secondary guides are available. (1991–96). Available from Zero Population Growth, 1400 16th St., NW, Ste. 320, Washington, DC 20036. (800) 767-1956 or (202) 332-2200. $19.95–$22.95

Books For Elementary Students

Herman and Marguerite: An Earth Story

by Jay O'Callahan describes how a friendship between an earthworm and a caterpillar helps to bring a broken-down orchard back to life. It discusses the role that invertebrates play in decomposition and soil enrichment. (Peachtree, 1996). $15.95

The Kids Environment Book: What's Awry and Why

by Anne Pedersen provides a clear, balanced discussion of the environmental problems we face today and are likely to face tomorrow. (John Muir Publications, 1991).

Of Things Natural, Wild, and Free: A Story About Aldo Leopold

by Marybeth Lorbiecki is an illustrated story about the conservationist Aldo Leopold. (Carolrhoda Books Inc., 1993). $19.93

Ride the Wind: Airborne Journeys of Animals and Plants

by Seymour Simon is a colorful story about the ways animals and plants migrate. (Harcourt Brace, 1997). $15.00

Welcome to the Green House

by Jane Yolen describes and illustrates a tropical rain forest house, where the trees are the walls, the canopy of leaves is the roof, and the inhabitants are fascinating creatures of all shapes and colors. (GP Putnam and Sons, 1993). $5.95.

Young Explorer's Guide to Undersea Life

by Pam Armstrong introduces children to sea lions, whales, sharks, moon jellies, and other sea life. (Monterey Bay Aquarium Press, 1996). $16.95

Books for Middle School Students

And Then There Was One: The Mysteries of Extinction

by Margery Facklam focuses on the process of extinction, both natural and human induced, rather than on the loss of one particular species. Concepts include evolution, adaptation, natural selection, speciation, and human factors of extinction. (Sierra Club, 1993). $7.95

Bats, Bugs, and Biodiversity: Adventures in the Amazonian Rain Forest

by Susan Goodman follows seventh and eighth graders on a trip to the rain forest. (Atheneum Books for Young Readers, 1995). $16.00

Beneath Blue Waters: Meetings with Remarkable Deep-Sea Creatures

by Deborah Kovacs and Kate Madin explores the diversity of life in the deep regions of the ocean. Young readers will be amazed at the number and beauty of the creatures living miles below the surface. (Viking, 1996). $16.99

Biodiversity

by Dorothy Hinshaw Patent discusses the interconnectedness of all living things and tells why it's important to save biodiversity. (Clarion Books, 1996). $18.00

Branch Out: A Book About Land

by Jill Wheeler examines how we are damaging our environment and

what we can do about it. The book introduces a variety of important topics, including deforestation, disappearing wetlands, and loss of topsoil. It also looks at what people can do about each. (Abdo & Daughters, 1993). $22.83

A Desert Scrapbook: Dawn to Dusk in the Sonoran Desert

by Virginia Wright-Frierson gives readers a tour of the Sonoran Desert. (Simon & Schuster, 1996). $16.00

Earthdance

by Joanne Ryder helps readers imagine that they are the Earth itself—home to myriad plants, animals, and cultures. An appreciation for the complexity of life on Earth is emphasized during this imaginary adventure. (Henry Holt & Co., 1996). $16.95

Earth Kids (Elementary–Middle School)

by Jill Wheeler tells empowering stories of individual children who have taken action to preserve the environment. Some of the children highlighted have helped to preserve forests, defend animals, and clean up beaches, water, and air. (Abdo & Daughters, 1993). $22.83

The Great Kapok Tree (Elementary–Middle School)

by Lynne Cherry describes how resident creatures of the rain forest save their kapok tree home from being chopped down by whispering in the ear of the man who has come to cut it down. (Harcourt Brace Jovanovich, 1990). $16.00

Insects (Elementary–Middle School)

by Laurence Mound helps readers discover the world of insects in a close-up look at their behavior, anatomy, and the important role they play in the Earth's ecology. (Eyewitness Books, Alfred A. Knopf, 1990). $7.95

Julie of the Wolves

(Middle School–Adult)

by Jean Craighead George describes the life of an Eskimo girl protected by a wolf pack. While lost on the tundra, the girl begins to appreciate the heritage and oneness with nature that is being destroyed in our modern world. (HarperCollins, 1972). $15.89

No More Dodos: How Zoos Help Endangered Wildlife

by Nicholas Nirgiotis and Theodore Nirgiotis describes the role of zoos in the preservation of wildlife diversity throughout the world. (Lerner, 1996). $23.95

Saving Planet Earth

(Middle School–Adult)

by Rosalind Kerven provides examples of worldwide conservation efforts. Identifying the Industrial Revolution as the starting point for many of our current environmental problems, the book looks at the way people around the world are meeting environmental challenges. (Watts, 1992). $17.00

Squishy, Misty, Damp, and Muddy: The In-Between World of Wetlands

by Molly Cone uses stunning photos and innovative text to demonstrate the diversity of life in wetlands. The book also explores threats to these habitats and ways to protect them. (Sierra Club, 1996). $15.95

Vanishing Animal Neighbors

by Geraldine Marshall Gutfreund provides readers with information and anecdotes about five endangered species. Filled with photos and sidebars, this book is written in a personal, engaging style. (Watts, 1993). $21.00

WOW!—A Biodiversity Primer

by World Wildlife Fund is a full-color, magazine-style primer for middle school students to help them understand biodiversity. This award-winning primer includes fiction and nonfiction stories and articles. (World Wildlife Fund, 1994). $3.00

Yuck! A Big Book of Little Horrors (Elementary–Middle School)

by Robert Snedden takes a big look at some small creatures. With bright color enlargements, the author highlights the diversity of life crawling around us everyday—on our beds, floors, food, and body. (Simon & Schuster, 1996). $15.00

Zoobooks: Exploring Ocean Ecosystems

(Elementary–Middle School)

contains eight student books of articles, photographs, illustrations, facts, games, and activities about animals, along with a curriculum guide. It takes a thematic approach to teaching science and the scientific process and encourages learning across the curriculum. Titles include *Dolphins and Porpoises, Penguins, Seals and Sea Lions, Sharks, Turtles, and Whales.* (1990–96). Available from Wildlife Education, Ltd., 9820 Willow Creek Rd., Ste. 300, San Diego, CA 92131-1112. (800) 477-5034. $29.95

Multimedia Resources

Audiotapes

"Life on the Brink" and "Oceans of Life" Kits (Middle School)

are part of the Radio Expeditions series produced by National Public Radio and National Geographic Society. Each kit includes a teacher's guide, a 30-minute audio cassette, and a map or poster. Hour-long CD recordings of Radio Expeditions shows are also available. "Life on the Brink" (1996) focuses on biodiversity on land, while "Oceans of Life" (1995) looks at marine biodiversity. To order kits, call (202) 414-2726. To order CDs, call (888) 677-3472. For more information, visit the Web site: <www.npr.org>.

Posters

Biodiversity— From Sea to Shining Sea

(Middle School–Secondary)

a poster kit produced by World Wildlife Fund, includes two 22-in. by 34-in. two-sided posters (the front features 12 photographs highlighting the amazing diversity of life on Earth; the back includes a map of the most threatened ecoregions in the United States) and a 12-page educator's guide with background information about biodiversity and suggestions for how to use the poster. (1996). Available from Acorn Naturalists, 17300 E. 17th St., #J236, Tustin, CA 92780. (800) 422-8886. $5.00

Diversity Endangered

(Middle School–Secondary)

a poster exhibition kit designed by the Smithsonian Institution Traveling Exhibition Service (SITES), includes 15 posters that feature issues and challenges related to biodiversity, ranging from the variety and interrelation of species to the complex habitats of tropical rain forests, wetlands, and coral reefs. The kit also includes a resource and programming guide. (1997). Available from SITES, 1100 Jefferson Dr., SW, Quad 3146, Washington, DC 20560. (202) 357-3168. $100.00

Eye of the Environment
(Middle School–Secondary)

by National Geographic Society features nine key global environmental issues, all linked to biodiversity. Each poster set includes three posters that give an overview of the problem, a focus on an individual issue, and a worldview offering possible solutions. A teacher's guide is also included. (1995). Available from National Geographic Society, Educational Services, P.O. Box 10597, Des Moines, IA 50340. (800) 368-2728. $45.95/set

Videos

Biodiversity! Exploring the Web of Life Education Kit
(Middle School)

introduces young people to biodiversity. This half-hour video explores the meaning of biodiversity, its status, and what people can do to protect it. An educator's guide is included in the kit along with *WOW! A Biodiversity Primer*, a full-color magazine for students. (1997). Available from Acorn Naturalists, 17300 E. 17th St., #J236, Tustin, CA 92780 (800) 422-8886. $24.95

Diversity Endangered
(Secondary)

covers a range of threats to biodiversity, including habitat loss, pollution, and climate change. This short video (nine minutes) was produced by the Smithsonian Institution Traveling Exhibition Service (SITES). (1997). Available from SITES, 1100 Jefferson Dr. SW, Quad 3146, Washington, DC 20560. (202) 357-3168. $10.00

The Diversity of Life (Secondary)

a 25-minute video that examines the incredible variety of plant and animal species, describes why it is important to preserve endangered species and habitats, and also explores possible solutions to the loss of biodiversity. (1993). Available from National Geographic Society, Educational Services, P.O. Box 98109, Washington, DC 20090-8019. (800) 368-2728. $99.00

Food Chains in the Biosphere (Middle School)

is an 18-minute video that looks at the Earth's biosphere with an emphasis on energy flow and food chains. (1993). Available from Hawkhill Associates, 125 E. Gilman St., Madison, WI 53703. (800) 422-4295. $79.00

Going, Going, Almost Gone! Animals in Danger Education Kit (Elementary)

includes an award-winning video about biodiversity, habitat loss, and the illegal trade of wildlife. The 28-minute video is produced by World Wildlife Fund and HBO. The kit includes an educator's guide and a color poster. (1997). $19.50. Also available is the *Science and Environmental Education Catalog* that describes science and nature education books, videos, and other materials for K–12 educators. Both available from Acorn Naturalists, 17300 E. 17th St., #J236, Tustin, CA 92780. (800) 422-8886.

Green Means
(Middle School–Secondary)

is a series of 32 short programs about ordinary people who are making positive contributions to the health of the planet. (1993). Available from Environmental Media, P.O. Box 1016, Chapel Hill, NC 27514. (919) 933-3003. $59.95

The Last Show on Earth
(Secondary–Adult)

celebrates the efforts of individuals who are struggling to save endangered species while it illustrates the causes of extinction. (1992). Available from Bullfrog Films, P.O. Box 149, Oley, PA 19547. E-mail: bullfrog@igc.apc.org. (800) 543-3764. $295.00 (public performance), $85.00 (rental); teacher discounts available

Life on Earth (Secondary–Adult)

is a compilation of programs from the television series of the same title hosted by David Attenborough. The four-hour video comes with an index for educators looking for segments pertinent to the subject they are teaching. (1986). Available from Delta Education, P.O. Box 3000, Nashua, NH 03061-3000. (800) 442-5444 or (603) 377-4740. $37.98

Spaceship Earth (Middle School)

is an Emmy Award-winning video that explores both the global links between nature, people, and technology, as well as many critical environmental issues, including rain forest destruction. It is narrated by young people. (1994). Available from Hawkhill Associates, 125 E. Gilman St., Madison, WI 53703. (800) 422-4295. $149.00

Understanding Biodiversity
(Middle School)

is a 27-minute video that covers the history of biodiversity, where it's found, why it's important, and the causes of its loss. It comes with a student page of video questions. (1996). Available from Educational Video Network, 1401 Nineteenth St., Huntsville, TX 77340. (409) 295-5767. $59.95

Variety and Survival
(Middle School)

explores why variation is important for the survival of communities and species, and examines overpopulation and species extinction. (1994). Available from Journal Films, 1560 Sherman Ave., Ste. 100, Evanston, IL 60201. (847) 328-6700.

The Video Project—Films and Videos for a Safe and Sustainable World
(Elementary–Adult)

an eco-video collection for schools, has themes related to biodiversity topics. Catalogues are available from The Video Project, 200 Estates Dr., Ben Lomond, CA 95005. (800) 4-PLANET.

Videodiscs

Animal Pathfinders
(Middle School–Secondary)

is an interactive program that simulates a field trip to study animals in their natural habitats. (1991). Available through Scholastic Software, Scholastic Inc., 2931 E. McCarty St., P.O. Box 7502, Jefferson City, MO 65102. (573) 636-5271. Item number 85234. $395.00

GTV: Planetary Manager
(Middle School–Secondary)

encourages students to investigate environmental problems and grapple with solutions. (1992). Available from National Geographic Society, Educational Services, 1145 17th St. NW, Washington, DC 20036. (800) 368-2720. Also available from Videodiscovery, 1700 Westlake Ave. North, Ste. 600, Seattle, WA 98109-3040. $395.00

Habitat and Dependence
(Middle School–Secondary)

looks at habitats, food chains, ecological interdependence, and how humans affect their surroundings. (1994). Available from Journal Films, 1560 Sherman Ave., Ste. 100, Evanston, IL 60201. (847) 328-6700.

STV: Rain Forest
(Middle School–Secondary)

gives students a close-up look at rain forests and why they're in trouble. (1991). Available from National Geographic Society, Educational Services, P.O. Box 10597, Des Moines, IA 50340. (800) 368-2728. $225.00

CD-ROMs

Cal Alive! (Middle School)

is a CD-ROM project created by the California Institute for Biodiversity to explore the plants, animals, and habitats of the state. The program uses photo images, computer animation, and satellite images in activities, games, and quizzes. Teacher resource guides, workshops, and titles for all age levels are planned. (1998). Available from California Institute for Biodiversity, 11 Embarcadero West, Ste. 120, Oakland, CA 94607. (510) 444-6629. $49.95

Earth's Endangered Environments (Middle School)

introduces rain forest and wetland ecosystems, deforestation, and pollution. The CD-ROM is compatible with Apple® Macintosh™ and Microsoft® Windows™ systems and includes a user's guide, student information, and classroom activities. (1994). Available from the National Geographic Society, Educational Services, P.O. Box 10597, Des Moines, IA 50340. (800) 368-2728. $45.95

Encyclopedia of U.S. Endangered Species
(Secondary–Adult)

contains reports on U.S. endangered species, maps, photos, and a glossary. Presentations (Windows™ compatible) by World Wildlife Fund, The Nature Conservancy, Threatened and Endangered Species Information Institute, and ABC News. (1996). Available from ZCI Inc. (800) 808-0623 ext. 101. $49.95

Eyewitness Encyclopedia of Nature (Elementary)

is an illustrated, interactive CD-ROM that is full of information and activities about plants and animals. Users can examine biodiversity through the *Web of Life* book, which includes subjects titled, "The Biosphere," "Natural Cycles," "Energy Flow," and "Living Together." (1997). Available from DK Multimedia, 95 Madison Ave., New York, NY 10016. (800) 356-6575. $49.95

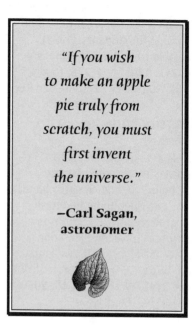

"If you wish to make an apple pie truly from scratch, you must first invent the universe."

—Carl Sagan, astronomer

The Great Ocean Rescue
(Middle School)

introduces students to marine ecosystems and marine biodiversity. The kit includes student reference booklets and a teacher's guide with lesson plans, worksheets, and activities. Windows™ or Macintosh™ compatible. (1996). Available from Tom Snyder Productions, Inc., 80 Coolidge Hill Rd., Watertown, MA 02172-9718. (800) 342-0236. Web site: <www.teachtsp.com>. $328.00

Rainforest Researchers
(Middle School)

engages students in projects about rain forest plants. Broader themes involve the study of biodiversity, tropical rain forest issues, and the management of biological resources. It includes student reference books, a teacher's guide, and a CD-ROM. Windows™ or Macintosh™ compatible. (1995). Available from Tom Snyder Productions, Inc., 80 Coolidge Hill Rd., Watertown, MA 02172-9718. (800) 342-0236. Web site: <www.teachtsp.com>. $199.95

Selected Internet Sites

Biodiversity and Biological Collections

provides biological data and links to journals, museums of natural history, and other reference materials for more advanced researchers. Web site: <kaw.keil.ukans.edu>.

Biological Resources Division of the U.S. Department of Interior

contains data, information, and links to all aspects of biodiversity in the public and private sector. It is useful as a springboard to state and local sites and also contains resources for educators and a kids' corner. Web site: <www.nbs.gov>.

Educational Resources Information Center (ERIC)

is designed to provide users with ready access to an extensive body of education-related literature. It allows access to the on-line ERIC Publications Catalog, the Question and Answer Service, the ERIC Database, links to other education sites, and numerous products to help access and use the information in the ERIC system. The site is supported by the U.S. Department of Education and the National Library of Education. Web site: <ericir.syr.edu>.

EE Link

connects educators with information and ideas that will help them explore the environment and investigate current issues with students. The site includes print documents and links to environmental education-related organizations and materials on the Internet. Web site: <eelink.net>.

Eisenhower National Clearinghouse (ENC) for Mathematics and Science Education

contains information on the ENC part of the U.S. Department of Education's continuing efforts to reform K–12 math and science education. It includes activities, journal articles, a listing of conferences and special events, and links to many other sites. Web site: <www.enc.org>.

Island Press

is a nonprofit organization that provides solutions-oriented resources that address the multidisciplinary nature of environmental problems. The site features on-line ordering of a wide range of books focusing on biodiversity and other environmental issues. Web site: <www.islandpress.com>.

JASON Project

provides an opportunity for teachers and students to take electronic field trips to explore the deep oceans. Teacher's guides and links to technology resources for the classroom can also be found on this site. Web site: <www.jasonproject.org>.

Second Nature

brings environmental education on-line with "Starfish." Starfish was developed to help educators infuse environmental and sustainability concepts into their teaching. Use "biodiversity" as the keyword to search databases on university-level courses offered by faculty members in over 35 disciplines, a bibliography with over 1,000 references, and comprehensive information on more than 20 innovative teaching and learning techniques. Web site: <www.starfish.org>.

Virtual Library of Ecology, Biodiversity, and the Environment

provides a wide range of ecology and biodiversity-related topics, including specific information on plant and animal species. The site links to periodicals and journals for more advanced research. Web site: <conbio.rice.edu/vl>.

Appendix G

Metric Conversions

When You Know	Multiply By	To Find
feet	.30	meters
yards	.91	meters
miles	1.61	kilometers
square feet	.09	square meters
square yards	.84	square meters
square miles	2.60	square kilometers
pounds	.45	kilograms
short tons (2,000 pounds)	.90	tonnes (metric ton)
gallons	3.79	liters

Student Book pages are in boldfaced type.

Activities Index

Unit Plans Index

*"Wherever we live and whatever we
do, we all make decisions and take actions
that directly impact on the structure and
quality of the world we inhabit."*

–Peter Martin, educator

FEEDBACK FORM

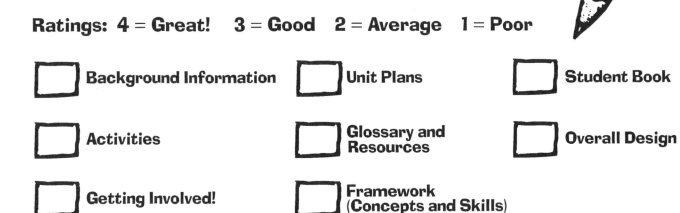

WHAT DO YOU THINK?

Please take a few minutes to give us feedback so that we can improve *Biodiversity Basics* when we reprint. We will also use your comments to help us develop other materials in the *WOW* family. Thanks for taking the time to give us your ideas!

RATING THE SECTIONS

Please use the numbers below to rate the sections in *Biodiversity Basics*. Feel free to add any specific comments on content, design, and usefulness.

Ratings: 4 = Great!   3 = Good   2 = Average   1 = Poor

☐ **Background Information**   ☐ **Unit Plans**   ☐ **Student Book**

☐ **Activities**   ☐ **Glossary and Resources**   ☐ **Overall Design**

☐ **Getting Involved!**   ☐ **Framework (Concepts and Skills)**

GENERAL COMMENTS:

What do you like best about *Biodiversity Basics*?

What recommendations do you have to improve *Biodiversity Basics*? If you have comments about specific activities, please list the activity title(s) along with your suggestions.

Other comments or ideas:

The following errors appear in Biodiversity Basics:

Page Number _____

Error (please describe clearly) _____

_____

_____

_____

Page Number _____

Error (please describe clearly) _____

_____

_____

_____

Page Number _____

Error (please describe clearly) _____

_____

_____

_____

Please complete (optional):

Name _____

School/Organization _____

Address _____

_____

Phone _____ E-mail _____

Please send this form to:

World Wildlife Fund
Environmental Education Dept.
1250 24th Street, NW
Washington, DC 20037
Fax: (202) 887-5293 or
        (202) 293-9211